PERVASIVE CARDIOVASCULAR AND RESPIRATORY MONITORING DEVICES

PERVASIVE CARDIOVASCULAR AND RESPIRATORY MONITORING DEVICES

MODEL-BASED DESIGN

MIODRAG BOLIC

School of Electrical Engineering and Computer Science, University of Ottawa, Ottawa, ON, Canada

ELSEVIER

ACADEMIC PRESS
An imprint of Elsevier

Academic Press is an imprint of Elsevier
125 London Wall, London EC2Y 5AS, United Kingdom
525 B Street, Suite 1650, San Diego, CA 92101, United States
50 Hampshire Street, 5th Floor, Cambridge, MA 02139, United States
The Boulevard, Langford Lane, Kidlington, Oxford OX5 1GB, United Kingdom

Notices
Knowledge and best practice in this field are constantly changing. As new research and experience broaden
our understanding, changes in research methods, professional practices, or medical treatment may become
necessary.

Practitioners and researchers must always rely on their own experience and knowledge in evaluating and
using any information, methods, compounds, or experiments described herein. In using such information or
methods they should be mindful of their own safety and the safety of others, including parties for whom
they have a professional responsibility.

To the fullest extent of the law, neither the Publisher nor the authors, contributors, or editors, assume any
liability for any injury and/or damage to persons or property as a matter of products liability, negligence or
otherwise, or from any use or operation of any methods, products, instructions, or ideas contained in the
material herein.

ISBN: 978-0-12-820947-9

For information on all Academic Press publications visit our
website at https://www.elsevier.com/books-and-journals

Publisher: Mara Conner
Acquisitions Editor: Carrie Bolger
Editorial Project Manager: Emily Thomson
Production Project Manager: Prasanna Kalyanaraman
Cover Designer: Matthew Limbert

Typeset by TNQ Technologies

Dedication

To my mom—who cared and loved so deeply and generously.

Contents

11. Conclusion and research directions: beyond wearable devices

Instructor Site URL:
https://educate.elsevier.com/book/details/9780128209479

Companion Site URL:
https://www.elsevier.com/books-and-journals/book-companion/9780128209479

Preface

Audience

Pervasive Cardiovascular and Respiratory Monitoring Devices is intended for senior undergraduate and graduate students in electrical and biomedical engineering as well as industry professionals. It is planned to become a textbook for courses on wearable devices and biomedical instrumentation. This book covers today's pervasive cardiac and respiratory devices used outside of the hospital as well as addresses emerging technologies such as noninvasive cuff-less monitoring of blood pressure.

Our approach

This book differs from other books on medical devices and biomedical instrumentation in several aspects:

- The focus is on understanding what parameters of the system (where the system consists of the human body, transducers, electronics, and algorithms) affect the performance of the device most. In order to determine that, the emphasis of the book is on model-based design. Engineers build models to understand and anticipate the technical choices they need to make. The reader will be able to simulate almost every design presented in the book using Matlab/Simulink by Math-Works Inc. However, the book is not about Matlab or Simulink. The models are provided on the book webpage, but the code was not explained in the book. It is assumed that the reader will have basic knowledge of Matlab.

- Another unique feature of the book is end-to-end models that involve the model of the interface between the transducer and the human body, the circuit, and the algorithm for processing the signal. In this way, the readers will understand the overall system and be able to analyze the effect of physiological parameters and/or parameters of the design on the performance of the device. In addition, we include noise and motion artifacts in the model to allow for the simulation of realistic situations.

- We showed that different design choices could be evaluated through uncertainty quantification and sensitivity analysis. This is unique compared to other books in this field that present and explain the design but do not focus on modeling aspects and analysis of design choices.

- This book will cover a smaller number of devices than traditional books on biomedical instrumentation and focus only on devices that monitor cardiovascular and respiratory signals that are used for monitoring at home. The fact that we cover a smaller number of devices allows us to go deeper into the functioning of the devices and their components. We focus on already pervasive technologies (electrocardiography, photoplethysmography, blood pressure) and on technologies still being researched. An example includes devices based on the fusion of different signals, such as cuff-less blood pressure monitoring.

- Current trends in industry and research trends are presented in most chapters. A general overview of new developments is also given in the last chapter.

The book follows a unified approach when describing devices in part II of the book. Normally, the chapter starts with a description of what the device measures. This requires a high-level description of a measurement technique and the physiological and physical aspects of the measurements. For example, when introducing photoplethysmography, a noninvasive optical technique for detecting blood volume changes in the blood vessels, we describe the optical properties of the tissues, light penetration, and choice of wavelength. Then we present the photoplethysmogram (PPG) signal, describe its morphology, and point out the type of noise and artifacts that are common for that signal. This is followed by performance metrics used in this field. All of these are extremely important because we need to understand the issues in measurements, the interference, and noise that are typically dealt with when designing the device, as well as the performance requirements before starting the design. Next, we present a physiological model that can be used to generate the signal. This is done while keeping in mind that the book is mainly intended for engineering students. Therefore, the models are less comprehensive than the ones found in books that focus on physiological models. However, they serve the purpose of showing the relationship between physiological variables and the morphology of the signal. Then, we present models of sensors used to acquire the signal and discuss typical sensor parameters. This is followed by the description of the front-end circuit and algorithms. A simulation of the whole system is presented, and the sensitivity or error analysis is performed, followed by a discussion of the results.

Organization

The book is divided into 2 parts and has 11 chapters. The first part introduces the topics, presents performance metrics, and introduces the basic electronic components of medical devices. Part II introduces pervasive devices classified based on the signal that they measure. For example, blood pressure devices are introduced through the oscillometric signal.

Chapter 1 introduces definitions and performance metrics. Then, uncertainty and sensitivity analysis are described in detail, followed by examples of how to perform sensitivity analysis.

Chapter 2 introduces transducers. We will discuss displacement transducers, including strain gauges, piezoelectric, and capacitive transducers. Also, we will discuss photodetectors and LEDs. Electrodes will be introduced as well.

Chapter 3 describes electronic components commonly used in biomedical devices. Conditioning circuitry, such as the Wheatstone bridge, is presented. We also describe operational amplifiers followed by differential and instrumentation amplifiers. Next, we presented analog filters. The next component discussed is the analog to digital converter. Digital to analog converters and pulse width modulation circuits are discussed as well. Several new components used in modern systems are introduced, including capacitance to digital and time to digital converters.

Chapter 4 is on the modeling and simulation of biomedical circuits and systems in general. It allows for connecting the concepts studied in Chapters 1–3. It is important not only to understand each component separately but also how they work as a part of a system. At the end of the chapter, we put all the components together and simulated a device that can generate an alarm when the breathing rate is too low. We simulated two types of devices: 1. the device in which the alarm is generated using mainly analog components without relying on algorithms/software, and 2. the system in which analog signal is converted into digital and processed in software. The first system represents the traditional "old-fashioned" type of system, while the second one represents the current devices in which electronics is used for data acquisition, and most processing is done in software.

Chapter 5 introduces blood pressure devices based on oscillometry.

Chapter 6 introduces devices based on PPG signals, including photoplethysmographs and pulse oximeters.

Chapter 7 describes the design of ECG systems.

In Chapter 8, we focus on devices based on a combination or fusion of data from sensors presented in the previous three chapters. Measuring the time difference between fiducial points of different signals is a foundation for several important concepts that are being researched nowadays, such as arterial stiffness and continuous blood pressure measurements—both of them are covered in Chapter 8.

In Chapter 9, we discuss the measurement of breathing rate. The extraction of the breathing signal using a breathing belt is described in Chapter 4. We will discuss other methods for estimating the breathing rate. Next, we discuss methods for measuring respiratory flow and volume that can be performed outside a hospital.

Chapter 10 discusses modern wearable devices. We start with commonly used sensor technologies not covered in previous chapters, including arterial tonometry, seismocardiography, and phonocardiography. Next, the chapter provides different classifications of wearable devices and then focuses on reducing power consumption.

Chapter 11 concludes the book. We point out research directions in the fields of sensors, electronics, processing, simulators, as well as in pervasive cardiac and respiratory monitoring in general. In addition, it introduces contactless monitoring systems that can perform monitoring of vital signs at distances of several cm to distances of several meters.

Potential course configuration

This book is intended to be used in upper-level undergraduate or graduate courses. The whole book can be covered in one semester. However, if the focus of the course is more on the devices and less on electronics, then the course's emphasis could be on part II of the book (Chapters 5—11). Course instructors are encouraged to use available course material, including slides, code, and video lectures. In addition, the book can be used as a part of a larger biomedical course. In this case, it could help cover parts related to electronics, cardiovascular, and respiratory devices, while other topics can be covered from other sources.

Conventions

The book uses a pale coloring scheme designed for color-blind readers. General block diagrams normally include pale blue, yellow, and green. Block diagrams that include circuit, modeling, or algorithm components have the following color scheme: light blue for sensors, analog, and digital processing blocks, pale gray for interfacing blocks and blocks not directly related to the processing of signals, pale yellow for feedback blocks, and white for algorithms/models. Databases and data sources are pale red. If the circuits are from Simscape, then the Simscape default coloring scheme was used.

The International System of Units (SI) notation was followed for assigning units of measurement to physical quantities. However, we used some non-SI units because they are commonly used in practice. For example, blood pressure is almost always given in millimeters of mercury (mmHg) instead of pascal (Pa). Likewise, heart and breathing rates are commonly expressed in beats/min or breaths per min instead of in hertz (Hz).

New terms are shown in italics letters. Interesting facts or applications and conflicting issues related to the material covered in this book are presented as separate text boxes.

Instructional aids

Questions and problems in the end of chapters include simple questions, computational problems, and simulation questions. Please note that questions that start with * refer to problems not covered in the book. They are added for the readers interested in exploring the material deeper or wider than what was covered here.

Models used for plotting figures in the book are uploaded on the book webpage and are available for download.

The modeling software Matlab/Simulink by MathWorks Inc. was selected because it is well known in the engineering community. In addition, it provides toolboxes covering a wide range of models, including circuits and algorithms. Also, it allows for performing different types of analysis, including parameter optimization and sensitivity analysis. Even though the author prefers open-source languages, no open-source software/simulator is available that is as popular as Matlab and has all the required toolboxes. In addition, a lot of open-source code was available in Matlab, and it was used in the book. The code is given in the form of notebooks (live editor in Matlab). Notebooks are suitable for learning because they present code with explanations so that students can read the material and then try the code in the same environment.

Students and researchers reading this book can rely on developed Matlab models, examples of using the models to perform different analyses, and pointers to research directions. We hope the book will be a useful resource for training researchers and developers in the biomedical field.

Acknowledgments

I express my sincere appreciation to my colleagues and graduate students who have helped me with their constructive comments and suggestions. Special thanks go to Dr. Hilmi Dajani for reviewing many chapters in the manuscript. I am also thankful to Dr. Sreeraman Rajan and Mohamad Hosein Davoodabadi Farahani for reviewing several chapters. I am also grateful to Shan He, who collected data used in several chapters of the book.

I am thankful to the University of Ottawa and Carleton University students who took my graduate course, Principles and Design of Advanced Biomedical Instrumentation, that I taught from 2013 to 2022. They pushed me to explain the material better, so I ended up writing this book in an attempt to do that.

I am grateful to many individuals and research groups who made their software and data available for others to use. Simulation in the book heavily relies on the available code.

I am grateful to the team at Elsevier, including Emily Joy Grace Thomson and Fiona Geraghty, for their guidance and patience with me even when I kept failing to submit the book in time to meet the deadlines for submission.

I'd like to thank my family for their patience and support while I was writing the book.

Overview of biomedical instrumentation and devices

Abbreviations and explanations

BCG	Ballistocardiogram
ECG	Electrocardiogram
EDA	Electro-dermal activity
EEG	Electroencephalogram
EMG	Electromyogram
OSA	Obstructive sleep apnea
PCG	Phonocardiogram
PPG	Photoplethysmogram
SCG	Seismocardiogram
Suffix-graph	The device (e.g., electrocardiograph) that measures a physical quantity and generates the record related to that physical quantity (e.g., electrocardiogram)
Suffix-gram	(e.g., electrocardiogram) The signal or the visual record that the device (e.g., electrocardiograph) produces

1. Introduction

The title of this book is *Pervasive Cardiovascular and Respiratory Monitoring Devices: Model-Based Design*. Let us first explain each term and concept in the title. We will provide details about each of these concepts later in the chapter.

Pervasive devices are devices that are used daily with the goal of continuous noninvasive monitoring of quantities of interest. Monitoring refers to data acquisition, measurement, estimation, and potential visualization of the quantity of interest over time. Let us consider an example of monitoring heart rate and arrhythmias using an electrocardiograph. An electrocardiogram (ECG) signal is continuously acquired, and the algorithms for heart rate estimation and detection of arrhythmias are executed continuously as well. Therefore, monitoring includes acquiring the biomedical signal and processing the signal to estimate/detect quantities of interest over time. In this book, cardiovascular monitoring includes monitoring cardiac activity as well as monitoring blood pressure. Respiratory monitoring is related to monitoring breathing rate and breathing rhythm as well as volume. Measurement of oxygen saturation is described as well. Many of these monitoring devices nowadays are designed to be wearable devices. Wearable devices are devices that a user wears potentially for a more extended period (for several days in some cases), containing technology for sensor data acquisition, processing, and communication with other devices or computers. We will discuss electronics that is a part of biomedical devices as well as several types of biomedical devices.

When discussing the design of a particular device, we will follow several steps, including understanding the problem and issues in measurements, analyzing signals and typical types of noise/interference, and understanding commonly used performance metrics, standard algorithms, and hardware. All these factors are complex and affect each other in different ways that are difficult to comprehend without implementing the device and performing measurements. Therefore, this book uses a model-based approach to focus on modeling system components and their interactions. Next, through simulations, we evaluate performance metrics and perform sensitivity analysis of the parameters to understand what parameters or design choices affect the performance of the device the most.

In this introductory chapter, we explain what pervasive healthcare is and the difference in designing devices for pervasive healthcare compared to the traditional design of medical devices. In addition, we introduce biomedical devices for cardiac and respiratory monitoring. We also explain the importance of modeling and simulation in engineering design. Components of biomedical devices will be introduced next. At the end of the chapter, we will describe the book's content.

2. Pervasive devices

2.1 Pervasive healthcare monitoring

In Varsney (2009), *pervasive healthcare* is defined as healthcare to anyone, anytime, and anywhere. In this book, we are mainly interested in pervasive healthcare for home healthcare monitoring and long-term monitoring, for example, in nursing homes.

Nowadays, healthcare is facing a number of challenges because obesity is growing, many people have a poor diet and are mainly sedentary during the day, the population is aging, and so on. A significant percentage of healthcare expenses is spent on chronic disease management, which opens immense possibilities for developing devices that can be used at home. Pervasive healthcare has the potential to:

- Improve preventive care because it will allow people to be continuously monitored
- Address the shortage of healthcare professionals by adding automation and continuous monitoring
- Reduce overall healthcare costs Varsney (2009)
- Actively involve patients as a part of the healthcare system where their use of technology has the potential to significantly improve adherence to monitoring, treatment, and medications (Vahdat et al., 2014).

Applications of pervasive healthcare include:

- Monitoring patients with chronic conditions
- Monitoring the elderly
- Monitoring patient behavior
- Stress monitoring
- Monitoring social activities
- Continuous monitoring of multiple physiological parameters

Pervasive healthcare will improve both wellness and disease management and provide support for independent and assisted living. We are interested in technologies that can be used at home and without supervision. The home environment is where people spend most of their time and where most chronic conditions are managed. Therefore, it makes sense to bring technology into that environment. Most homes already have some medical devices, such as blood pressure monitors. We will discuss traditional devices, such as blood pressure monitors, but we will also present newer technologies, such as wearable patches for ECG monitoring.

Several definitions related to health monitoring are introduced next.

"*Preventive care* is any medical service that defends against health emergencies. It includes doctor visits, such as annual physicals, well-woman appointments, and dental cleanings" (Amadeo, 2020).

"*Acute conditions* are severe and sudden in onset. This could describe anything from a broken bone to an asthma attack. A *chronic condition*, by contrast, is a long-developing syndrome, such as osteoporosis or asthma" (Vorvick, 2019).

Outpatient care or *ambulatory care* includes care that does not require hospitalization. *Ambulatory monitoring* is the monitoring of patients done outside of the hospital. Therefore, we are interested in devices used for ambulatory monitoring. Biomedical devices in this book will include both medical and wellness devices.

2.2 Medical devices

A medical device is any instrument, appliance, or software intended for people to diagnose, prevent, monitor, treat, or alleviate disease, injury, or similar (Medical Device Overview, 2018). Medical devices are used in ambulatory monitoring to help patients monitor their health variables. Fitness and lifestyle wearable devices are for individuals that wish to keep track of their activity levels and vital signs. These devices are less focused on providing high accuracy in monitoring and they should not be used to diagnose or manage health conditions.

Based on recent FDA guidelines, the main difference between medical and *wellness devices* is in their intended use. "A general wellness product has (1) an intended use that relates to maintaining or encouraging a general state of health or a healthy activity, or (2) an intended use that relates the role of healthy lifestyle with helping to reduce the risk or impact of certain chronic diseases or conditions and where it is well understood and accepted that healthy lifestyle choices may play an important role in health outcomes for the disease or condition" (General Wellness, 2019).

As the main difference between wellness and medical devices for monitoring vital signs and other relevant quantities of interest is in their use, in this book, we will describe and model devices that can be therefore used for both applications.

2.3 Pervasive healthcare monitoring devices versus traditional medical devices

The major differences in designing devices for pervasive healthcare use in comparison with designing traditional medical devices include:

1. Pervasive devices need to address unique challenges related to how and where the monitoring is done, including motion artifacts, unsupervised data collections, and non-trained users of the devices.
2. Pervasive devices are built for use by a nonspecialist and therefore have simple interfaces and are easy to use. They are also designed to require minimal maintenance, and therefore, the focus is on extending battery life, small form factor, and robust design.
3. Wearable devices are commonly battery-operated, and therefore there is a reduced risk of a hazard compared to AC-operated devices.
4. Accuracy requirements are commonly less stringent with pervasive devices than with traditional medical devices.
5. If the pervasive healthcare monitoring device belongs to the lifestyle monitoring device, then it requires fewer approvals than medical devices.

2.4 Issues in pervasive health monitoring

Some of the issues in pervasive monitoring include unsupervised data collection and privacy.

Unsupervised data collection means that the data is collected by the patient himself or herself. Some devices require following specific procedures before and during the measurements, which the patient might not be aware of or forget to follow. An example is a blood pressure device that requires that the patient has been still for several minutes before the measurements. Therefore, unsupervised data collection can result in low-quality data. As such, devices and systems that rely on unsupervised data collection should include algorithms that assess the quality of raw data and processed data, learning algorithms that can identify and quantify noise, artifacts, activities, and algorithms to enhance noisy signals. In this book, we will mention some criteria for determining signal quality. However, machine learning and signal enhancement algorithms will not be part of this book.

Privacy is a significant issue in pervasive computing. As the system collects more information about the subject being monitored, there is a danger of misusing this information against the subject. Subjects need to be able to trust the manufacturers of pervasive devices that their privacy will be protected and not to turn off their devices as soon as their tasks are completed (Varsney, 2009).

3. Biomedical devices for cardiac and respiratory monitoring

3.1 Definitions related to measurements and signals

In this section, we will provide several definitions of the terms used in the book. The definitions are mainly related to measurements, signals and systems.

3.1.1 Measurements

In Gupta (2012), the quantity is defined as: "the property of phenomenon, body or a system to which a number may be assigned with respect to the reference. The reference can be a unit of measurement, a measurement procedure, or a reference material."

For example, the value of systolic blood pressure is 120 mmHg. Here, systolic blood pressure is *the quantity*, and the *value* of this quantity is 120 mmHg. The unit of this physical quantity is "mmHg" in this example.

Next, we will introduce several definitions related to measurements. *Measurement* is a process of experimentally obtaining the values of a physical quantity. In Gertsbakh (2003), the measurement is defined as "the assignment of a number to the measurand, using special technical means (measuring instruments) and a specified technical procedure." So this means that to perform the measurement, there needs to be a phenomenon or system in which property we are interested in measuring, the measurement instrument or device, and the measurement procedure or the protocol. The m*easurand* is the quantity that we intend to measure. So, the measurand does not represent the measurement value. The *result of the measurement* or *measured value* is the estimate of the value of the measurand. An *estimate* is defined as a calculated approximation of the value of a physical quantity. The caret ˆ will be used to denote for "estimate of." For example, $\hat{x}$ is used to represent the estimate of x. *Direct measurement* is the measurement where the quantity of interest is measured directly. An example of direct measurement is measuring a length of an object using a ruler. *Indirect measurements* allow for estimating the variable of interest that cannot be assessed or measured directly based on measuring other quantities and using a known relationship between the quantities to estimate or compute the variable of interest. The measurement model needs to be defined to estimate the value of the measurand. The model can be a mathematical expression, or it can be in the form of an algorithm. So, when we estimate blood pressure, our measurand is the arterial pressure that we intend to measure. We measure the pressure of the deflating cuff directly. The blood pressure is then indirectly estimated by applying an algorithm or a model for calculating the blood pressure.

The accuracy is defined as the degree of closeness of a measurement or estimate to its true value. The *error* is then defined as the difference between a computed estimate or measured value and its true value. *Measurement* can also be defined as an act of estimating the true value of the measurand (Willink, 2013).

Calibration is the process of determining the deviation of a measurement from an established standard of known accuracy.

3.1.2 Signals and systems

In this book, a *signal* is defined as a time-varying physical quantity $x(t)$. The patterns, including time and amplitude of variations, give us information about the behavior or nature of some phenomena. *A processed signal* is a signal obtained from one or more signals which have been processed by some mathematical algorithm. *A filtered signal* is obtained from another signal processed by a *filter*. The filter is a component or a process that removes some unwanted features from the input signal.

We use the term pulse in two ways. In the biomedical field, the *pulse* is defined as arterial palpation that results from the heart's beating. In signal processing, the *pulse* is defined as a signal with rapid change from the baseline and rapid return to the baseline. A commonly used pulse is a rectangular pulse. Many physiological signals, such as PPG, can also be seen as a pulse during one cardiac cycle since there is a rapid rise from and then drop to the baseline. The *baseline* represents the voltage level from which the pulse is initiated.

We also use the term *wave* in the book in two ways. One is related to a signal that contains a series of electric or electromagnetic pulses, such as a sine wave. We will also see that pseudo-periodic pulses of some biomedical signals are also called waves, such as the P wave of an ECG signal. These waves occur during each cardiac cycle and repeat with some possible variations. We will also use the term wave to define the signal propagated through the medium. An example includes the pressure wave propagated from the heart through the arterial tree.

The *system* is defined as a set of interacting elements, principles, and/or procedures that act according to a set of rules to perform a certain task. The system has a different meaning depending on the scale we consider it. The cardiovascular system is an example of a physiological system. In signal processing, the system is commonly defined as a process in which the input signal is transformed into the output signal. Similarly, for an electric circuit, the system is commonly defined as a circuit in which the input signal is transformed into the output signal. When discussing modeling the physiological system and the device, the system will include both the device and the part of the physiological system being monitored.

Waveform represents a graphical representation of a varying quantity over time. A *recording* is a process of capturing physical information and storing it digitally for offline processing or analysis.

A *continuous-time signal* $x(t)$ is defined for every value of time t. A *discrete-time signal* $x(n)$ is defined only at discrete times that are multiples of the sampling period. The *sampling period* is the period at which the continuous-time signal is sampled. *Sampling* is a conversion of the continuous signal $x(t)$ into a discrete signal $x(n)$. A discrete-time signal $x(n)$ that takes values from a finite set of real numbers is known as a *digital signal*. We can say that the digital signal is the quantized discrete signal.

A *continuous-time system* is a system that transforms a continuous-time input signal $x(t)$ into a continuous-time output signal $y(t)$. Continuous-time systems are physically implemented using analog electronic circuits, like resistors, capacitors, inductors, and operational amplifiers. The physical implementation of a continuous-time system is known as an *analog system*.

A system that transforms a discrete-time input signal $x(n)$ into a discrete-time output signal $y(n)$, is called a *discrete-time system*. In practice, a discrete-time system is implemented as a *digital system* using hardware and/or software.

3.1.3 Definition of parameters and indices

This book uses the term *index* to represent an equation expressing the relationship of multiple values or properties. Indices define composite measures that are either used as features for machine learning or shown to the medical specialist for analysis.

The term *parameter* is used as a constant or a coefficient in a mathematical equation. In physiologic models, the values that remain constant over measurements such as age, gender, weight, height, and lung capacity may be considered parameters of the model. These parameters are unique for a patient or a specific patient population. They are collected or estimated once and then used in the model. We assume that the parameters we use in the model are true—for example, the patient's age is entered correctly.

Based on Squara et al. (2015), a *variable* refers to a value of a measurable quantity or attribute of an individual or a system measured by a device. Examples of physiological variables are heart rate and blood pressure.

The term fiducial means that something is accepted as a reference. We will use the term *fiducial points* to indicate specific points of interest on a signal.

We will try to avoid terms commonly used elsewhere, such as physiological parameters, to point to a specific quantity being measured or estimated because the parameters are defined in this book as coefficients in equations or values that remain constant during measurements.

3.2 Biomedical signals of interest

The physiological or biomedical signal is collected by a sensor or transducer, which is an element of a measurement system directly affected by a phenomenon or substance carrying a quantity to be measured (Squara et al., 2015). Signals of interest for cardiorespiratory measurements include ECG, PPG and oxygen saturation, oscillometric, ballistocardiogram (BCG), seismocardiogram (SCG), arterial tonometry signal, bioimpedance signals, and phonocardiogram (PCG). These signals are presented in Table 1. These signals are obtained using various sensors that operate on different physical principles and will be described in this book.

The ECG is a measure of heart electrical activity. This is the only electrical signal in Table 1. A PPG is a signal related to the pulsatile volume of blood in tissue and is measured using optical sensors. PPG signal obtained from at least two photodiodes can be used to infer the oxygen saturation levels. The device that measures oxygen saturation is called a *pulse oximeter. The oscillometric* signal is obtained using cuff-based blood pressure measurement devices, and it represents the signal obtained from a pressure sensor during the cuff deflation, during which the volume of the occluded artery changes from zero to its nominal value. The BCG is a signal obtained by a noninvasive measurement of the body motion generated by the ejection of the blood in each cardiac cycle. As BCG is a mechanical signal, it is normally measured using pressure or accelerometer sensors. The SCG signal records cardiac motion transmitted to the chest wall, and it is normally measured using accelerometers. *Radial tonometry* is based on pressing and partially flattening the radial artery on the wrist. The force exerted by the arterial pressure on a pressure sensor represents the arterial pressure signal. *Phonocardiogram* PCG is the signal related to the acoustic vibrations of the heart as well as blood circulation, and it is commonly measured using microphones.

Vital sign monitoring provides insight into the following: heart rate, blood oxygen content, respiratory rate, blood pressure, and body temperature.

TABLE 1 Common cardiac and respiratory signals and operating principles.

Signal	Transduction element or sensor	Operating principle of the transducer	Signal or action observed	Typical measurement site	Technique
ECG	Biopotential electrodes	Ionic conduction → electric conduction	Electrical activity of the heart	For example, a wrist for pervasive measurements	Attaching electrodes
PPG	Photoelectric	Light → current	Blood volume changes in the blood vessels	Finger	Attaching PPG probe
Oscillometric	For example: resistive	Displacement → resistance	Oscillometric pressure signal	Upper arm	Requires the cuff to be inflated to prevent the blood from going through the artery, and measure the pressure while the pressure in the cuff is being released.
BCG	For example: piezoelectric	Displacement → pressure	Motion due to ejection of blood	A sensor under the mattress	Requires the sensor to be placed on or under the person.
SCG	Accelerometer	Displacement → electrical signal that corresponds to acceleration	Mechanical heart motion measured at the surface of the chest	Chest	An accelerometer is placed on the chest
Arterial tonometry	For example: piezoelectric	Pressure → electric potential	Arterial pressure pulse	Wrist	An artery is pressed, and the pressure is observed using a pressure sensor
PCG	For example: capacitive (microphone)	Sound → electrical current	Acoustic vibrations of heart and blood circulation	Chest	The microphone normally touches the chest
Bioimpedance	Bioimpedance electrodes	Current → voltage	Can measure breathing rate	Chest	Current is injected into tissue, and voltage is measured

3.3 Physical examination of the patient with suspected cardiovascular or respiratory disease

Expert clinicians often share the adage that 'the patient's diagnosis is found within their history.' Listening to the patient is key to diagnosing episodic, chronic, urgent, and/or emergent problems. The patient history is often the predominant determinant of a diagnosis; it typically comprises 80% of the diagnosis being made by a clinician. If a comprehensive thorough history is taken, the likelihood of making a correct diagnosis is extremely high. The physical assessment most often just confirms what a clinician is thinking as a result of a thorough history. (*Gawlik et al., 2020*)

If the patient history is the leading determinant in making the diagnosis, then the diagnosis can be made much more objectively by providing the physician with much richer and more objective historical data through continuous monitoring using pervasive devices at home. It is very important to do that at home, where people spend most of their time. In this way, the physician would have a much better picture of the patient's conditions than in cases where data is collected only during doctors' appointments. However, of course, pervasive devices would not be able to measure all the variables that the physician would like to know about the patient.

Here is an example of what the physician might evaluate from a patient with known or suspected cardiovascular disease. Major historical signs of the cardiac disease include chest discomfort, shortness of breath, fatigue, palpitation (the feeling that the heart is pounding or racing), and syncope. In addition, limb pain, edema (excess of fluid), and skin discoloration indicate a vascular disorder (Fang and O'Gara, 2020). In the case of congestive heart failure, it was shown that the combination of the symptoms, such as shortness of breath together with physical examination findings, results in a high predictive value in determining heart failure (Fang and O'Gara, 2020). Some of these signs and symptoms, but not all, could be measured non-invasively using pervasive devices at home.

3.3.1 What signals are used to monitor or diagnose several common cardiac and respiratory diseases

Table 2 shows several chronic conditions and signals that need to be measured to diagnose or monitor these conditions. The table shows only the methods and devices that are easily accessible by the patient. For example, evaluating the severity of *obstructive sleep apnea* requires going into the sleep lab to perform *polysomnography* which is the study of sleep patterns and the body's response to the various stages of the sleep cycle. Examples of cardiac and respiratory signals measured during sleep apnea include the oxygen saturation level and breathing rate. Several other conditions are shown in Table 1, including hypertension, congestive heart failure, stress level monitoring, and chronic obstructive pulmonary disease, together with examples of cardiac and respiratory signals that are being monitored. Please note that this book is about modeling biomedical devices, and therefore, we will not focus on particular chronic conditions in the book.

3.4 Classification of biomedical devices

Types of biomedical devices based on the type of measurements can be classified as:

- Direct versus indirect
- Invasive versus noninvasive
- Monitor versus actuator
- Permanent versus induced
- Continuous versus intermittent

Indirect and direct measurements were defined in Section 3.1. The majority of measurements in the biomedical field are indirect measurements. For example, the systolic and diastolic blood pressure values are inferred from the oscillometric signal obtained indirectly by measuring cuff pressure when the cuff is placed on the arm. Therefore, systolic and diastolic pressures are not directly measured.

TABLE 2   Some chronic conditions and signals that are commonly monitored using pervasive health devices.

Chronic conditions	Definition	Indices and signals monitored or inferred using biomedical instrumentation	An example of monitored cardiovascular signals	An example of monitored respiratory signals
Hypertension	Hypertension is a medical condition in which the blood pressure in the arteries is elevated.	Blood pressure	Oscillometric signal	—
Congestive heart failure	"Heart failure is present when the heart is unable to pump blood forward at a sufficient rate to meet the metabolic demands of the body ..." (Lilly, 2016)	Weight increase, specific respiratory patterns, fluids in the limbs, elevated jugular venous pressure	Observing the height of the jugular vein on the neck	The respiratory signal from the breathing belt
Obstructive sleep apnea	"Sleep apnea resulting from collapse or obstruction of the airway with the inhibition of muscle tone that occurs during REM sleep." (Dorland, 2012)	Polysomnography includes EEG, pulse oximetry, temperature and pressure sensors to detect nasal and oral airflow, respiratory belts, ECG, electromyogram (EMG) sensors to detect muscle contraction in the chin, chest, and legs.	PPG signal	Nasal and oral airflow
Monitoring stress	"Chronical stress has negative health consequences, such as raised blood pressure, bad sleep, increased vulnerability to infections, decreased performance, and slower body recovery." (Segerstrom and Miller, 2004)	Respiratory rate, ECG, EMG	ECG	The respiratory signal from the breathing belt
Respiratory disorders	"Chronic obstructive pulmonary disease (COPD) is any disorder characterized by persistent or recurring obstruction of bronchial airflow, such as chronic bronchitis, asthma, or pulmonary emphysema." (Dorland, 2012)	Lung capacity	—	Volume

Invasive biomedical devices normally require implantable sensors or electrodes in the body. On the other hand, *noninvasive* devices allow for measurements from the surface of the body or even without contact. An example of a noninvasive measurement device is the electrocardiograph.

Monitoring devices monitor or detect chemical, electrical, or physical variables. An *actuator* delivers external energy to the subject's body via or without direct contact.

Biomedical devices might require *injection* or application of electric current or other applied energy to be able to acquire the signal. For example, bioimpedance is measured by

TABLE 3 Classifications of pervasive devices based on different parameters.

	Type	Examples
Availability	Always available	Smartwatch
	Only when in the field or when required	Cameras for breathing rate monitoring
Requires patient's action	No	Monitoring breathing using cameras
	Minimal	Wearable heart rate monitoring (requires recharging the battery and wearing the device)
	Requires proper placement and following measurement procedure	Blood pressure monitoring
Distance	Touching the user (zero distance)	Devices based on electrodes, displacement transducers
	Small distance	Capacitive transducers
	Larger distance	Cameras, radars
Placement	Wearable	Blood pressure device
	Smart spaces/contactless monitoring	Heart rate estimation using cameras
Medical application	Yes	Traditional medical devices
	No	Fitness/wellness and non-medical applications

injecting current into tissue and then measuring the voltage generated in the tissue to respond to the injected current. These signals are called *induced signals,* and they exist roughly for the duration of the excitation. *Permanent signals* exist without any external impact, energy or trigger. The source of the permanent signal is already in the body. An example of such a signal is an ECG. Devices can be classified as devices that measure either induced or permanent signals.

Regarding the frequency of measurements, the systems can be *continuous* and *intermittent.* Intermittent measurements mean that the physiologic variables are measured at particular time instants. Continuous measurements are performed continually normally at the fixed sampling rate.

Several other classifications based on availability, requirements of patient's actions, distance and placement, type of devices, real-time decision making, and type of medical services are shown in Table 3. Examples of devices for each classification are shown in the last column. Please note that even though many parameters are considered, the classification is not comprehensive.

3.4.1 Wearable versus contactless devices

Regarding the placement of the device, the devices that are placed directly on the patient, that are lightweight and can be worn for some time, are *wearable devices.* They are an emerging technology that enables continuous ambulatory monitoring of vital signs during daily life (during work, at home, during sports activities, etc.) or in a clinical environment, with the advantage of minimizing discomfort and interference with normal human activities.

Wearable devices are not suitable for monitoring that takes longer periods, such as weeks or months, because people forget to wear them, the devices become uncomfortable to wear, and their batteries run out and need to be recharged. These issues resulted in many people stopping using their wearable devices within several months of the first use. To address these issues, a new research direction on *contactless monitoring* is gaining popularity. Contactless monitoring allows for monitoring displacements on the human body due to respiration, heartbeat, or blood circulation or monitoring other variables on the skin using cameras, thermal imaging, radars, sonars, or other devices. These devices can measure physiological variables from several millimeters to several meters. They are less accurate in general than wearable devices and can measure fewer variables (e.g., cannot measure the electrical activity of the heart at distances larger than several millimeters). However, they do not require any patient compliance or device maintenance like changing/charging batteries. This book is focused on the devices that are in contact with the patient. Contactless devices are briefly described in Chapter 11.

4. Modeling and simulation

A *model* is a representation of a system, phenomenon, or process using mathematical concepts. The purpose of modeling is to represent the system and its components to evaluate its behavior when subjected to certain inputs and for different values of model parameters. Computer programs are used to describe models and the inputs, conditions, and parameters that are used to set up the virtual experiment. The *simulation* is the examination of the model or the problem by using a computer program without direct experimentation. The simulation starts when the computer program is executed, and the output of the model is calculated for the given inputs, parameters, and conditions. Simulations allow us to determine or predict the behavior of the system under different conditions.

In designing biomedical devices, several types of experiments need to be performed. They include feasibility and pivotal clinical studies. An exploratory or feasibility study is conducted in the early stages of medical device development to evaluate the safety and effectiveness of the device. It usually includes a small number of subjects (less than 40). Pivotal studies are larger studies performed to demonstrate the device's safety and effectiveness within a specific patient population. When mentioning the *study* in the book, we will refer to a clinical study performed to evaluate the device's effectiveness on human subjects without distinguishing between the types of studies.

4.1 Model-based engineering

Traditionally, the design of biomedical devices is not model-based—it is based on the experience of engineers and researchers who design the devices. In this case, the following issues might emerge: there might be a gap between specified performance and performance measured in realistic conditions, data collection is expensive, especially if one needs to cover difficult cases such as different patients' conditions, and the overall system is often not well understood. Sometimes, it is difficult, expensive, and time consuming to run real experiments or perform clinical studies. In addition, it is difficult to run experiments to cover all possible cases and evaluate different situations. Here are some unique challenges in running experiments in the biomedical engineering field:

1. Running studies on human subjects takes a long time and requires obtaining ethics approval, agreements of the patients, collaboration, and interest of the medical personnel taking care of the patients.
2. Very often, the outcomes of the studies need to be evaluated and labeled by an expert, and this evaluation can be very expensive. In some cases, it is necessary to have two or three physicians evaluate the results to quantify the measurement agreement.
3. Some situations and conditions are very difficult to observe and happen rarely. An extreme example is syncope or losing consciousness.
4. Patients are different, and we often do not know why some patients react in different ways to the same stimulus.
5. True reference value might be unknown, and therefore, it is not easy to estimate the accuracy of a device. An example is blood pressure measurement, where the real arterial blood pressure is not known unless the invasive intra-arterial blood pressure measurement is performed.

For all these reasons, it is very important to model the system, where the system includes both the physiological subsystem being evaluated and the device. In addition, it is important to know the morphology of the signal and types of noise in the system to understand what parameters affect the performance of the device.

In the end, the device is designed and evaluated using real experiments. However, the results of the experiments could be better explained if the whole system is well-understood through modeling and simulation.

The model should simplify concepts and allow for predicting outcomes for varying conditions and situations. *Model-based engineering* can be defined as Hart (2015): "An approach to engineering that uses models as an integral part of the technical baseline that includes the requirements, analysis, design, implementation, and verification of a capability, system, and/or product throughout the acquisition life cycle."

4.2 What will be modeled in this book?

Biomedical devices include hardware and software, and we will model both. However, they are not designed to work in isolation. The measurement system includes the patient, environment as well as the device itself. Modeling all of these will allow us to analyze how different parameters of the system affect the accuracy and precision of the measurement. Parameters of the environment include temperature, interference from other devices or AC interference, motion artifacts, and so on. The device's parameters are related to the sensors and components used in designing the device. We will also look at how different physiological parameters affect the device—for example, how very dry skin affects the quality of ECG.

4.3 Simulation tools

A commonly known simulation software is Matlab/Simulink by MathWorks Inc. Simulink models represent mathematic relationships. In other words, the signal flow lines that we draw between each Simulink block transmit variables from the output of the block to the input of the next block. The calculations are done in the blocks themselves as algorithms. As such, Simulink models represent the real environment as a set of algorithms. This is a "white-box approach" in which the known physics of the model is explicitly described. Simscape is a part of Simulink used for representing the physical system by using components that are already prebuilt and have physical meaning. These components are, for example, an operational amplifier or a transducer. In Simscape, all physical parameters are accessible within the model, and connections between components normally have physical meaning (for example, current or voltage in an electrical circuit). Simscape allows for modeling multiple simultaneous physical phenomena, including heat transfer, deformations, acoustics, and similar. All circuits in this book will be modeled using Simscape.

Besides Simulink and Simscape, some programs will be written in Matlab. In this book, no code will be provided. However, the Matlab code and Simulink and Simscape models for every figure in the book resulting from the simulation are posted on the book website.

5. Design cycle

The design cycle followed in Part II of the book, where we will introduce several biomedical devices, is shown in Fig. 1. The design cycle is divided into three stages: analysis, design, and evaluation.

5.1 Analysis stage

The analysis stage includes several important steps. First, it is important to understand the problem well. This requires modeling and simulation of the physical phenomenon and the interaction between the device and the tissue. An example is modeling ECG electrodes, the electrical properties of the tissue, as well as the interaction between the tissue and the electrodes. This model allows us to explore situations such as sweating or motion artifacts.

The next step is to understand the requirements and performance measures. For several devices that we will introduce in Part II of the book, there are standards or guidelines that define performance requirements as well as specify gold standards for comparison. For other devices, there are only recommendations but no standards, and for some that are still in the research phase, there are no well-defined performance requirements. We introduce standards and recommendations for well-established devices in Part II.

We need to understand the scenario in which the device will be used. This brings us to the third step in the analysis stage, which is understanding the signal and types of noise. The type of noise will depend on the application scenario. For example, if we would like to use dry ECG electrodes on the smartphone, we should expect a larger effect from motion artifacts than in cases where we use gel electrodes. Biomedical devices are designed with a specific signal in mind, which determines what kind of processing to apply, what amplifiers and

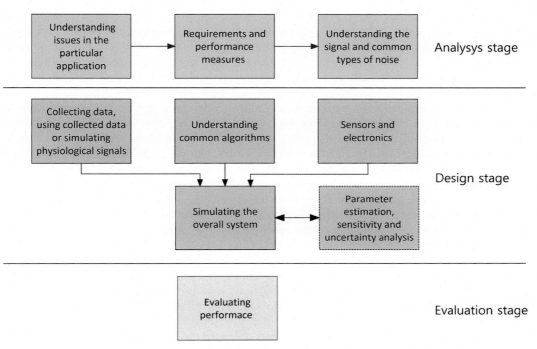

FIGURE 1   The model-based design cycle that will be followed in the book.

filters to design in hardware, etc. For example, for an ECG signal, it is important to know its amplitude and the DC baseline drift range to design appropriate amplifiers and filters. Next, it is important to know the frequency range of the signal of interest. For example, the frequency range of the ECG signal in clinical applications is between 0.05 and 150 Hz.

5.2 Design stage

The design stage involves the design (in this book, modeling) of hardware and software. In this book, we will model hardware in Simscape. The current trend in biomedical instrumentation and instrumentation, in general, is that the devices have a minimal number of analog components needed to convert analog sensor signals into digital data that can be further processed in software on a processor or a microcontroller. Therefore, we will model transducers, conditioning electronics, and analog-to-digital (A/D) converters. Hardware should also be designed with the algorithms/software in mind. If some specific features from the signal need to be extracted in software, the hardware should be designed to facilitate that. For example, if we are interested in the time difference between the specific points on the signals, then the A/D converter and the antialiasing filters must be designed to support a high sampling rate to provide enough resolution for measuring the time difference. This is why it is important to understand algorithms and how the parameters and features of the algorithms interact with the hardware before the hardware is designed. In software, algorithms will be implemented in Matlab. The physiological system, together with the hardware and algorithms, will be modeled. Data will be generated from the physiological model and used to estimate the system's parameters in the design stage. All of these blocks will be simulated together.

Understanding the applications, requirements, data generation process through modeling and simulation, signal morphology, type of noise, algorithms for data processing as well as hardware and its limitations are all necessary steps to design a device. However, biomedical devices are complex and include many parameters that need to be tuned. This is done in the next block in the design cycle, which includes parameter estimation, sensitivity analysis, and uncertainty quantification. This block is bounded by a dotted line meaning that we do not always apply it when describing devices in Part II of the book. *Parameter estimation* is the process of adjusting unknown parameters of the model to reduce the error between the estimates and the reference measurements. It involves tuning values of the parameters of the model so that its output matches the reference values. To perform parameter estimation, optimization needs to be performed to minimize the error or optimize some other parameter of interest (e.g., the frequency response).

Sensitivity analysis answers the question of what the most important parameters in the model or system are by looking at how the change of each parameter affects the output. Sensitivity analysis for complex nonlinear systems that include hardware components and algorithms is normally done using Monte Carlo simulations. The analysis is done by following these steps: identify the probability distribution of each parameter of interest, generate samples from these distributions, and compute the output value of the model based on these parameters. After this procedure is repeated many times, we can measure the association between the output and each parameter. Sensitivity analysis allows us to reduce the dimension of the model by fixing unimportant parameters and to focus the design effort on parameters and parts of the system that affect the output the most.

Uncertainty propagation allows us to provide estimates of the quantity of interest (e.g., the accuracy of the device) that are not just single numbers (point estimates) but also includes uncertainty quantification through confidence intervals or some other statistical measure. Uncertainty propagation normally involves (1) characterization of the sources of uncertainty, (2) propagation of the uncertainties from the model input through the model itself, and (3) analysis of the model outputs.

To better understand the difference between sensitivity analysis and uncertainty propagation, let us consider a design of a blood pressure device. The model of the system can include the physiological model of tissues and arterial blood flow, a model of the cuff, and a model of dynamic changes in the volume of the artery during cuff inflation/deflation. In addition, the pressure sensor and conditioning electronics, as well as the algorithm, are modeled. These sub-models have many parameters that need to be tuned. Tuning of the parameters is done through parameter estimation. Sensitivity analysis determines the most important parameters in the algorithm and hardware. Uncertainty propagation allows us to compute systolic and diastolic blood pressure with confidence intervals.

5.3 Evaluation stage

After the device is modeled and the parameters are estimated, we need to test the device's performance. Normally, data will be collected during pilot or clinical studies. However, in this book, we will generate new data from our model under different conditions and evaluate the performance of the system in that way. Evaluation of performance can be done under different effects, including the effects of physiological parameters as well as the effects of respiration, posture, and motion on the accuracy of the device.

6. Biomedical device components and design issues

6.1 Components of the system

A biomedical device is shown in Fig. 2. A *transducer* is a device that converts one energy source into another. For example, a strain gauge converts displacement (mechanical energy) into resistance change. The transducer in wearable devices is the component that is normally in contact with the subject. *Sensors* convert the output of the transducer into current or voltage signals. For example, the strain gauge is followed by the Wheatstone bridge to convert the resistance change into the voltage change. The sensor that can produce an analog signal at its output (usually in volts) is called an *analog-output sensor*, while the signal that produces a digital signal at its output is called a *digital-output sensor*. Sometimes, engineers use the term digital sensor for the digital-output sensor. However, this term is not accurate because there is normally a lot of analog processing inside the sensor, and therefore, we will refer to these sensors as digital-output sensors.

The analog signal from the transducer/sensor usually needs to be preprocessed (conditioned) before it can be converted into the digital domain. Analog processing includes amplification, filtering, antialiasing filtering, and in clinical devices, signal isolation. *Amplifiers* provide high input impedance that is normally needed when connecting transducers to the conditioning circuits as well as they amplify the signal so that it can cover the input range of the A/D converter. Various types of amplifiers will be covered that are suitable for different transducers. For example, transimpedance amplifiers are commonly used to amplify the current from the photodiode and perform the current-to-voltage conversion. *Analog filters* limit the frequency range of the signal. One of the filters is an antialiasing low pass filter used for preventing aliasing effects when the signal is sampled. Filters are used for removing

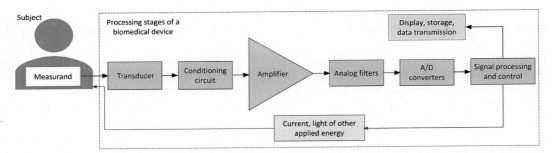

FIGURE 2   Processing stages of a biomedical device. The blocks shown in *blue* represent typical blocks found in monitoring devices. The block in *yellow* is related to providing energy to the circuit. Blocks that are light grey are related to storage, communication, power consumption, and all other operations not related to processing or measurements.

baseline drift (high pass filters), removing 50 Hz or 60 Hz interference (notch filters), and so on. Signal isolation is needed if the device is connected to AC power to protect the patient and the device in some cases. We will not discuss signal isolation in this book as we assume that the devices are battery-operated.

The next component is the *analog-to-digital (A/D) converter* that converts an analog signal into a digital that a microcontroller or a processor will process. Nowadays, the majority of processing, including signal processing and machine learning algorithms, is done in software on a processor. The processor is also responsible for the control of the system, storing the data, communication, and interface.

In Fig. 2, blocks that provide energy to the tissue or feedback to the user are colored yellow and might not be present in some biomedical devices. Some devices need to apply energy to the tissue to induce the signal and measure the response. For example, to measure bio-impedance, the current is injected into the body through the electrodes and then the voltage is measured over the same electrodes or electrodes placed nearby. Optical sensors are based on LED-photodetector pair where the LED emits light to the tissue, and the photodetector detects the tissue response. Some devices also provide a feedback to the patient in the form of a signal that can stimulate or alert the patient.

Power-related circuitry includes battery management, power management circuitry, and the battery. Battery management includes a battery charger and possibly a battery protector. Power management blocks are voltage regulators that can convert the voltage from the battery into voltage levels needed for the operation of the processor, sensors, and other components. A graphical interface is a display for presenting the results. Devices contain memory for storing the data and some parameters.

6.2 Multimodal monitoring devices

Unlike traditional devices that monitor one signal, nowadays, devices that can measure multiple signals are becoming pervasive. Some of these devices are used to measure and present different variables separately, like most smartwatches nowadays. In other cases, sensor fusion is performed to develop a better estimate or increased accuracy in monitoring some quantities of interest.

A modern multimodal monitoring device is shown in Fig. 3. Two types of sensors are shown: sensors that measure environmental conditions and sensors that measure physiological signals. Sensors shown in Fig. 3 include PPG, accelerometer, bioimpedance sensor, ECG, and microphones.

Communication between digital-output sensors as well as peripheral components with the microcontroller can be done through some of the most common serial, such as the Serial Peripheral Interface (SPI) or the Inter-Integrated Circuit (I^2C) protocol. A microcontroller is responsible for interfacing with sensors and reading sensor data at the specified sampling rate.

The other part of the device (grey blocks in Fig. 3) includes components that are not directly related to sensing and monitoring. Communication with the personal computer or smartphone is commonly done through Bluetooth communication protocol. Other components include memory, power, and battery management unit, as well as the battery, USB interface, and touchscreen display.

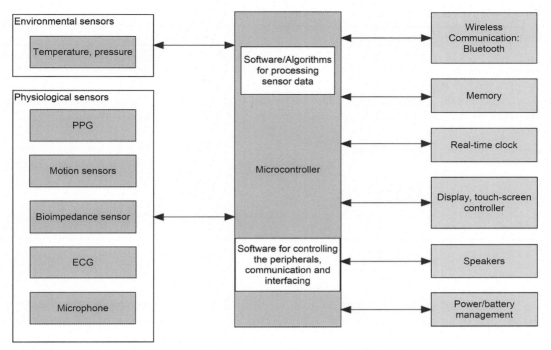

FIGURE 3   Multimodal monitoring device with multiple sensors and microcontroller. Light blue blocks are directly involved in measurement and processing while the light grey blocks are responsible for other operations including communication, storage, and power management. White blocks inside the microcontrollers are related to software processing.

6.3 Issues in designing biomedical devices

Issues that one encounters when designing biomedical devices are listed in Table 4. Normally, measured signals have a very low amplitude. They are also corrupted by noise and subject to baseline drift. Baseline drift is a low-frequency variation of a signal caused by interference or motion artifacts that results in varying baseline voltage. Therefore, the signal must be first filtered to remove the noise and the baseline drift and then amplified. Interference from other organs affects the signal of interest. For example, ECG signal amplitude is modulated by respiration. External interferences include power line interference, interference from other devices, and ambient interference. Power line interference results in 50 Hz or 60 Hz signal that is added to the signal of interest. Sometimes, this interference signal has an even larger amplitude than the signal of interest. Commonly it is suppressed using an instrumentation amplifier with a large common mode rejection ratio and 50 Hz or 60 Hz notch filter. Ambient light changes can affect the PPG signal. Physiological variability means that physiological signals and parameters vary over time. For example, systolic blood pressure can vary from minute to minute, and therefore, it is difficult to estimate it accurately. Motion artifacts affect the quality of all biomedical signals causing noise bursts and baseline drifts

TABLE 4   Issues in designing biomedical devices.

Issue	Description	Example
Very low amplitude of the signals	The amplitude of the signal can be very low, and it depends on the measurement site and the type of sensors.	The amplitude of the PPG signal can be very small, especially when the PPG probe is not placed on the finger
Interference from other organs	When measuring signals on the body, it is common for one physiological signal to affect the other	The breathing signal modulates ECG or PPG and appears as a baseline drift
External interference	Electromagnetic interference	Can induce interference at the power-line frequency (50 Hz or 60 Hz)
Physiological variability	Signals vary in time	Blood pressure can vary significantly from minute to minute
Motion artifacts	Motion causes baseline changes and/or bursts of noise	Movement of a finger with PPG probe attached

that are difficult to filter out. Some devices or configurations are more sensitive to motion artifacts than others. For example, dry ECG electrodes are more sensitive to motion artifacts than gel electrodes.

7. Organization of the book

7.1 Topics covered

The book is divided into two parts and has 11 chapters in total. The organization of the chapters and their relationship is shown in Fig. 4. Each block represents one chapter. The introduction and conclusion are in light yellow; chapters in Part I are in green and Part II in blue.

Part I of the book deals with components of biomedical devices and concepts related to performance evaluation and sensitivity analysis. Chapter 1 introduces performance requirements, sensitivity analysis, and uncertainty quantification. We provide examples of applying these concepts to electrical circuits in Chapter 4. Chapter 2 introduces displacement transducers such as strain gauges, capacitive and piezoelectric transducers, photodetectors, and electrodes. Chapter 3 introduces conditioning circuits, including amplifiers and filters needed to remove the noise and baseline drift and amplify the signal to utilize the full range of the A/D converter. A/D and D/A converters are presented as well in Chapter 3.

Transducers and electronic components are put together in Chapter 4 to understand how they interact and be able to simulate the device and evaluate its performance. Chapter 4 starts by describing ways to obtain data to be used to evaluate the devices. Also, it introduces modeling techniques that would allow us to model the interaction between the physiological system and the device to understand what parameters of the model of the physiological

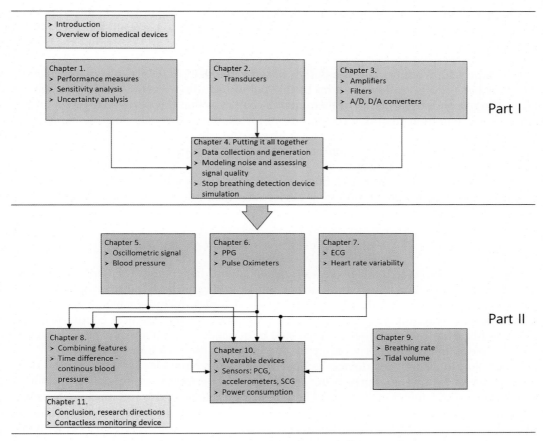

FIGURE 4 Topics covered in the book and associated chapters.

system affect the performance of the device the most. Next, we present models of several circuits together with their sensitivity analysis and uncertainty quantification starting from a relatively simple model of a transducer and ending with the complete system for generating alarms in case the breathing rate is below a certain range.

Part II presents devices for cardiac and respiratory monitoring. The blue arrow that points from Part I to Part II means that there is a direct connection between each chapter in Part I with the chapters in Part II. We decided to divide chapters based on the signals being measured/processed. Therefore, Chapter 5 describes the oscillometric signal used for blood pressure measurements. Chapter 6 introduces the PPG signal and describes the design of PPG devices and pulse oximeters. Chapter 7 introduces ECG systems. In Chapter 8, we show how one can utilize multiple physiological signals in one device. It shows how physiological signals introduced in Chapters 5–7 are related to each other temporally and how one can take advantage of their temporal relationships in designing biomedical devices. We will give details about continuous blood pressure monitoring in Chapter 8.

Chapter 9 introduces monitoring respiratory signals, including breathing rate and volume. Breathing rate based on bioimpedance measurements is discussed as well. We also show how we can use cardiovascular signals to extract the breathing signal and determine the breathing rate.

In Chapter 10, we introduce the design of wearable devices. We also present several signals that have not been described up to this point, including arterial tonometry, SCG, and PCG. The reason for placing all of them in one chapter is that the devices based on these signals are less common than devices based on, for example, the oscillometric signal. Understanding these signals allows us to address the design of a wearable multimodal monitoring system. We describe components such as power management circuits and explain basic low-power design concepts.

The book finishes with a section that summarizes the material covered in the book. It also briefly discusses contactless monitoring devices and methods as they will also become pervasive in the future.

7.2 Scope

The scope within which this book was written is introduced next. The book focuses on the modeling and design of only several cardiovascular and respiratory monitoring systems used pervasively. This book touches on several medical topics, including blood pressure, heart rate, heart rate variability, arterial stiffness, and heart failure. Medical conditions are not explained in detail—we are mainly interested in biomedical signals that need to be acquired and processed to diagnose or monitor the medical conditions of interest.

Processing in all modern devices is done in software. Nowadays, many medical devices have a suite of digital signal processing algorithms that perform noise reduction and filtering, detection, and parameter estimation in real time. Many devices also run machine learning algorithms to detect and classify patient conditions such as arrhythmias and other conditions. Even though signal processing and machine learning have become necessary components of pervasive biomedical devices, the topics of machine learning and signal processing are so vast that we decided not to introduce and cover them in this book. A Matlab notebook with some practical signal-processing algorithms can be found on the book website. We present only elementary signal processing algorithms for processing biomedical signals in Chapters 5—10.

Other limitations of this book include:

- Certifications and regulatory issues are not addressed at all.
- Safety issues are not discussed as we will assume that the pervasive devices are battery-operated.
- Only a limited number of biomedical devices have been introduced. In addition, even within these devices, we do not show many different implementations. This is done because we want to present the whole design cycle as explained in Section 5, where hardware implementation represents only one of the design steps. Therefore, to describe all the steps needed to be understood when designing a system, we had to limit our presentation to several devices and their most common implementations.

- The book figures based on simulation are obtained from Matlab and Simulink by Math-Works. However, the code is not included in the book. Instead, all the details about the code are given on the book website. MATLAB and Simulink are trademarks of Math-Works, Inc. MathWorks does not warrant the accuracy of the text or exercises in this book. This book's use or discussion of MATLAB and Simulink software or related products does not constitute endorsement or sponsorship by the MathWorks of a particular pedagogical approach or a particular use of the MATLAB and Simulink software.

8. Summary

This chapter introduced the topics of pervasive computing for health applications and different classifications of biomedical instrumentation systems. Then we presented major signals of interest used for cardiac and respiratory monitoring. As this book is model-based, we provided motivations for performing modeling and simulation of biomedical instrumentation systems. The design cycle that includes analysis, design, and evaluation stages is also presented. Components of biomedical devices were discussed as well as issues in designing biomedical devices. The chapter ends with the organization of the book and the boundaries within which the book is written.

The book website includes a step-by-step guide on using Simscape to build a simple circuit model as well as several notebooks that provide an introduction to digital signal processing that will be useful in later chapters.

9. Problems

9.1 Simple questions

1. Describe the differences between wellness and medical devices.
2. What are the significant differences between pervasive and traditional medical devices?
3. SCG signal is caused by (select one answer).
   (a) Mechanical heart motion measured at the surface of the chest.
   (b) Sound vibrations of heart and blood circulation.
   (c) Force due to the motion of blood.
4. Bioimpedance measurement requires current to be injected into the body.
   (a) True
   (b) False
5. Indirect measurements allow for (select one answer.)
   (a) Directly observing the phenomenon of interest.
   (b) Inferring the physiological variables of interest based on the measurements.
   (c) Measuring on the surface of the body only.
6. A transducer is defined as (select one answer).
   (a) A device that converts one source of energy into another.
   (b) A device that converts the physical measurand into an electric output.

7. What are the objectives of uncertainty quantification?
8. What are the objectives of sensitivity analysis?
9. Define the following terms related to:
   (a) Signals: permanent, induced, continuous-time, discrete-time, digital, waveform.
   (b) Systems: analog, digital.
   (c) Measurements: measurand, error, accuracy, direct, indirect, estimate.
   (d) Types of cardiorespiratory signals: electrocardiogram, photoplethysmogram, oscillo-metric signal, ballistocardiogram, seismocardiogram, arterial tonometry signal, bio-impedance signals, phonocardiogram.

9.2 Problems

10. Give an example of a medical device that is, based on the classification provided in Section 3.4, an indirect, noninvasive, monitoring, continuous biomedical device.
11. *Find the research paper that describes the development of a device of your interest and analyze the major design steps and major components of the device.

10. Further reading

The author assumed that the readers have some background at the undergraduate level in circuit design, signal processing, and modeling. In addition, even though this book is not about simulation tools, it relies on Matlab and Simscape. Therefore, below are several resources that the readers can consult to refresh their knowledge and/or learn about the simulation tools. Please note that this list is not comprehensive and that many excellent and relevant resources might have been omitted unintentionally.

Learning Matlab: There are many excellent resources for learning Matlab. Gilat (2014) provides an introduction to programming in Matlab. The study of Kapshe and Jain (2016) introduces Matlab and Simulink and then focuses on toolboxes used for signal processing and machine learning.

Learning Simulink/Simscape: The book (Liebgott, 2016) is recommended to learn Simulink and Simscape. Chapter 3 of Liebgott (2016) guides students on building and evaluating simple electrical and multiphysics models.

Circuit design: From the book of Nahvi and Edminister (2003), please review Chapters 1, 2, 4, 5, 6, 8, and 11.

Modeling: Excellent resource for modeling and simulation that discusses different modeling techniques and provides examples in Matlab is Chaturvedi (2010). For modeling physiological systems and processes, please refer to van Meurs (2011).

Digital signal processing: Very comprehensive signal processing reference is Oppenheim and Schafer (1998).

References

Amadeo, K., "Why Preventive Care Lowers Health Care Costs," Updated May 28, 2019, https://www.thebalance.com/preventive-care-how-it-lowers-aca-costs-3306074, Accessed on January 6, 2020.

Chaturvedi, D.K., 2010. Modeling and Simulation of Systems using Matlab® and Simulink®. CRC Press.

Dorland's Illustrated Medical Dictionary, thirty second ed., 2012. Elsevier.

Fang, J.C., O'Gara, P.T., 2020. The history and physical examination: an evidence-based approach. In: Evidence-Based Physical Examination: Best Practices for Health and Well-Being. Springer.

Gilat, A., 2014. MATLAB®: An Introduction with Applications, sixth ed. Pearson.

Gupta, S.V., 2012. Measurement Uncertainties: Physical Parameters and Calibration of Instruments. Springer.

General Wellness: Policy for Low Risk Devices Guidance for Industry and Food and Drug Administration Staff, September 27, 2019. Food and Drug Administration Document number, p. 1300013.

Gertsbakh, I., 2003. Measurement Theory for Engineers. Springer.

Gawlik, Kate, Mazurek Melnyk, Bernadette, Teall, Alice M., 2020. Approach to Evidence-Based Assessment of Health and Well-Being. In: Evidence-Based Physical Examination: Best Practices for Health and Well-Being. Springer.

Hart, L.E., July 30, 2015. Introduction To Model-Based System Engineering (MBSE) and SysML. In: Delaware Valley INCOSE Chapter Meeting.

Kapshe, S., Jain, S., 2016. Modeling and Simulation Using MATLAB®—Simulink®: For ECE. Wiley.

Liebgott, I., 2016. Modeling and Simulation of Multi-Physics Systems with MATLAB®—Simulink® for students and engineers, second ed. https://github.com/Ivan-LIEBGOTT/Livre_2018a_English/blob/main/Book.pdf. (Accessed 30 March 2022).

Lilly, L.S., 2016. Pathophysiology of Heart Disease: A Collaborative Project of Medical Students and Faculty, sixth ed. Wolters Kluwer.

Medical Device Overview, Food and Drug Administration (FDA)," 09/14/2018, https://www.fda.gov/industry/regulated-products/medical-device-overview, Last accessed on April 24, 2021.

van Meurs, W., 2011. Modeling and Simulation in Biomedical Engineering: Applications in Cardiorespiratory Physiology. McGraw-Hill.

Nahvi, M., Edminister, J.A., 2003. Theory and Problems of Electric Circuits, Schaum's Outline Series, fourth ed. McGraw-Hill.

Oppenheim, A.V., Schafer, R.W., 1998. Discrete-time signal processing, second ed. Prentice Hall.

Segerstrom, S.C., Miller, G.E., 2004. Psychological stress and the human immune system: a meta-analytic study of 30 years of inquiry. Psychol. Bull. 130, 601—630.

Squara, P., Imhoff, M., Cecconi, M., 2015. Metrology in medicine: from measurements to decision, with specific reference to anesthesia and intensive care. Anesth Analg 120 (1), 66—75.

Vahdat, S., Hamzehgardeshi, L., Hessam, S., Hamzehgardeshi, Z., 2014. Patient involvement in health care decision making: a review. Iran Red Crescent Med J 16 (1), e12454.

Varsney, U., 2009. Pervasive Healthcare Computing: EMR/EHR, Wireless and Health Monitoring. Springer.

Vorvick, L.J., "Acute vs. Chronic Conditions," Review Date 2/7/2019, https://medlineplus.gov/ency/imagepages/18126.htm, Accessed on January 6, 2020.

Willink, R., 2013. Measurement uncertainty and probability, Cambridge.

Concepts in performance evaluation and uncertainty analysis

Pervasive Cardiovascular and Respiratory Monitoring Devices
https://doi.org/10.1016/B978-0-12-820947-9.00008-8

1

Variables used in this chapter include:

y_j	the output of a function, model, or a device
y_{REF}	the reference measurements
x	the sample from a random variable X or the input to the function f.
$x^{(i)}$	i^{th} sample drawn from a random variable X during Monte Carlo analysis.
$y^{(i)}$	the output obtained after computing $y^{(i)} = f(x^{(i)})$
x	in bold represents a vector
μ_x	the mean of random variable X
$\bar{x}$	the sample mean
e	the random error
σ_x^2	the variance of random variable X
s	the sample or experimental standard deviation
σ_e^2	the error variance
D_i	the measurement difference
S	sensitivity
p_X	probability density function
F	cumulative distribution function
f	any function
$q_U(x)$ *and* $q_L(x)$	the upper and the lower confidence intervals
z	standard normal random variable
ρ	Pearson correlation coefficient
u	standard uncertainty
u_c	combined uncertainty
U	extended uncertainty
N_p, N_n	the total number of actual positive and negative samples, respectively
n	the number of measured values

1.1 Introduction

This chapter will introduce the sensor model and focus on performance measures. Any measurement aims to obtain the true value of the measurand. If the true value of the measurand is not known, which usually is the case, then the error is also not known. This is the case, for example, with blood pressure measurement, where the true value of blood pressure inside the person's artery is not known. Therefore, we need to look at the agreement between the results obtained from two or more measurement procedures or come up with metrics that can be used to quantify the uncertainties in the measurement results.

Uncertainty quantification is studied in this chapter. *Measurement uncertainty* is defined as a probable interval around the measured value where we believe that the true value is with some probability. Uncertainty differs from error which is defined as a difference between the measured and the true value. To understand the concept of uncertainty, we will review basic concepts in probability.

We will also introduce a framework for modeling and analyzing the models, including model parameter estimation, uncertainty quantification, and sensitivity analysis. Sensitivity analysis will allow us to analyze how the parameters affect the output of the system.

Biological systems are complex, and it is often difficult to measure the quantity of interest directly. Biomedical devices are also complex and include a combination of hardware and software. Therefore, we will represent the medical device as a system with the input $x(t)$ and the output $y(t)$. Their relationship is presented as $y(t) = f(x(t))$ where f is a function that describes the system. An example is a blood pressure measurement system where $x(t)$ represents the pressure signal from the cuff and $y(t)$ represents the estimates of systolic and diastolic blood pressure. Except in rare cases, function f is a highly nonlinear function, and it is difficult to describe it mathematically. Therefore, to analyze the uncertainty and sensitivity of the system, we describe inputs and model parameters as random variables and then use Monte Carlo sampling to sample data points from these distributions. *Monte Carlo* methods are computational algorithms that rely on repeated random sampling to perform a numerical experiment. Using Monte Carlo sampling allows for evaluating the uncertainty in complex systems that include hardware and software components that cannot be described completely using mathematical formulas.

1.2 Basic probabilistic concepts

In this section, we introduce the concept of cumulative and probability distribution functions and define normal, uniform, and Student t-distributions. Next, we introduce concepts of the central limit theorem and confidence intervals.

1.2.1 Distribution functions, mean, and variance

The probability that the random number X is smaller than or equal to some given number x is denoted as $P(X \leq x)$. The cumulative distribution function of random variable X is the probability that $X \leq x$:

$$F(x) = P(X \leq x)$$

The derivative of the cumulative distribution function is called the probability density function:

$$p_X(x) = \frac{dF(x)}{dx}$$

In case X is a discrete random variable, the probability distribution function will also be discrete, and it is called the probability mass function. The probability mass function has non-zero values p_i only at discrete points.

The expected value, or the mean of a random variable X, is defined as

$$E(X) = \int_{-\infty}^{+\infty} x p_X(x) dx$$

In the case of discrete random variables $E(X) = \sum_i p_i x_i$. $E(X)$ is often represented using the symbol μ representing the mean.

For a random variable X, $X - \mu$ represents the deviation from the mean. The average square deviation from the mean is called variance:

$$\sigma_X^2 = E\left[(X - \mu)^2\right] = E(X^2) - (E(X))^2$$

If x_i, for $i = 1, \ldots, n$, are independently and identically sampled, then the variables $\bar{x}$ and s^2 represent sample mean and variance, respectively. They are defined as

$$\bar{x} = \frac{1}{n} \sum_{i=1}^{n} x_i$$

$$s^2 = \frac{1}{n-1} \sum_{i=1}^{n} (x_i - \bar{x})^2$$

Please note that $s/\sqrt{n}$ is called the *standard error*.

1.2.2 Several probability distributions

1.2.2.1 Gaussian distribution

The *Gaussian* or *normal distribution* has the probability distribution function defined as

$$p_X(x) = N(x; \mu, \sigma) = \frac{1}{\sigma(2\pi)^{\frac{1}{2}}} e^{-\frac{1}{2}\left[\frac{(x-\mu)}{\sigma}\right]^2} \tag{1.1}$$

It is entirely determined by two parameters μ and σ that represent the mean and standard deviation of the random variable X.

If x_i, for $i = 1, \ldots, n$, are independently and identically sampled from the normal distribution with the mean μ and standard deviation σ, then the random variable

$$z = \frac{\bar{x} - \mu}{\frac{\sigma}{\sqrt{n}}}$$

has a *standard normal distribution* $N(z; 0, 1)$.

The normal distribution is used to model events that occur by chance, such as variation of dimensions of mass-produced items during manufacturing, experimental errors, and variability in measurable biological characteristics such as people's height or weight. In addition, the normal distributions can be applied in situations when the random variable is the aggregate of several other independent variables in the system, which is often the case in engineering.

1.2.2.2 *Student t-distribution*

If the number of random samples is small ($n < 30$), then the Student *t*-distribution, rather than the normal distribution, should be used. Student *t*-distribution is used to compute confidence intervals and model data with outliers.

If x_i, for $i = 1, ..., n$, are independently and identically sampled from the normal distribution, then the random variable

$$t = \frac{\bar{x} - \mu}{\frac{s}{\sqrt{n}}}$$

has the Student *t*-distribution $t(\mu, s, v)$ where s is the sample standard deviation and $v = n - 1$ is the degrees of freedom. We will show that the Student *t*-distribution is used in computing confidence intervals. When estimating the mean with the small number of samples, the distribution of the sample mean can be approximated by the Student *t*-distribution. Student *t*-distribution should not be used with small samples from populations that are not approximately normal.

1.2.2.3 *Uniform distribution*

The uniform probability distribution applies to both continuous and discrete data whose outcomes have equal probabilities.

The probability density function for the discrete case where X can assume values $x_1, x_2, ..., x_k$ is given by

$$U(x; k) = \frac{1}{k}$$

with mean

$$\mu = \frac{\sum_{i=1}^{k} x_i}{k}$$

and variance

$$\sigma^2 = \frac{\sum_{i=1}^{k} (x_i - \mu)^2}{k}$$

For random variables that are continuous over an interval (c, d), the probability density function is

$$U(x) = \frac{1}{d-c}$$

with mean

$$\mu = \frac{c+d}{2}$$

and variance

$$\sigma^2 = \frac{(d-c)^2}{12}$$

Some applications include:

- Uniform random number generation.
- Sampling from an arbitrary distribution: Sampling from the uniform distribution is the first step in sampling from arbitrary distributions when the inverse transform sampling method is used (Papoulis and Pillai, 2002).
- Quantization error: During analog-to-digital conversion, a quantization error occurs. This error is either due to rounding or truncation, and it is typically modeled using the uniform distribution. The same is with the reading results of digital instruments.

1.2.3 Central limit theorem

Let us consider a case where we have independent random variables X_i, $i = 1, ..., n$. Please note that the variables do not need to be identically distributed. Let us form a new variable X: $X = \sum_i X_i$. Random variable X has the mean $\mu = \sum_i \mu_i$ and variance $\sigma^2 = \sum_i \sigma_i^2$. The central limit theorem states that the distribution of X is *normal* with mean μ and variance σ^2 if n is sufficiently large (Papoulis and Pillai, 2002). Fig. 1.1 shows the empirical distribution

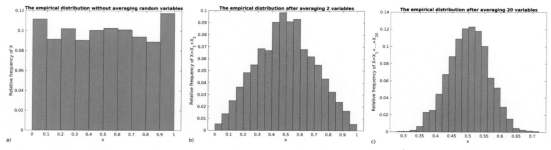

FIGURE 1.1   The empirical distribution of a random variable X obtained after averaging (A) $n = 1$, (B) $n = 2$, (C) $n = 20$ uniform random variables. 2000 samples are drawn for each random variable from the uniform distribution over an interval [0, 1].

obtained after sampling 2000 uniform random numbers with $n = 1$ shown in Fig. 1.1A, $n = 2$ shown in Fig. 1.1B, and $n = 20$ shown in Fig. 1.1C. As expected, the uniform distribution is observed in Fig. 1.1A. However, the distribution of the sum of 20 random variables already resembles the normal distribution.

Noise in biomedical systems is a sum of many different noise sources that are not necessarily normally distributed themselves. If the contribution of the different sources is linear, we can assume that the noise in the system is additive Gaussian noise due to the Central limit theorem. This is a common assumption in many biomedical systems.

1.2.4 Confidence intervals

Several texts pointed out that the measurement result is incomplete without including a statement of the measurement uncertainty or the margin of errors (Willink, 2013). Therefore, it is important to present the measurement results together with the confidence intervals.

Let us define functions $q_L(x)$ and $q_R(x)$ such that, for a variable of interest q is bound as: $q_L(x) < q < q_R(x)$ where x represents samples $x = \{x_1, ... x_n\}$ of a random variable X. Here subscripts L and U mean the lower and upper bounds, and q can be any statistical parameters such as mean or standard deviation. The random interval $[q_L(X), q_U(X)]$ is termed as an *interval estimator*. The interval estimator, in combination with a *confidence coefficient* $1 - \alpha$ is commonly called a *confidence interval*. The term $(1 - \alpha)100\%$ is commonly referred to as the *confidence level*. The confidence level $(1 - \alpha) \cdot 100\%$ can be interpreted as the frequency of times, in repeated sampling, that the interval will contain the target variable q:

$$P\left[q_L(X) \leq q \leq q_U(X)\right] = 1 - \alpha$$

Confidence intervals of the mean of samples that follow normal distribution are calculated like this:

confidence bounds = point estimate $\pm$ (measure of how confident we want to be) $\times$ (standard error)

Here, the confidence bounds are $\{q_L(X), q_U(X)\}$, the point estimate is the variable q, and the measure of how confident we want to be depends on the sampling distribution of the statistics and the confidence level. We will show next how to obtain that measure in two special cases. The standard error of the mean has already been defined in Section 1.2.1, and it represents the estimate of the standard deviation of the sampling distribution scaled by the square root of the sampling size.

For example, let us consider a situation where we would like to estimate confidence intervals of the mean of a normal random variable X with a 95% confidence level. So, our parameter of interest $q = \mu$. Suppose that $x = \{x_1, ..., x_n\}$ are samples from Normal distribution $x_i \sim N(\mu, \sigma^2)$. Empirical mean $\bar{x} = \sum_{i=1}^{n} x_i/n$ can be computed and the standard deviation σ is known and given. To determine information about the unknown mean, we consider the sample mean $\bar{x}$. 95% of the area of a normal distribution lies within two standard deviations of the mean:

$$P\left(\bar{x} - 2\frac{\sigma}{\sqrt{n}} < \mu < \bar{x} + 2\frac{\sigma}{\sqrt{n}} \right) \approx .95$$

The confidence bounds are $\bar{x} \pm 2\,\sigma/\sqrt{n}$. Here, 2 (exactly 1.96) is a coverage factor. Coverage factor k represents the measure of how confident we want to be. The value of the coverage factor of 2 ($k = 2$) defines an interval having a level of confidence of approximately 95%, and $k = 3$ defines an interval having a level of confidence greater than 99%. Fig. 1.2 presents the normal distribution with a mean of 0 and a standard deviation of 1. The gray area shows the 95% area under the normal curve that corresponds to the coverage factor of 2.

If the mean and the standard deviation are not given but computed as the sample mean and standard deviation and samples are drawn from the normal distribution, we should not compute the confidence intervals using the normal distribution. When the sample size is small, there tends to be an underestimation in the standard deviation, resulting in a lower level of confidence. When σ is estimated from the sample standard deviation s, the sampling distribution of $(\bar{x} - \mu)/(s/\sqrt{n})$ follows a Student t-distribution with $v = n - 1$ degrees of freedom. Therefore, we rely on the Student t-distribution to compute the confidence intervals. The Student t-distribution has thicker tails for smaller sample sizes, which means that confidence intervals are wider than those of a normal distribution for the same confidence levels. We denote by $t_{n-1,1-\alpha/2}$ to reflect the $n - 1$ degrees of freedom and probability $1 - \alpha/2$. Let us assume that we are interested in estimating the mean with 95% confidence. In this case:

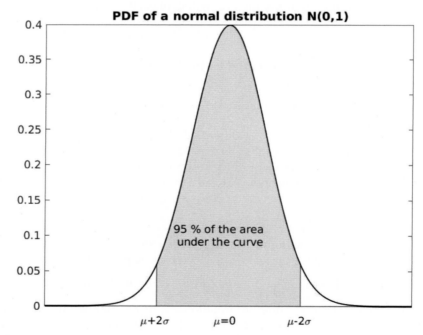

FIGURE 1.2   The probability density function of the normal distribution. The gray area represents a 95% area under the normal curve with confidence intervals $[\mu - 2\sigma, \mu + 2\sigma]$.

$$P\left(-t_{n-1,0.025} < \frac{\bar{x} - \mu}{s/\sqrt{n}} < t_{n-1,0.975} \right) = 0.95$$

This can be further written as:

$$P\left(\bar{x} - \frac{t_{n-1,0.025}s}{\sqrt{n}} < \mu < \bar{x} + \frac{t_{n-1,0.975}s}{\sqrt{n}} \right) = 0.95$$

Therefore, the confidence intervals for the mean at 95% confidence levels are $\bar{x} - t_{n-1,0.025} \cdot s/\sqrt{n} < \mu < \bar{x} + (t_{n-1,0.975} \cdot s)/\sqrt{n}$. In traditional (non-Bayesian) statistics that we discuss here, a 95% confidence interval means that the true parameter q is contained in 95% of estimated confidence intervals. It does not mean that the probability that the true parameter q lies between $q_L(x)$ and $q_U(x)$ is 95% because the traditional statistics requires repeated experiments.

Example 1.1. The mean value of a quantity of interest is measured $n = 50$ times. From these measurements, we obtain the sample mean and standard deviation: $\bar{x} = 2.018$ V and $s = 0.483$ V, respectively.

(a) What are the confidence intervals at 95% confidence levels?
(b) What are the confidence intervals at 95% confidence in case we use the normal instead of Student t-distribution?

Solution:

(a) We will use the approach based on Student t-distribution because both the mean and the standard deviation are estimated from the population. Here, $\alpha = 1 - 0.95 = 0.05$. In Matlab, we can find the confidence bounds for the 95% confidence interval for Student t-distribution. The values are $t_{49,0.025} = -2.01$ and $t_{49,0.975} = 2.01$. The lower confidence bound can be computed as $\bar{x} + (t_{49,0.025}\, s)/\sqrt{n} = 1.882$ V. Similarly, the upper confidence bound is 2.155 V.
(b) When using a normal distribution (instead of Student t-distribution since the number of samples is relatively high), $k = 2$ for a 95% confidence interval. Then, the lower confidence bound is $\bar{x} - 2s/\sqrt{n} = 1.883$ V, where we used s instead of σ. The upper confidence bound is 2.154 V.

1.3 Performance metrics

1.3.1 The model of a sensor or a device

A generic model of a sensor is shown in Fig. 1.3. This sensor model allows us to simulate sensors and/or devices that have linear or nonlinear transfer functions, additive noise, are bandlimited, and can provide discrete and/or analog outputs. The left part of the model (between input and Transducer output) shows the transfer function and the additive noise—we

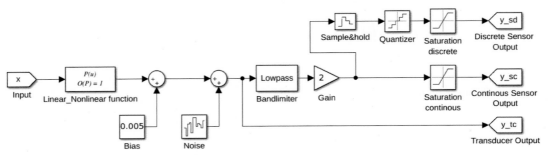

FIGURE 1.3   The generic sensor model.

will call this part the transducer model. In this model, $x(t)$ is sensor input at time step t, $y_{sd}(t)$ is the sensor discrete output at time t, $y_{sc}(t)$ is the sensor continuous output at time t and $y_{tc}(t)$ is the transducer output at time t which is also continuous. The other blocks are explained next.

A Linear or nonlinear function block allows us to introduce any polynomial transfer function into the model. The linear transfer function is $\beta_1 x(t)$. At the output of the transducer, the linear model is $y_{tc}(t) = \beta_0 + \beta_1 x(t) + e(t)$, where β_0 represents the bias that adds a constant value to the input signal. This bias can be temperature-dependent and affected by random variations—however, we will assume that it is constant in this model. The noise $e(t)$ represents white additive noise. We will see in Chapter 4 that we can also include other types of noises. We can also model any other polynomial using this model. For example, the third-order polynomial can be modeled as: $\beta_1 x(t) + \beta_2 x^2(t) + \beta_3 x^3(t)$. Please note that the bias term is not included in the linear or nonlinear function, but it is kept as a separate block in the model.

To limit the bandwidth, we used the low-pass filter. It is designed to take any desired values for the passband and the stop-band frequencies. It models the low-pass behavior of the device itself and can be used to model the antialiasing filter if the sensor with the discrete output is used. The saturation block is used to set the sensor full-scale or the measurement range of the output. The gain is used to amplify the signal. The output of the gain block is typically given in V. Sample and hold, and quantization blocks convert the signal from continuous to discrete.

The continuous and discrete sensor outputs are shown in Fig. 1.4 for the case where the input signal is $x(t) = 1 + 0.1 \sin(2\pi f t)$, where $f = 1$ Hz. The model is linear: $y_{tc}(t) = 0.005 + 1.01 x(t) + e(t)$, where the noise $e(t)$ has a mean zero and the variance of 10^{-6}. The sampling rate of the sample and hold unit is set to 100 Hz, and the number of quantization levels is set to 2^8. The saturation range is between 0 and 5 V. We can see the discrete nature of the signal in Fig. 1.4B due to sampling and quantization.

1.3.2 Continuous variables

1.3.2.1 Error metrics

Assume that the quantity of interest y_i, for $1 < i < n$, is obtained using n measurements with the same medical device. Assume that we can measure the true (reference value) using

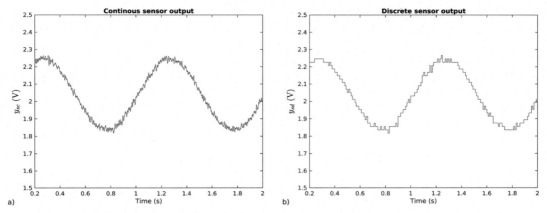

FIGURE 1.4 The continuous (A) and discrete (B) signals at the output of the model shown in Fig. 1.3.

another method and obtain paired measurements: for each measurement y_i, we obtain one reference measurements y_{REFi}. Then, we can define several performance metrics:

- Mean error: $ME = \sum_{i=1}^{n} \left(y_{REFi} - y_i\right)/n$

- Sum of squared errors: $SSE = \sum_{i=1}^{n} \left(y_{REFi} - y_i\right)^2$.

- Mean square error: $MSE = \sum_{i=1}^{n} \left(y_{REFi} - y_i\right)^2/n$. An advantage of using this metric is that it is easy to use. Disadvantages include that it does not consider the direction of error and is sensitive to outliers.

- Root mean square error: $RMSE = \sqrt{MSE}$. The advantage is that the error has the same units as the quantity of interest.

- R^2: $R^2 = 1 - MSE/(\text{var}(\hat{y}))$

- Mean absolute error (MAE) or deviation (MAD): $MAD = \left(\sum_{i=1}^{n} \left|y_{REFi} - y_i\right|\right)/n$. The advantage of this error is that it is robust to outliers, while the disadvantage is that no direction is considered.

- α-trimmed MSE: discard $\alpha\%$ of the largest residuals and then compute MSE.

There is no standard way in biomedical instrumentation to present and compute the measurement error. In addition, measurements from other devices are commonly taken as reference values when computing the measurement error of a new device. For example, it is common to present the mean absolute deviation and the standard deviation between paired measurements made with the new device and the reference device in blood pressure measurements. In pulse oximetry, root mean square error is commonly used to report the error in oxygen saturation measurements of pulse oximeters.

1.3.2.2 Accuracy and precision of linear systems

The random error causes a random variation in output values during the repeated measurements of a variable. *The systematic error* causes an offset between the value of the mean of the output of the measurements and its true value.

To define accuracy and precision, let us consider a simple linear measurement model that would allow for defining these metrics quantitatively. Assume the following model for the paired measurements (y_{i1}, y_{i2}) obtained using devices 1 and 2 (or methods 1 and 2 using the same device) (Choudhary and Nagaraja, 2017):

$$y_{i1} = x_i + e_{i1} \tag{1.2}$$

$$y_{i2} = \beta_0 + \beta_1 x_i + e_{i2} \tag{1.3}$$

where $i = 1, 2, \ldots, n$, and x_i is the true unobservable measurand. The experimenter would collect one measurement from both devices simultaneously and then repeat these *paired measurements* n times. We are interested in observing the difference between the measurements. Parameter β_0 represents the fixed bias of the measurement of the second method (offset) and β_1 is the proportional bias or scale factor of the second method. In (1.2) and (1.3), e_{i1} is the random measurement error of first measurement method and e_{i2} is the random error of second measurement method.

We assume that measurement one is more accurate than measurement 2 and can be considered a reference measurement. The reference measurement device is the device that provides the *gold standard* measurements. The gold standard measurements represent the established standard method of measurement that is not free from errors. We assume that device one is calibrated and, therefore, there is no systematic bias, including proportional and fixed bias, and the random noise has a smaller variance than the noise of device 2. We will also assume that x, e_1 and e_2 are normally distributed, where $x \sim N(\mu_x, \sigma_x^2)$ and $e_1 \sim N(0, \sigma_{e_1}^2)$ and $e_2 \sim N(0, \sigma_{e_2}^2)$. Measurements y_{i1} and y_{i2} are, of course, correlated because we are measuring the same unobservable measurand x. The measurand x does not need to be considered as a random variable—in that case, $\sigma_x^2 = 0$. In the case of indirect measurements, x can come from a previous processing or measurement stage and, therefore, contains an error. We will see an example of that in Chapter 8, where we will model blood pressure as a function of pulse arrival time. The pulse arrival time is not measured directly but is calculated from other measurements and should be modeled as a random variable.

The model is flexible and allows for introducing several ways to represent the errors. For example, $\beta_1 = 1$ can be used if we assume that the methods have the same scale. The disadvantage of this model is that it is a linear model and assumes that all the parameters follow the normal distribution. These assumptions are too restrictive for many real problems—however, they allow for deriving analytic expressions for the metrics of interest.

From the above equations, the measurement difference D_i can be written as

$$D_i = \beta_0 + (\beta_1 - 1)x_i + e_{i1} - e_{i2} \tag{1.4}$$

This difference is used, for example, to compute the MSE introduced in Section 1.3.2.1. We can see that the difference between the two measurements depends on several factors, including the value of the true measurand x_i, as well as the bias, scale factor, and measurement errors.

1.3.2.2.1 Precision

The precision of a measurement method is linked to the size of the random errors in measurement. The error variances σ_e^2 measures the variability in the observed variable y around its error-free value. Therefore, σ_e^2 is the measure of the precision of the method/device. Method 1 is more precise than method 2 if $\sigma_{e1}^2 < \sigma_{e2}^2$. *Random Error* is defined as an error due to unpredictable and unknown variation in the experiment that causes the randomness of the measurement results.

1.3.2.2.2 Accuracy

In the model presented by Eq. (1.3), β_0 is a constant independent of the true value of the measurand. If $\beta_0 = 0$, the fixed bias is the same in methods 1 and 2. β_1 is the amount of change detected by the device if the unobserved variable x changes by 1 unit and represents the difference in scales of the devices. If $\beta_1 = 1$, same scale is observed in the device. The fixed bias β_0 and the scale factor β_1 cause systematic errors in measurement. The smaller the systematic error is, the better the *accuracy of the device* is. The expected value of the difference (*the systematic error*) is:

$$E[D] = \beta_0 + (\beta_1 - 1)\mu_x, \tag{1.5}$$

and the variance is

$$\text{var}(D) = (\beta_1 - 1)^2 \sigma_x^2 + \sigma_{e2}^2 + \sigma_{e1}^2 \tag{1.6}$$

We can consider the following situations.

- If $\beta_1 = 1$ and $\beta_0 = 0$, then $E[D] = 0$ and $\text{var}(D) = \sigma_{e2}^2 + \sigma_{e1}^2$. This means that there is no bias and that the error of measurements depends only on random errors. We can say that this measurement is *accurate*. If the random errors are small, the measurement is also *precise*.
- If $\beta_1 = 1$ and $\beta_0 \neq 0$, then $E[D] = \beta_0$ and $\text{var}(D) = \sigma_{e2}^2 + \sigma_{e1}^2$. This means that there is a fixed bias and that there are both systematic and random measurement errors. If the bias is large, then the measurement is inaccurate. This situation is shown in Fig. 1.5, where 100 data points y_{i1} and y_{i2} are presented for the case where $\beta_1 = 1$, $\beta_0 = 0.01$, $\sigma_{e1} = 0.001$, $\sigma_{e2} = 0.01$ and $\sigma_x = 0$.
- If $\beta_1 \neq 1$ and $\beta_0 \neq 0$, then $E[D]$ and $\text{var}(D)$ can be computed using (1.5) and (1.6). The systematic error depends on the data through its mean. The measurement error variance depends both on the variance in data and on random errors.

1.3.2.2.3 Sensitivity

Sensitivity is the ability of a method or a device to distinguish small changes at the input. Sensitivity refers to how much the change in any parameter of the system induces the change in the output:

$$S = \frac{\partial y}{\partial x}$$

If the parameter that we consider is an input signal, then sensitivity relates changes in the input to changes in the output. If the system is linear, sensitivity is the same as the scale factor β_1. Also, a steeper slope implies a higher sensitivity. In circuits, sensitivity is affected by several parameters, such as temperature and supply voltage.

1.3.2.3 Hysteresis error

Sometimes, it is of interest to observe the behavior of the system when the input of the system changes randomly or when the input is sequentially increased or decreased over the range of input values. The *range* is defined as the difference between the maximum and the minimum values of the signal. These limits define the operating range of the system. Therefore, both input and output ranges need to be specified.

In a *sequential test*, the input to the device is changed in fixed steps sequentially over the desired range, and the output is observed. The change can be done by increasing (the upscale sequential test producing the output $y_{us}(x)$) or decreasing (downscale sequential test with the output $y_{ds}(x)$) the value at the inputs. In a *random test*, the input values are generated/applied randomly, and the output is again observed. The random test has advantages since it removes the effects of hysteresis and observation errors by assuming that each input value is independent of the previous value (Figliola and Beasley, 2012).

Hysteresis error is defined as a difference in the outputs for the upscale and downscale sequential test for the same input value $e_h(x) = y_{us}(x) - y_{ds}(x)$. Hysteresis error in data sheets is normally reported as the maximum hysteresis error $e_{h_{max}}$ divided by the output range:

$$\%e_{h_{max}} = \frac{e_{h_{max}}}{y_R} \cdot 100\%,$$

where y_R is the full-scale output range.

1.3.2.4 Linearity error

Linearity error $\%e_{L_{max}}$ is reported as the maximum deviation from a linear curve divided by the output range. We define two linearity errors based on how the linear curve is obtained. In the *terminal method*, a straight line is drawn between the endpoints of the *transfer curve*. In the *best-fit straight-line* method, a tool like Matlab is used to find a linear curve that fits the transfer curve with the smallest mean square error. Therefore, we can define terminal line linearity error and best-fit straight-line linearity error depending on which straight line is used as a reference.

Example 1.2. We will use the sensor model shown in Fig. 1.3 to define linearity errors. The following nonlinear function is used: $y_{tc}(t) = 0.1 + x(t) + 0.08x^2(t) - 0.015x^3(t)$. No noise is added and the gain is set to 1.

(a) Compute the nonlinearity errors
(b) Plot the curves and the errors

Solution: Fig. 1.6 shows the transfer curve of the potential sensor (blue line called output). It is clear that the curve is nonlinear, and it intersects the y-axis at 0.1, which corresponds to the fixed bias of 0.1. The terminal line (red line) is obtained by connecting the endpoints of the transfer curve. The difference between each point of the transfer curve and the terminal line is computed. The linearity error is calculated for the maximum difference. It is shown as a vertical yellow line in Fig. 1.6. The terminal line linearity error is $\%e_{Lt_{max}} = 4.8\%$.

The line of best fit was obtained using the Matlab function for fitting polynomials. It is shown as a purple line in Fig. 1.6, and the maximum difference between the transfer curve and the best-fit straight line is shown as a green vertical line. The best-fit (BF) straight-line linearity error is $\%e_{LBF_{max}} = 4.2\%$. The best-fit straight-line linearity error is normally smaller than the terminal line linearity error.

1.3.3 Categorical variables

In some cases, the devices need to provide a discrete output where the discrete values can represent different classes. If there are only two classes, then this problem is called binary classification. Therefore, the output does not represent a continuous but a categorical variable. Examples include the classification of breathing types based on the respiratory signal, classifications of arrhythmias based on the ECG signal, etc. Let us consider only the binary classification problem in which we refer to classes as positive or negative. The number of times we correctly classified the positive class is referred to as true positive (TP), while the number of times the negative class is correctly classified is a true negative. Misclassifications are false positives (FP) or false negatives (FN). Common metrics include:

1. Recall, sensitivity or true positive rate: $Recall = \frac{TP}{TP+FN} = \frac{TP}{N_P}$, where N_P is the total number of actual positive samples. This metric has important applications in clinical diagnosis. We will use the term recall instead of sensitivity because the term sensitivity is used for a different purpose in this book, as explained in Section 1.3.2.2.
2. Specificity: $Spec = \frac{TN}{TN+FP} = \frac{TN}{N_N}$, where N_N is the total number of actual negative samples.
3. Positive predictive value (PPV), precision: $PPV = \frac{TP}{TP+FP}$.
4. Negative predictive value (NPV): $NPV = \frac{TN}{TN+FN}$.
5. Accuracy in classification is defined as: $Acc = \frac{TP+TN}{N}$, where $N = TP + FN + TN + FP = N_P + N_N$ is the total number of samples.

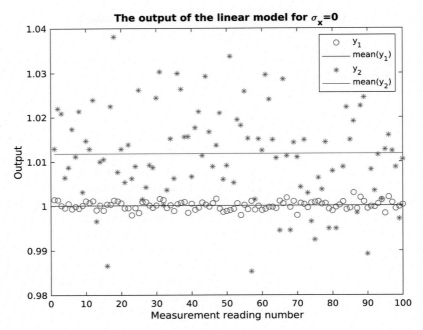

FIGURE 1.5 The repeated measurements y_1 (blue circles) and y_2 (red stars) versus the measurement number for the case where $\sigma_x = 0$. The straight blue line presents the true value, while the straight red line represents the bias in the measurement of y_2.

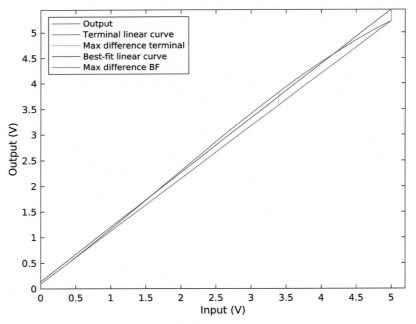

FIGURE 1.6 Linearity error for the generic sensor shown in Fig. 1.3. The sensor model is modified to include a nonlinear transfer function. The blue line represents the continuous sensor output, the red line represents the straight terminal line, and the purple line represents the best-fit straight line. The maximum difference between the transfer curve and the terminal line is shown as a yellow line, while the maximum difference between the transfer curve and the best-fit straight line is shown as a green line.

1.3.4 Metric for relationship

1.3.4.1 Pearson correlation coefficient

When the quantity of interest y is continuous, the Pearson correlation coefficient can be used to measure the relationship/correlation of two measurement methods or devices. The Pearson correlation coefficient measures the linear relationship between two random variables. Suppose we have paired data (y_{i1}, y_{i2}), $i = 1, \cdots, n$. The Pearson correlation coefficient $\hat{\rho}$ can be estimated as

$$\hat{\rho} = \frac{s_{12}}{s_1 s_2}$$

where s_{12} is the sample covariance between y_1 and y_2 defined as

$$s_{12} = \frac{1}{n-1} \sum_{i=1}^{n} (y_{i1} - \bar{y}_1)(y_{i2} - \bar{y}_2),$$

s_1^2 is the sample variance of y_1

$$s_1^2 = \frac{1}{n-1} \sum_{i=1}^{n} (y_{i1} - \bar{y}_1)^2 \text{ and}$$

s_2^2 is the sample variance of y_2 defined as

$$s_2^2 = \frac{1}{n-1} \sum_{i=1}^{n} (y_{i2} - \bar{y}_2)^2.$$

In addition, $\bar{y}_1 = \frac{1}{n} \sum_{i=1}^{n} y_{i1}$ and $\bar{y}_2 = \frac{1}{n} \sum_{i=1}^{n} y_{i2}$

If the variances are known, then the Pearson correlation coefficient can be computed as

$$\rho = \frac{\sigma_{12}^2}{\sigma_1 \sigma_2}.$$

Pearson correlation coefficient has the following properties (Choudhary and Nagaraja, 2017).

1. $\rho = 1$ means y_1 and y_2 show perfect positive, linear correlation.
2. $\rho = -1$ means y_1 and y_2 show perfect negative, linear correlation.
3. $\rho = 0$ means that y_1 and y_2 are not correlated. It does not mean y_1 and y_2 are independent because they may be nonlinearly related.
4. If y_1 and y_2 follow the normal distribution, then $\rho = 0$ indicates they are independent.

1.4 Measuring agreement

In this section, we consider paired measurements between the new measurement device and the reference or gold standard measurement device. Here, we cannot observe the true value of the quantity of interest, and therefore even the gold standard measurements might contain random and systematic errors. If two measurement methods show a high level of agreement, then the new method can be used interchangeably with the old one. Achieving a high level of agreement between the new and the reference device is important when developing a new device that is less expensive, less intrusive, or has better power efficiency for battery-operated devices. The perfect agreement between the two measurement devices with the outputs y_1 and y_2 correspond to the conditions: $\mu_1 = \mu_2, \sigma_1^2 = \sigma_2^2, \rho = 1$ where μ_1 (μ_2) and σ_1 (σ_2) are the mean and standard deviation of y_1 (y_2), respectively. Therefore, it is not enough to observe the correlation coefficient to determine the agreement between the two methods. If y_1 or y_2 gets scaled up or down, the correlation coefficient would not change, but the variance would change, so there would be no perfect agreement. To evaluate agreement, graphical and statistical methods can be used. Graphical methods include the Bland−Altman plot and the scatter plot. Statistical methods include concordance correlation coefficient (CCC), total deviation index, and others (Box 1.1).

1.4.1 Concordance correlation coefficient

The CCC is defined as

$$CCC = \frac{2s_{12}}{s_1^2 + s_2^2 + \left(\bar{y}_1 - \bar{y}_2\right)^2}$$

CCC has the following properties (Choudhary and Nagaraja, 2017).

BOX 1.1

Calibration versus measurement of agreement

It is important to understand that comparing the results of two measurements for the purpose of evaluating the agreement between them is very different from comparing the results of the measurements with the goal of calibrating the device. During calibration, measurement results are obtained from a highly accurate method/device as well as from a device that needs to be calibrated in order to develop an equation or a model that converts the measurements from the new device into the more accurate measurements. In contrast, when measuring the agreement, the devices or methods that are being compared have already been calibrated. The model can be developed as well to transfer the measurement results to achieve better agreement.

1. CCC $= 1$ means y_1 and y_2 perfectly agree.
2. CCC $= -1$ implies a perfect negative agreement between y_1 and y_2.
3. CCC $= 0$ implies uncorrelated measurements.
4. $0 \leq |CCC| \leq |\rho| \leq 1$.

1.4.2 Scatter and Bland−Altman plot

A *scatter plot* visualizes paired observations y_1 in the abscissa and y_2 in the ordinate. We expect data to be scattered symmetrically around the identity line in case of good agreement. The *identity* or *equity line* is the line that forms a 45 degrees angle with the abscissa.

Bland−Altman plot shows the average, $\overline{y}_i = (y_{i1} + y_{i2})/2$, versus the difference $D_i = y_{i2} - y_{i1}$, for $i = 1, 2, ..., n$ (Choudhary and Nagaraja, 2017). Bland−Altman plot visualizes the mean $\mu = E(D_i)$ and the standard deviation of the difference between the measurements $\sigma = std(D_i)$ presented in Eqs. 1.5 and 1.6. *Limits of agreement* at 95% confidence are $[\mu - 2\sigma; \mu + 2\sigma]$. *Outliers* are the points outside of the range $[\mu - 2\sigma, \mu + 2\sigma]$.

In a Bland−Altman plot, we can observe the following.

- If the points are close to the zero line, the methods agree well.
- The large limits of agreements indicate low agreements between the two methods and possibly large random errors.
- The mean of differences $\mu \neq 0$ indicates a systematic error between the measurements.
- The linear trend along the average horizontal line could be caused by (1) unequal scale factors or (2) unequal precision between the two measurement methods.

1.4.2.1 Heteroscedasticity

Heteroscedasticity refers to the change of variance of the difference in measurements D as the average $\overline{y}$ increases. It indicates the variability of D with the magnitude of the measurement. In such cases, the Bland−Altman plot shows a shape where the spread of the points increases with increasing the mean (Choudhary and Nagaraja, 2017).

Example 1.3. Plot the agreement of systolic blood pressure measurements made simultaneously by an observer and an automatic blood pressure measurement device.

Data is provided at https://www-users.york.ac.uk/~mb55/datasets/bp.dct. Even though more data per observer and the device are provided, we only used the first measurements by the observer and the device. There are 80 measurements performed in total.

Solution: The scatter plot is shown in Fig. 1.7A. As can be seen from the plot, data points are mainly on one side of the identity line. Data points are far from the identity line, and also their trend is not parallel to the equality line. These observations indicate that there might be a different scale factor between the measurement methods.

Bland-Altman plot is shown in Fig. 1.7B. The mean difference in methods is significant, and it is 16.3 mmHg indicating a large bias. The standard deviation is 19.6 mmHg. We can see many outliers that represent the points outside of the range $[\mu - 2\sigma; \mu + 2\sigma]$. In addition, it seems that data are scattered more with increased blood pressure indicating the effect of heteroscedasticity. In conclusion, the agreement between the observer's and the device's measurements is very low.

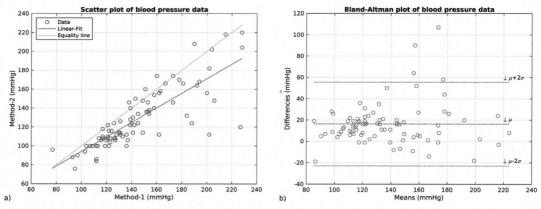

FIGURE 1.7 (A) Scatter plot of blood pressure measurement data from (Bland and Altman, 1999). The line of equality and the line of best fit are also presented. (B) Bland–Altman plot of the same data.

1.5 Uncertainty in measurements

The goal of uncertainty quantification (UQ) is to characterize the impact of limited knowledge on the quantities of interest. We have limited knowledge of every point in our system, including the input data, model and model parameters, physical parameters (such as tolerances of the resistors), and estimations. The limited knowledge can be described at the level of observations, predictions, and model parameters. Regarding observations, we are unable to observe the phenomenon of interest accurately due to.

- The limited accuracy of measurement instruments,
- The high cost of performing accurate experiments and/or measurements,
- The lack of accurate models.

Predictions or estimations result from a combination of observations, modeling, and computations, and each introduces its uncertainties. Models are based on certain levels of approximation to make them manageable. Therefore, they capture only the limited knowledge and never completely represent the phenomenon of interest.

An uncertainty u is an estimate of an interval $\pm u$ that should contain an error. Uncertainty can be presented in multiple ways, as shown in the next example. Measurements results that include a standard error or confidence intervals can be presented as follows for the example of estimating the heart rate:

- Standard error: For example, heart rate estimates are presented as (60 ± 4) beats/minute (mean $\pm$ standard error)
- Confidence intervals: For example, 95% confidence interval for the heart rate is [52, 68] beats/minute.
- Probability distributions can be estimated from data, and then confidence intervals can be derived from the probability distribution.

1.5.1 The workflow of uncertainty quantification

The workflow is divided into three main steps, as shown in Fig. 1.8 (Baudin et al., 2015). Different colors and arrows are used to represent different elements of the system. The blue horizontal line pointing from left to right shows the computation done from the input to the output, while the yellow horizontal line that points from right to left shows that step B represents the inverse problem. In the inverse problem, inputs and outputs to the model are provided, and we need to estimate the parameters of the model. This is a common problem while designing biomedical devices. The following steps for uncertainty quantification are presented:

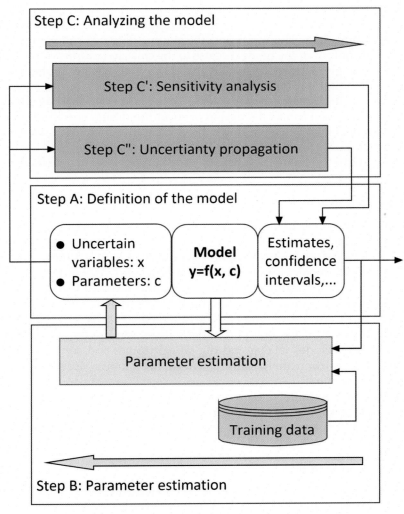

FIGURE 1.8   The workflow of uncertainty quantification.

Step A: The step includes model definition, identification of inputs and outputs, and the parameters of the model. Model is presented as a function $y = f(x, c)$, where the output y depends on input data x and fixed parameters c. Besides the estimates of y, the confidence intervals and different performance metrics can be calculated. Also, one needs to determine the probability distribution over the set of input variables and model parameters. Some variables might be correlated, and this needs to be captured using multidimensional probability distributions. Dependencies between inputs themselves and the parameters need to be modeled as well. Some models contain tens or hundreds of parameters, so expert knowledge in required to assign the appropriate probability distribution and determine the relations among the parameters. If the input data is available, fitting of probability distributions can be performed using statistical or machine learning methods.

Step B: Parameter estimation is performed based on training data. Parameter estimation is the process of learning unknown model parameters to reduce the error between the model output and the output data. The input and the output data are collected, and parameters are fitted so that the output of the model is close to the collected output values for given input data based on some pre-defined criteria. Approaches that can be applied here belong to parameter estimation and machine learning fields.

Step C: In *uncertainty propagation* step (step C''), the aim is to quantify the output uncertainties. It can also be used to aid the design to meet the design specification of the device. With *sensitivity analysis* in step C', we study the effect of the changes in the system's parameters on the output.

After all these steps, the model is functional and can be used in a device or a system.

1.5.2 Uncertainty propagation

Propagation of uncertainty quantifies the effect of the uncertainty in parameters and input variables on the output. For example, we would like to estimate the confidence intervals at the output of the device if we know the confidence intervals in measurements of all of its variables. If we can model our device or system as a function f, where $y = f(x)$, then we can propagate the uncertainties from the input to the output through the function f. Two main methods are used to propagate uncertainties: the perturbation method based on Taylor series expansion and Monte Carlo-based method. The first method can be applied if the function f is presented analytically and if the first partial derivatives can be computed. However, in the case of a system that includes analog and digital electronics and software, it is not easy to describe the system's function analytically. Then, we need to use Monte Carlo-based method for uncertainty propagation.

Uncertainty propagation allows us to compute the estimates or predictions with confidence intervals and compute the uncertainty budget. The uncertainty budget represents the combination of all uncertainties in the system, and it is commonly used to meet the specifications of the device.

1.5.2.1 Perturbation method

The perturbation method is based on the Taylor series expansion of the function f around the mean value of the input parameters. We will consider K dimensional vector x with elements $x = [x_1, x_2, ..., x_K]^T$, where superscript T refers to transpose. The Taylor series expansion of the model $f(x)$ around some point x_0 is

$$f(x) = f(x_0) + \sum_{i=1}^{K} \frac{\partial f}{\partial x_i}\Big|_{x=x_0}(x_i - x_{0i}) + \frac{1}{2}\sum_{i=1}^{K}\sum_{j=1}^{K} \frac{\partial^2 f}{\partial x_i \partial x_j}\Big|_{x=x_0}(x_i - x_{0i})(x_j - x_{0j}) + ...$$

Let us consider a special case where $x_0 = E[x]$. We obtain $E[f(x)]$ as

$$E[y] = E[f(x)] \approx f(x_0) + \sum_{i=1}^{K} \frac{\partial f}{\partial x_i}\Big|_{x=x_0}E[(x_i - x_{0i})] + \frac{1}{2}\sum_{i=1}^{K}\sum_{j=1}^{K} \frac{\partial^2 f}{\partial x_i \partial x_j}\Big|_{x=x_0}E[(x_i - x_{0i})(x_i - x_{0j})]$$

We know that $E[(x - x_0)] = 0$ and $E[(x_i - x_{0i})(x_i - x_{0j})] = C_{ij}$ is the covariance of x_i and x_j, so

$$E[y] \approx f(x_0) + \frac{1}{2}\sum_{i=1}^{K}\sum_{j=1}^{K} C_{ij} \frac{\partial^2 f}{\partial x_i \partial x_j}\Big|_{x=x_0}$$

If we ignore the second degree of the Taylor series expansion, we get $E[y] \approx f(x_0)$. The variance of y is then computed as follows:

$$Var[y] = E[(y - E[y])^2]$$

The variance can be approximated as (Pavese and Forbes, 2009)

$$Var[y] \approx \sum_{i=1}^{K}\sum_{j=1}^{K} C_{ij} \frac{\partial f}{\partial x_i}\Big|_{x=x_0}\frac{\partial f}{\partial x_j}\Big|_{x=x_0} \tag{1.7}$$

If $C_{ij} = 0$ for $i \neq j$, we can get that

$$Var[y] \approx \sum_{i=1}^{K} \left(\frac{\partial f}{\partial x_i}\Big|_{x=x_0}\right)^2 \sigma_{x_i}^2 \tag{1.8}$$

The problem with this method is that derivatives can be challenging to compute, and the covariances need to be estimated. In addition, if we cannot describe the overall model analytically (some subsystems are black boxes), we cannot use this technique.

1.5.2.2 Monte Carlo method

Monte Carlo methods are computational algorithms used for inference in machine learning, optimization, numerical solution of integrals, propagation of uncertainties, and several other applications. The idea behind the algorithms is to sample from a pre-defined or estimated probability distribution repeatedly and then use these samples for estimation, simulation, or other purposes.

The Monte Carlo method for uncertainty propagation is shown in Fig. 1.9. The major components of Monte Carlo Carlo-based methods are:

- The uncertain variables and parameters need to be described by probability distribution functions. These probability distribution functions need to incorporate the correlation between variables and parameters.
- A random number generator or a sampling method needs to be used to allow for sampling from the probability distribution functions.
- Scoring the outcomes needs to be defined to provide the metrics such as quantiles, confidence intervals, etc.
- The last step is an estimate of the statistical error as a function of the number of trials which will not be discussed in this book.

The idea behind Monte Carlo sampling is that one can generate a sample $x^{(i)}$ from a probability distribution $p(x)$ as described in step A of Section 1.5.1, and then computes one output value as $y^{(i)} = f(x^{(i)})$. This procedure is repeated M times where $i = 1, 2, ..., M$. After M itterations, we can estimate the *empirical distribution* of y. Having empirical distribution allows for computing various metrics such as standard deviation, percentiles of interest, etc.

Monte Carlo sampling applies to complex and nonlinear models as well as models not described implicitly using analytical functions. The latter models include software algorithms

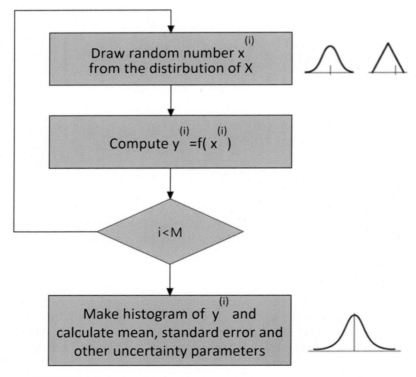

FIGURE 1.9   Monte Carlo sampling steps for uncertainty propagation.

and electrical circuits. Disadvantages include that the probability distributions of the input variables need to be determined and that the method converges slowly.

To obtain metrics from the samples of $y^{(i)}$, one can do the following:

- To get $P(a < y < b)$ let N be the number of $y^{(i)}$ such that $a < y^{(i)} < b$ and then compute $P(a < y < b) \approx \frac{N}{M}$.

- To get σ_Y, compute the sample mean $\bar{y} = \frac{1}{M}\sum_{i=1}^{M} y^{(i)}$ and $\sigma_Y \approx \sqrt{\frac{1}{M-1}\sum_{i=1}^{M}(y^{(i)} - \bar{y})^2}$

- Compute the qth quantile of the empirical distribution of y.

This book mainly uses Monte Carlo-based sampling for uncertainty propagation and sensitivity analysis.

Example 1.4. Let us define a ratio R of AC currents obtained from a photodetector after emitting red and infrared light as:

$$R = \frac{I_R}{I_{IR}}$$

Assume the AC currents at the photodiode obtained after emitting red (I_R) and infrared (I_{IR}) light are measured with the standard deviation of 0.5 μA. The values of I_R and I_{IR} are $I_R = 10\ \mu A$ and $I_{IR} = 20\ \mu A$. Let us assume that the measurements are correlated and that the correlation coefficient is $\rho = 0.7$. What is the variance of the ratio R? Let us assume that the currents I_R and I_{IR} follow the normal distribution. The distribution of R obtained after dividing two normal random variables is also normal.

Solution: Method 1: Perturbation

We will apply Eq. (1.7) for only two inputs:

$$\text{Var}[y] \approx \left(\frac{\partial f}{\partial x_1}\right)^2 \sigma_{x1}^2 + \left(\frac{\partial f}{\partial x_2}\right)^2 \sigma_{x2}^2 + 2\frac{\partial f}{\partial x_1}\frac{\partial f}{\partial x_2}C_{12}$$

Here, $R = y$, $I_R = x_1$ and $I_{IR} = x_2$. We can also compute $\partial R/(\partial I_R) = 1/I_{IR}$ and $\partial R/(\partial I_{IR}) = -I_R/I_{IR}^2$. The covariance between I_R and I_{IR} is $C_{R\_IR} = \rho\sigma_{I_R}\sigma_{I_{IR}}$

$$u_R^2 = \text{Var}[R] = \left(\frac{1}{I_{IR}}\right)^2 \sigma_{I_R}^2 + \left(-\frac{I_R}{I_{IR}^2}\right)^2 \sigma_{I_{IR}}^2 - 2\frac{I_R}{I_{IR}^3}C_{R\_IR}$$

We obtain that $\text{Var}[R] = 0.0185$.

Method 2: Monte Carlo method

The same result $\text{Var}[R] = 0.0185$ is obtained using Monte Carlo sampling. First, we need to sample from two-dimensional multivariate Gaussian distribution and obtain correlated samples $I_R^{(i)}$ and $I_{IR}^{(i)}$, for $i = 1, ... M$, where M is the number of Monte Carlo samples. We are using $M = 100,000$ samples. Then, these samples are passed through the function $R^{(i)} = I_R^{(i)}/I_{IR}^{(i)}$.

Values $R^{(i)}$ are fitted to a normal distribution, and the confidence intervals are computed. The solution in Matlab is shown on the book webpage.

1.5.2.3 *Guide to the expression of uncertainty in measurement*

The Guide to the expression of uncertainty in measurement (Evaluation of measurement data, 2008), commonly referred to as GUM, establishes directions for uncertainty propagating and expression of the uncertainties in measurements in general. GUM divides the uncertainties into two types based on the type of analysis performed on the observations.

- Type A: "method of evaluation of uncertainty by the statistical analysis of series of observations" from (Evaluation of measurement data, 2008),
- Type B: "method of evaluation of uncertainty by means other than the statistical analysis of series of observations" from (Evaluation of measurement data, 2008).

We introduce the following several definitions of the types of uncertainties in the context of GUM. They include (Evaluation of measurement data, 2008):

- Standard uncertainty u is the uncertainty of the result of a single type of measurement, including Type A and/or Type B uncertainties,
- Combined standard uncertainty u_c is used to express the uncertainty of multiple measurement results,
- Expanded uncertainty U is the standard or combined standard uncertainty multiplied by a coverage factor k: $U = ku_c$

In general, the value of the coverage factor k is chosen based on the desired level of confidence to be associated with the interval, as discussed in Section 1.2.4. Typically, k is 2 or 3.

The procedure for measuring uncertainty and uncertainty propagation based on GUM is described next.

If the expanded uncertainty is given at a certain level of confidence for variable x, then the coverage factor needs to be determined based on that level of confidence. Then, the standard uncertainty can be computed as $u_x = U_x/k$. When the uncertainty is given as a standard deviation with no information about the sample size, then: $u_x = \sigma_x$. When the uncertainty is given as an empirical standard deviation s_x and the sample size n is also given, then the expanded uncertainty is computed based on Student t-distribution: $U_x = \frac{t_{n-1,1-\alpha/2}s_x}{\sqrt{n}}$

Combining several standard uncertainties can be done separately to find the standard uncertainty for a single variable of interest. The method recommended in the GUM is to first categorize all the uncertainties as Type A or Type B. The Type A and Type B combined standard uncertainties are then calculated as follows:

$$\text{Type A combined standard uncertainty} = u_A = \sqrt{\sum_i u_{A,i}^2}$$

$$\text{Type B combined standard uncertainty} = u_B = \sqrt{\sum_i u_{B,i}^2}$$

The combined standard uncertainty for the measurement is then determined as: $u_c = \sqrt{\{u_A^2 + u_B^2\}}$.

A summary of steps needed to determine the uncertainty of measurement based on GUM is given below. The steps are.

1. Determine the standard uncertainty for each variable, u.
2. Combine the standard uncertainties separately for the Type A and Type B uncertainties.
3. Combine the total Type A and Type B uncertainties to get the total standard uncertainty in a measurement.
4. Multiply the standard uncertainty in measurement by the coverage factor to get the expanded (total) uncertainty of measurement U for a pre-defined level of confidence.

GUM was expanded in 2008 by providing the expression of uncertainty based on Monte Carlo sampling (Evaluation of measurement, 2008). It follows the methods already described in Section 1.5.2.2.

It is important to consider guidelines when propagating uncertainties and presenting the results. The following items should be typically reported (Evaluation of measurement, 2008):

a) An estimate y of the value of interest;
b) The standard uncertainty u_y associated with y;
c) The confidence level $100(1 - \alpha)\%$ (e.g., 95%);
d) Any other relevant information.

Example 1.5.

(a) For a transducer with the following calibration curve, $y = bx$, estimate the expanded uncertainty for $x = 5.00$, if $b = 1$ with $U_b = 0.01$ and $U_x = 0.05$ at 95% confidence. Assume that all variables follow the normal distribution. Please note that we did not introduce units here and assume that all the values are relative.
(b) Compute the expanded uncertainty at 99% confidence.

Solution:

(a) Let us first compute the uncertainty using the perturbation method. We can follow the steps listed below:
1. Standard uncertainties are computed for $k = 2$ (since the confidence is 95%) as $u_x = U_x/k = 0.025$ and $u_b = U_b/k = 0.005$
2. The combined expanded uncertainty for y at 95% confidence is as follows:

$$u_y = \sqrt{\left(\frac{\partial y}{\partial b}\right)^2 u_b^2 + \left(\frac{\partial y}{\partial x}\right)^2 u_x^2}$$

$$u_y = \sqrt{x^2 u_b^2 + b^2 u_x^2}$$

$$u_y = \sqrt{5.00^2 0.005^2 + 1^2 0.025^2}$$

$$u_y = 0.035$$

$$U_{y_{95\%}} = 0.07$$

(b) U_y at 99% confidence is calculated as:

$$U_{y_{99\%}} = k_{99\%} u_y = 3 \cdot 0.035 = 0.105$$

The same results can be obtained using Monte Carlo sampling—please check the code on the book webpage.

1.5.3 Sensitivity analysis

Sensitivity analysis represents block C' in Fig. 1.8. Sensitivity analysis investigates the relationship between changes in the parameters and the changes of the output of a model or a system. Important properties of sensitivity analysis include:

- It does not need input data and can be conducted based on purely mathematical analysis,
- It identifies the most important parameters in the model in terms of their effect on the output, which allow us to focus our design effort on those parameters, and
- It helps reduce the dimension of the model by fixing unimportant parameters.

Local sensitivity analyzes the effect of x_i to the output while all other input variables x_j, $j \neq i$, are kept at the nominal values. The sensitivity is then determined by computing the partial derivatives of the output functions with respect to the input parameters

$$S_{x_i} = \frac{\partial f(x)}{\partial x_i}$$

Sensitivity is given in physical units. To present the sensitivity in the percentage of change of the output value, one can define normalized sensitivity as:

$$\%S_{x_i} = \frac{\dfrac{\partial f(x)}{\partial x_i}}{\dfrac{f(x)}{x_i}} \, 100\%$$

Normalized sensitivity allows us to observe the change in the output in percentage if the parameter x_i varies by, for example, 1%. Local sensitivity analysis should not be used when the model is nonlinear and various input parameters are affected by uncertainties.

Global sensitivity analyzes the effect of x_i on the output while all other input variables or parameters x_j, $j \neq i$, are varied as well. Parameter space exploration, in this case, is commonly done using Monte Carlo simulations (Saltelli et al., 2008). Global sensitivity analysis allows us to detect how interactions between multiple inputs affect the outputs. Global sensitivity analysis is performed nowadays using variance decomposition methods (Smith, 2014). However, coverage of that method is outside of the scope of this book.

Using sensitivity analysis, one can rank the contributions of the model parameter to the output of the model. In that way, we can better understand how these parameters affect the model outputs and find the most effective way to improve our model/system.

1.5.3.1 *Sensitivity analysis of a cost function*

When doing sensitivity analysis, one can observe the contributions of the parameters not only to the output of the system but also to a user-defined cost function. It involves the following steps.

- Generate samples from a parameter space
- Define a cost function
- Compute the cost function for each combination of parameter values using Monte Carlo simulations and evaluate the results.
- Analyze statistical relations between the parameters and the cost function and determine what parameters affect the cost function the most
- Use the results to modify or improve the model

The cost function can be related to any quantity of interest that we would like to optimize in our design. For example, cost functions can be related to the properties of the signal in the time domain, including the minimum of the signal, the maximum of the signal, the mean, the variance and so on. It can also be related to step response characteristics of the system in the time domain, including the rise time and the settling time.

After the sensitivity analysis is performed, it is useful to analyze what parameters affect the output significantly and what parameters do not affect it. Therefore, one needs to fix unimportant parameters and not consider them for future analysis.

Example 1.6. Uncertainty propagation and sensitivity analysis of a differential amplifier

A differential amplifier is a circuit that amplifies the difference between the input voltages. It is shown in Fig. 1.10. We will discuss differential amplifiers more in Chapter 3. At this stage, it is enough to know that, for the selected values of resistors, the output of this circuit is 10 times larger than the difference between the inputs. Resistors have the following values: $R_1 = 10\,k\Omega$, $R_2 = 100\,k\Omega$, $R_3 = 10\,k\Omega$ and $R_4 = 100\,k\Omega$. Tolerance of each resistor is 1%. For the input voltage difference of 0.2 V, the output is 2 V.

(a) Perform uncertainty propagation, assuming that the resistor values follow a uniform distribution.
(b) Perform sensitivity analysis and present the scatter plot of the output V_O versus the values of the resistors.

Solutions: We generate 1000 random values of the resistance {R1, R2, R3, R4}, assuming that their values follow the uniform distribution. The empirical distribution of the output V_O is

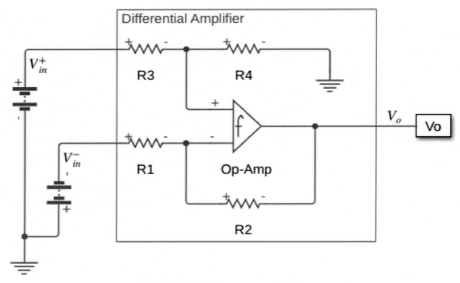

FIGURE 1.10 Differential amplifier.

shown in Fig. 1.11. The mean is $\mu = 1.9780$ V and the standard deviation is $\sigma = 0.0152$ V and the distribution resembles the normal distribution. Interval estimation with 95% confidence is approximately $[\mu - 2\,\sigma, \mu + 2\,\sigma] = [1.9476, 2.0084]$.

  Sensitivity analysis is performed in Simscape for the differential amplifier shown in Fig. 1.10. One hundred values of resistances {R1, R2, R3, R4} are generated for sensitivity analysis. The scatter plot showing the relationship between the amplifier output and the values of the resistors is shown in Fig. 1.12. Correlation coefficients computed between the values of each resistor and the output are: [−0.76, 0.76, −0.04, 0.07]. We can see that the output of the system is much more affected by changes in the resistance of resistors R1 and R2 than by changes in the resistance of resistors R3 and R4. Therefore, it is more important to select resistors R1 and R2 to have low tolerances because large tolerances in these resistors will result in large variations in the output. Resistors R3 and R4 can be chosen to have larger tolerances since they affect the output voltage less.

1.6 Summary

  This chapter introduced performance measures for both continuous and categorical data. We discussed the concept of measuring agreement since this is very important in biomedical instrumentation, where the true reference values are commonly unknown. We also discussed uncertainty propagation and sensitivity analysis, which are necessary to be performed when analyzing how different parameters of the systems affect the output.

  Many important aspects have not been covered in this chapter. For example, the detailed computation of the uncertainty budget is not explained. Modern sensitivity analysis based on

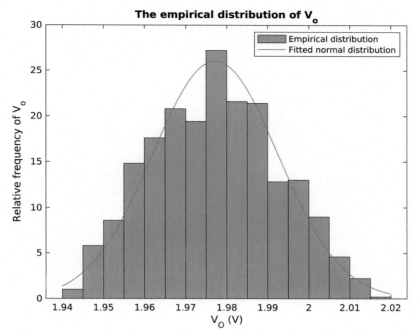

FIGURE 1.11 Empirical distribution of the output voltage in case it is assumed that the tolerance of each resistor is 1%.

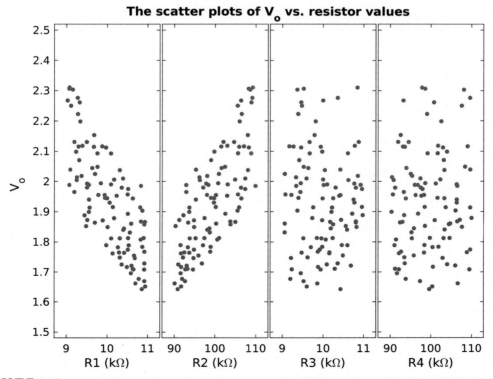

FIGURE 1.12 Scatter plot that shows the output versus values of the resistors of the differential amplifier.

variance decomposition was also not covered. In addition, the optimization of parameters (step B in Fig. 1.8) and estimating the uncertainty of model parameters were not explained. We refer the reader to the Further Reading Section for more information about these topics.

1.7 Problems

1.7.1 Questions

1.1. Answer the following questions:
  (a) Is the sensitivity constant for linear systems? Why?
  (b) When comparing the agreement between a new device and a device commonly used in clinical practice, relying only on a correlation coefficient is not recommended. List several disadvantages of using the correlation coefficient in this case.
  (c) How would one present the agreement between two devices graphically?
  (d) What can be used to quantify the agreement between two devices analytically?
  (e) If the new device agrees with the old device, what can the manufacturer of the new device claim (why would they even care if the agreement with the old device is achieved and instead focus on the accuracy of the new device)?
1.2. Several error metrics are defined in Section 1.3.2.1. Which metrics would one use in case:
  (a) There are large outliers in the output.
  (b) There is a possibility of having a systematic error.
  (c) One needs to use the same units to quantify the error as the units of the quantity of interest?
1.3. Assume that the variance of device one in Eq. (1.2) is negligible $\sigma_{e1}^2 = 0$. Fill Table 1.1 by doing the following:
  (a) Indicate reasonable values of the parameters β_1, β_0, σ_{e2}^2 in Eq. (1.3) for four combinations of accuracy and precision.
  (b) Draw the following four diagrams for the different combinations of accuracy and precision. Each diagram should show the distribution of y_{i2} (draw it as a normal distribution) and it should indicate the mean of y_{i2} and the true value of x. Label each diagram to indicate the systematic and random errors.
  (c) Simulate these four cases explained in question b) in Matlab and present the results.

TABLE 1.1 Accuracy versus precision table.

		Accuracy	
		Accurate	Not accurate
Precision	Precise		
	Not precise		

1.4. Consider the uncertainty quantification framework presented in Section 1.5.1.
    (a) Describe the steps of uncertainty quantification.
    (b) Describe terms such as inverse modeling, parameter estimation, uncertainty quantification, and local and global sensitivity analysis.

1.7.2 Problems

1.5. Let us consider a case where we have independent random variables $X_i, i = 1, ..., n$. Let us form a new variable X: $X = \sum_i X_i$. Show that the random variable X has the mean $\mu = \sum_i \mu_i$ and the variance $\sigma^2 = \sum_i \sigma_i^2$.

1.6. Let us analyze Example 1.1

    (a) What are the confidence intervals at 99% confidence?
    (b) Comment on increasing/decreasing confidence intervals with decreasing the sample size n?
    (c) Is there a difference in confidence intervals computed using Student-t and normal distribution when the number of samples is (1) 5, (2) 15, (3) 50?
    (d) What would be a necessary number of samples to achieve the uncertainty in the mean value due to data variation to within 0.01 V at 95% confidence?

1.7.3 Simulations

1.7. Fig. 1.2 shows 95% of the area under the normal curve. Draw the area under the normal curve for 66% and 99% confidence levels.
1.8. Plot normal distribution and Student t-distribution with 2 and with 9 degrees of freedom for the same parameters. Comment on the "heavy tails" of Student t-distribution and show that the confidence intervals are wider for Student t-distribution than those of a normal distribution for 95% confidence levels.
1.9. For the data shown in Fig. 1.5, do the following:
    (a) Compute mean error, the sum of squared errors, mean square error, root mean square error, R^2 and mean absolute error.
    (b) Next, compute all these errors but make β_0 negative. Comment on your results. Which errors are insensitive to the direction (sign) of β_0?
    (c) Introduce the standard deviation in x (e.g., $\sigma_x = 0.01$) and repeat step (a).
1.10. For the data shown in Fig. 1.5, do the following:
    (a) Show and discuss the Bland-Altman plot.
    (b) Next, change the standard deviation in x to be 0.01 (or higher) and show the Bland-Altman plot. Comment on the bias and the variation in data.
1.11. For the model described using Eqs. (1.2) and (1.3),
    (a) Modify the code accuracy.mlx to estimate the correlation coefficient.
    (b) Increase the variance σ_{e2}^2 and σ_x^2 and observe changes in the correlation coefficient.
    (c) Compute the concordance correlation coefficient.

1.12. For the linear sensor model presented in Fig. 1.3, perform uncertainty propagation. Assume that both bias and the scaling factor are represented as normal distributions where their mean is equal to their nominal values and the standard deviation is 1% of their nominal values. Next, assume that the input x is the normal random variable with a mean of one and a standard deviation of 0.01. What are the mean values and standard deviation at y_{sc} and y_{sd}?

1.13. For the linear sensor model presented in Fig. 1.3, perform sensitivity analysis. Assume that both bias and the scaling factor are represented as normal distributions where their mean is equal to their nominal values and the standard deviation is 1% of their nominal values. Assume that the gain can vary based on uniform distribution between 1.95 and 2.05. Next, assume that the input $x = 1$ and it is constant. Plot the scatter plot that shows the relationship between the parameters and the error $y_{sc} - gain \cdot x$?

1.14. Consider Example 1.6, where the confidence intervals at the output are computed using Monte Carlo sampling. Instead of computing confidence intervals, compute the limits at the output using the worst-case analysis. This would require to choose the maximum or minimum values of the resistances so that the output is maximum (or minimum). What are the maximum and minimum values of V_o for resistors that have 1% tolerance?

1.8 Further reading

Additional resources related to different topics covered in this chapter are listed below. Please note that this list is not comprehensive, and it is selected mainly based on the availability of the books by the author.

For measuring agreement between the methods, the author strongly recommends (Choudhary and Nagaraja, 2017).

Uncertainty propagation for sensors and measurement systems is well covered in Figliola and Beasley (2012).

Excellent theoretical discussion about uncertainty quantification is covered in Smith (2014). There are also chapters on sensitivity analysis and uncertainty propagation. One of the main references in the field of global sensitivity analysis is (Saltelli et al., 2008).

Modeling and parameter estimation is covered in van der Heijden et al. (2004).

Some uncertainty quantification-related simulators include the following:

- OpenTurns, http://www.openturns.org/
- UQLab, http://www.uqlab.com/
- MIT Uncertainty Quantification Library (MUQ)

References

Baudin, M., Dutfoy, A., Iooss, B., Popelin, A.-L., 2015. OpenTURNS: An Industrial Software for Uncertainty Quantification in Simulation. arXiv:1501.05242v2 [stat.CO] 5 Jun.

Bland, J.M., Altman, D.G., 1999. Measuring agreement in method comparison studies. Statistical Methods in Medical Research 8, 135–160.

Choudhary, P.K., Nagaraja, H.N., 2017. Measuring Agreement: Models, Methods, and Applications. John Wiley & Sons, Inc., New York.

Evaluation of measurement data — Supplement 1 to the "Guide to the expression of uncertainty in measurement" — Propagation of distributions using a Monte Carlo method. JCGM 101, 2008.

Evaluation of measurement data — Guide to the expression of uncertainty in measurement. JCGM 100, 2008.

Figliola, R.S., Beasley, D.E., 2012. Theory and Design for Mechanical Measurements, fifth ed. Wiley.

Papoulis, A., Pillai, S.U., 2002. Probability, *Random Variables and Stochastic Processes*, fourth ed. McGraw-Hill.

Pavese, F., Forbes, A.B., 2009. Data Modeling for Metrology and Testing in Measurement Science. Springer.

Saltelli, A., et al., 2008. Global Sensitivity Analysis. The Primer. John Wiley and Sons.

Smith, R.C., 2014. Uncertainty Classification: Theory, Implementation and Applications. SIAM.

van der Heijden, F., Duin, R., de Ridder, D., Tax, D.M.J., 2004. Classification, Parameter Estimation, and State Estimation: An Engineering Approach Using MATLAB. Wiley.

Willink, R., 2013. Measurement Uncertainty and Probability. Cambridge University Press.

OUTLINE

Acronyms and explanations

LED light-emitting diode
MEMS micro-electromechanical system

Variables used in this chapter include:

ρ resistivity
L wire length, capacitance plate gap
A area
ε strain
δ stress
Y Young's modulus of elasticity
μ Poisson ratio
G gauge factor
D diameter of the wire
q charge
F force
k_p piezoelectric constant
k piezoelectric coupling coefficient
ε_0 permittivity of vacuum
ε_r relative permittivity
S sensitivity
P power
A surface area
E irradiance is the light power that irradiates a surface area A.
F fluence rate refers to the irradiance, which is incident from all angles onto a small region of space.
Φ luminous flux: the amount of light being emitted by a source in all directions

2.1 Introduction

A transducer is a device that converts one energy source into another. For example, a photodiode converts light energy into current. A *sensor* converts the physical measurand into an electric output that can be a current or a voltage signal. This chapter will cover transducers but not sensors since sensors normally require electronics to convert the signal into the electrical signal, and the electronics components are covered in Chapter 3. Sensors will be briefly introduced in Part II of the book when discussing particular devices.

A *self-generating transducer* is a transducer that operates without using an external power source. An example is a piezoelectric transducer, where the conversion between displacement caused by compression or tension and electric potential is done without the external power source. On the other hand, an *externally powered transducer* converts one form of energy into another form with the help of an external power source. An example includes strain gauges that need to be connected to the external power source to detect the resistance change of the strain gauge. Please note that sometimes these transducers are called active and passive, but we will avoid that terminology here.

TABLE 2.1 Transducers and their typical use in sensors and devices.

Transducer	Sensors based on these transducers	Devices
Resistive	Pressure	Blood pressure, respiratory belts
Capacitive	Pressure, accelerometers	Blood pressure
Piezoelectric	Microphones, ultrasound, pressure, force	Blood pressure, arterial tonometry, respiratory belts
Optical	Photodetectors	PPG, pulse oximeters
Electrodes	Biopotential electrode, bioimpedance	ECG, breathing belt based on bioimpedance measurements

In this chapter, we will consider the following transducers.

- *Resistive* transducers change resistance with *displacement* due to applied force or pressure. An example is a strain gauge.
- *Capacitive* transducers change their capacitance with displacement.
- *Piezoelectric* transducers react on displacement caused by force and produce electric potential. Piezoelectric transducers can only measure dynamic changes (displacement or pressure) but not DC values. They have very high sensitivity in comparison with other types of displacement transducers.
- *Optical* transducers convert light energy into current.
- *Electrodes* convert a biopotential into an electric potential. *Biopotentials* are "voltages generated by physiological processes occurring within the body" (Macy, (n.d)). An example includes ECG electrodes.

Table 2.1 presents transducers covered in this chapter, sensors based on these transducers, and devices where these transducers/sensors are used. Please note that the list of sensors and devices shown in Table 2.1 is not comprehensive and that we showed only typical examples.

2.2 Strain gauges

This section will analyze displacement transducers based on piezoresistive effect—resistive displacement transducers or strain gauges (Box 2.1). *Strain gauges* are transducers that change their resistance with applied stress. These transducers are used in various applications to measure flow and pressure indirectly.

Strain gauges are attached to objects that get elongated or compressed, causing the resistance of the transducer to increase or decrease, respectively. When attached to a conditioning circuit, like the Wheatstone bridge, the resistance change is converted into the voltage change of the bridge.

BOX 2.1

Piezoresistive and piezoelectric effects

Prefix piezo is from Greek word piezein which means press. Therefore, *piezoresistive* effect means the ability of metals and semiconductors to change resistance in response to applied mechanical stress. Similarly, *piezoelectric* effect refers to the ability of metals and semiconductors to generate voltage in response to applied mechanical stress.

Next, we will analyze the sensitivity of the strain gauge. Let us assume that a strain gauge is composed of a single wire, whose cross-sectional area is A, the length is L, and a resistivity constant is ρ [Ωm]. The resistance of the strain gauge can be computed as follows:

$$R = \rho \frac{L}{A}$$

A change in the length of the wire due to applied force over the total length is called the *strain* $\varepsilon = dL/L$. Strain is positive for the elongation process and negative for the shortening process. Even though the strain is unitless, it is commonly presented in the units of microstrain, parts per million (ppm) or m/m. Here $1\ \mu\varepsilon = 1$ ppm. *Stress* δ [Pa] is defined as the force F applied to the material with the surface area A

$$\delta = \frac{F}{A}$$

Many materials exhibit a linear relationship between stress and strain up to a certain point, referred to as Hooke's Law. The slope of the stress–strain curve is referred to as Young's modulus of elasticity $Y = \delta/\varepsilon$. Tensile stresses are positive, while compressive stresses are negative.

Ideally, the resistance change of the resistive displaced sensors would be proportional only to the change in the length of the wire. The gauge sensitivity is defined in terms of the characteristics called the gauge factor that relates the change of the resistance with strain.

$$G = \frac{\dfrac{dR}{R}}{\dfrac{dL}{L}}$$

However, this assumption that the strain will cause only changes in resistance is not satisfied in practice. The elongation or compression of the wire results in changing the diameter D of the wire as well. This change is quantified through the Poisson ratio μ:

$$\mu = -\frac{\dfrac{dD}{D}}{\dfrac{dL}{L}} = -\frac{\dfrac{dA}{A}}{2\dfrac{dL}{L}}$$

If we consider straight wire placed horizontally, then the Poisson ratio represents the negative ratio of strains in the transverse direction which is perpendicular to the applied force, and in the axial direction, which is in the direction of the applied force. The diameter of the wire will reduce with the elongation and vice versa, and therefore the Poisson ratio is always positive.

To obtain the sensitivity of the strain gauge, we can find the partial derivative of R as follows:

$$dR = d\left(\rho\frac{L}{A}\right) = \frac{\rho}{A}dL - \frac{\rho L}{A^2}dA + \frac{L}{A}d\rho$$

Plugging this into the definition of the gauge factor, we obtain the following:

$$G = 1 + 2\mu + \frac{\dfrac{d\rho}{\rho}}{\dfrac{dL}{L}}$$

So, the gauge factor depends on the Poisson ratio and the piezoresistive effect $(d\rho/\rho)$. Strain gauges are mainly made of metals or semiconductors. The gauge factor of a metal strain gauge is generally small (around 2) and is mainly a function of the Poisson ratio. Therefore, for a range of strains, the change in resistance is proportional to the strain with a factor of 2. On the other hand, semiconductor strain gauges have large gauge factors mainly due to the piezoresistive effect and, therefore, can be used to measure a smaller strain. However, they are normally more sensitive to temperature changes.

Strain gauges are designed for specific nominal resistance, for example, 120 Ω or 350 Ω. The resistance is specified with the resistance tolerance—for example, 0.3%.

Example 2.1. Strain of 0.001 m/m (1000 ppm) is measured using a strain gauge with a gauge factor of 2. What is the resistance change if the resistance of the strain gauge is 120 Ω?

Solution. $dR = RG\frac{dL}{L} = 0.24\ \Omega$. Therefore, the change of resistance is relatively small.

A large length of the strain gauge is required to achieve a significant change in resistance. A typical strain gauge is fabricated as a conductive strip folded in a grid pattern multiple times, as shown in Fig. 2.1. This allows longer wire (strip) to be packed onto the smaller area. Regarding sensitivity change, the strain for the overall strain gauge is the same as the strain of a single strip. Therefore, the gauge length is given as the length of a single trace. The strain gauge length is commonly several millimeters, even though it can be as small as 0.1 mm. The strain range is typically given in a datasheet, for example, 3% or 5%. Some parameters and their typical values are shown in Table 2.2.

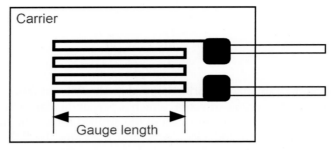

FIGURE 2.1   Typical strain gauge with a grid pattern.

TABLE 2.2   Characteristic parameters of strain gauges.

Parameters	Typical values
Strain range	±3%
Strain gauge length	3.2 mm
Temperature range	−20°C to 100 °C
Gauge factor	For metal 2 ±5%

The conductive strip is bonded to a thin layer called the carrier—see Fig. 2.1. The carrier is attached directly to the object whose displacement needs to be measured. The strain experienced by the object is transferred directly to the strain gauge resulting in its resistance change.

In biomedical instrumentation, strain gauges are used in many applications, including measuring breathing and pressure. The measuring element can be a diaphragm with a strain gauge attached to it that will deform under pressure.

2.3 Piezoelectric transducers

Piezoelectric transducers are used in various biomedical applications and are components of pressure transducers, accelerometers, microphones, ultrasound systems, and other systems and sensors. Piezoelectric transducers induce charge in response to the applied force, and in this way, they represent self-generated transducers. Also, when an electric field is applied, piezoelectric materials can be used as actuators because they generate strain.

Let us assume that the piezoelectric transducer acts as a parallel-plate capacitor of capacitance C. The applied force F on the material is directly proportional to the induced charge, q

$$q = k_p F$$

where k_p is the *piezoelectric constant* ([C/N] or [m/V]). The constant k_p is typically presented in pC/N, and it is about 2.3 pC/N for quartz and 140 pC/N for barium titanate. Therefore, the open-circuit voltage can be computed as follows:

$$V = \frac{q}{C_p} = \frac{k_p \cdot F}{C_p} = \frac{k_p FL}{\varepsilon_0 \varepsilon_r A} \tag{2.1}$$

where $C_p = \varepsilon_0 \varepsilon_r A / L$, A is the area of the plates of the capacitor, L is the plate separation, ε_0 is the permittivity of the vacuum while ε_r is the relative permittivity.

Example 2.2. A piezoelectric transducer with dimensions of 3 mm $\times$ 1.2 mm $\times$ 1.6 mm is used. After applying the force due to 1 g weight, compute the voltage at the output if the piezoelectric material used is (a) quartz and (b) barium titanate.

Solution. $F = mg$ where g is the gravity acceleration and $\varepsilon_0 = 8.854 \times 10^{-12}$ C/Vm. ε_r for quartz is 4.5, and for barium titanate is 1255. After plugging the numbers into Eq. (2.1), we obtain that the open-circuit voltage at the output is (a) 0.25 V and (b) 0.055 V.

One of the disadvantages of these devices is that, although their materials have high resistance, the finite value of the resistance leads to charge leakage through the resistor when a static deflection, L, is applied. The charge generated is proportional to the deflection with a proportionality constant K (Webster, 2010):

$$q = KL$$

A piezoelectric sensor can be modeled using the circuit shown in Fig. 2.2 (Webster, 2010). R includes the total leakage resistance of the piezoelectric crystal together with the input impedance of the amplifier. The leakage resistance of the piezoelectric crystal is in the order of $G\Omega$ and the input resistance of the connected amplifier is in the order of $M\Omega$. Typical capacitance of the transducer is in the order of nF. In Fig. 2.2, $R = 10 \ M\Omega$ and $C = 1nF$. Please also note that i and v represent ammeter and voltmeter. Therefore, the time constant of the transducer/amplifier system is 10 ms, and the voltage at the output drops to zero at about 50 ms, as shown in Fig. 2.3B. The current source that is externally controlled was used. This circuit

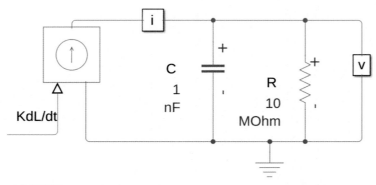

FIGURE 2.2 Simplified equivalent circuit of a piezoelectric transducer.

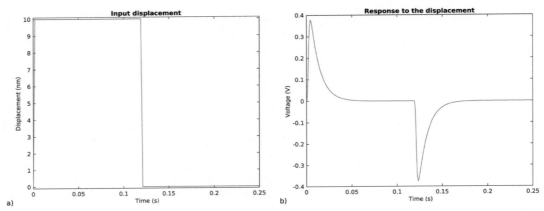

FIGURE 2.3 (A) Step displacement of the piezoelectric transducer, (B) the response of the piezoelectric transducer based on the model shown in Fig. 2.2.

does not have any resistance in series with the capacitor, and therefore, DC signal of a piezoelectric transducer is zero.

The piezo film can be used as a sensor. Piezoelectric films can be bent similarly to strain gauges producing a voltage. Their thickness is between 28 μm and 110 μm. Piezoelectric films are commonly covered with a metal such as silver ink. Different alloys can be used as well. The piezo film transducers can generate about 10 mV per microstrain, which is about 60 dB higher than foil strain gauges. The piezoelectric elements can be very small and with a rapid response to changing pressure. Some properties of piezoelectric film transducers used for continuous blood pressure measurements are shown in Table 2.3 (Wang and Lin, 2020). The excellent sensitivity of 2 mV per mmHg can be achieved.

The sensitivity of the circuit in Fig. 2.2 is then computed as

$$\frac{V_o(j\omega)}{L(j\omega)} = \frac{K_s j\omega\tau}{j\omega\tau + 1} \tag{2.2}$$

TABLE 2.3   Characteristic parameters of piezoelectric sensors used as a pressure sensor (HK−2000B, Hefei Huake Electronic Technology Research Institute, Hefei, China).

Parameters	Typical values
Measurement range	Up to 300 mmHg
Sensitivity	2 mV/mmHg
Sensitivity temperature coefficient	$1 \times 10^{-4}/°C$
Hysteresis	0.5%
Accuracy	5%

where $K_s = K/C$ (sensitivity in V/m) and $\tau = RC$ (time constant). The derivation can be found in Appendix 2.8.1. More comprehensive models and their comparisons are shown in Staworko and Uhl (2008).

The efficiency of piezoelectric material as a transducer is measured by the *piezoelectric coupling coefficient* or factor (Vijay, 2013). It indicates the ability of a piezoelectric to convert electrical energy to mechanical energy and vice-versa. It is represented as k, and defined as $k^2 = (\text{converted energy}) \div (\text{input energy})$. The k value for quartz is about 1%. Synthetic piezoelectric materials can have k values of 30%–75%.

2.4 Capacitive transducers

Capacitance transducers are widely used in microphones and for non-contact measurements. They measure the displacement due to force or pressure. Some non-biomedical applications include detecting a presence and capacitive touchscreen displays. One of the most important applications of multiplate capacitors is for accelerometers. Capacitive transducers produce no current when the capacitor plates are not in motion. This is adequate for microphones that need not operate at lower frequencies. Here, we will consider parallel plate capacitive transducers to measure the displacement.

The variation in capacitance of a capacitive transducer can be achieved by changing.

- The distance between the capacitor plates,
- The effective area between the plates, or
- The amount or properties of the dielectric.

When the displacement is large, the capacitance variations due to changes in the area are preferred.

A capacitance transducer comprises parallel plates of area A separated by a distance L (gap). The capacitance also depends on the permittivity of the vacuum, ε_0 and that of the insulator, ε_r:

$$C = \varepsilon_0 \varepsilon_r \frac{A}{L}$$

For the capacitor that allows for changing the distance between the plates due to the applied force or pressure, the sensitivity S is dependent on the plate separation L, as shown below:

$$S = \frac{dC}{dL} = -\varepsilon_0 \varepsilon_r \frac{A}{L^2} = -\frac{C}{L}$$

Thus, it is possible to increase the transducer's sensitivity by decreasing the plate separation. The change in the distance is non-linearly related to the capacitance. However, for very small changes in the distance that exists, for example, in MEMS (micro-electromechanical system) devices, the relation between the changes in the distance between the electrodes and the capacitance is almost linear. In addition, this non-linearity due to gap variation can be overcome if the three-plate capacitive transducer is used. Let us assume that the outer plates are fixed and that the middle plate that can move due to displacement is at an equal distance L from both external plates. The movement of the middle plate will reduce the distance to one of the plates by ΔL and increase the distance to another plate also by ΔL. The differential form of the capacitance for the typical case when $\Delta L^2 \ll L^2$ is proportional to the change in the distance ΔL (Aezinia, 2014):

$$C_d = \varepsilon \frac{A}{L - \Delta L} - \varepsilon \frac{A}{L + \Delta L} \approx 2\varepsilon \frac{A}{L^2} \Delta L$$

Therefore, the current phasor through this capacitor $I = j2\pi f C_d V$, where f is the frequency of operation and V is the voltage phasor, is proportional to the displacement and can be converted into the voltage using current-to-voltage converters.

Multiplate or multielectrode capacitors, also called *interdigitated* parallel plates, have n plates placed in a comb-like structure making $n - 1$ capacitors. Alternate plates have the same voltage and are connected in parallel with the total capacitance $C = (n - 1)\varepsilon A / L$, where L is the gap between each pair of plates (Aezinia, 2014).

Capacitive accelerometers consist of two sets of plates, one fixed and stationary, and one moving, fixed to a proof mass. The changes in the capacitance of the plates as a result of motion are converted into a voltage signal (Xie et al., 2008).

2.5 Optical sensors

The photodiode converts the light energy to the current generated when light is incident on the device. The light-emitting diode (LED) converts the current to light energy. Diodes are semiconductor devices that act as a switch in an electrical circuit. They can be forward-biased, allowing a large current to flow through them, or reverse-biased, preventing the current flow.

2.5.1 Radiometry and photometry

This section describes parameters related to the intensity or amount of light emitted by a source or irradiated to a surface. *Radiometry* is concerned with quantities related to radiant energy, while *photometry* is concerned with radiant energy only in the visible spectrum (Chan, 2016; Jacques, 2020).

Radiometric parameters include radiant power, radiant intensity, irradiance, fluence rate, and surface reflectance. The power output of a source is described by its *radiant power P* [W]. The power P from a source that is directed into a particular direction along the center of a cone encompassing a solid angle Ω [steradians] or [sr] is called the *Radiant Intensity, I* [W/sr]. If the beam angle θ is given in degrees, then the solid angle can be computed as $\Omega = 2\pi(1 - \cos(\theta\pi/360))$. A sphere has 4π steradians. The power P [W] that irradiates a surface area A [cm^2] is called the Irradiance E [W/cm^2]. The *Fluence Rate F* [W/cm^2] refers to the irradiance, which is incident from all angles onto a small region of space. It is defined as the radiant power incident on a small sphere divided by the cross-sectional area of that sphere. *Surface reflectance* is the fraction of photons reaching the detector surface per cm^2.

The *visible spectrum* is between 380 and 780 nm. Luminous flux and intensity are related to the light coming from an emitter. Illuminance is related to the light falling on a surface. The amount of light emitted by a source in all directions is called *luminous flux φ*[lm] measured in lumens. The luminous flux Φ is equal to the product of power P and luminous efficacy η [lm /W], $\Phi = P\eta$. *Luminous intensity I*[cd] is defined as luminous power emitted by a point light source in a particular direction per unit solid angle, given in candelas $I = \Phi/\Omega$. It measures the directionality of the energy radiated by the light source. If an emitter emits 1 lumen and the optics of the emitter are set up to focus the light evenly into a 1 steradian beam, then the beam would have a luminous intensity of 1 cd. *Illuminance* is defined as the luminous flux per unit area, and it is a measure of power that irradiates the area A.

2.5.2 LED

Red and near-infrared light is commonly used in pulse oximetry (see Box 2.2 for other applications). *LEDs* are light sources with a narrow emission spectrum. They are ubiquitous low-cost devices and, as such, suitable for PPG and pulse oximetry. The physics of LEDs is described in multiple textbooks and will not be covered here (Webster, 2010). For more information, please see Chapter 10 of Dakin and Brown (2017).

BOX 2.2

Other applications of LEDs

Besides using LED in photoplethysmography, LEDs are used in a wide range of applications including providing monochromatic light for imaging and optocouplers for contactless signal transmission and for light-based therapies.

Photobiomodulation-based LED therapies are safe, nontoxic, and non-invasive and are used to promote wound healing and reduce inflammation, act on lymph nodes to reduce edema and inflammation, and promote muscle relaxation. In imaging, LED panels can be used to generate light of specific wavelengths (Dong and Xiong, 2017).

Important features of LEDs include forward voltage and current, power dissipation, reverse current, switching time, and intensity vs. wavelengths. The forward voltage of LEDs is normally higher than the forward voltage of ordinary silicon diodes. LED's forward voltage is between 0.9 and 2.5 V, while the silicon diode's forward voltage is about 0.7 V. The forward current is defined as the current flowing through the LED in the direction from the anode to the cathode. With sufficient current, an LED will emit light. Fig. 2.4 shows an LED connected in series with a voltage source and a resistor. The resistance selected in this case is about 100 Ω.

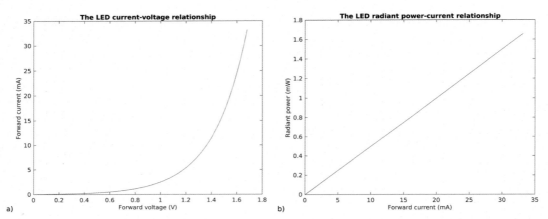

FIGURE 2.4   Simple circuit that includes a voltage source, LED, and resistor. Here, components named i, v, and W mean ammeter, voltmeter, and power meter. The power meter measures the radiant power of the LED.

Fig. 2.5A presents the relationship between the forward current and forward voltage. The results are obtained by changing the input voltage from 0 V to 1.85 V. Fig. 2.5B shows the radiated power versus. forward current. The radiated power of an LED is measured in mW. The typical radiant power of both the red and IR LEDs used in pulse oximetry is in the order of 1 mW at 20 mA direct current. Normally, the radiant power of LEDs does not exceed 10 mW in wearable electronics.

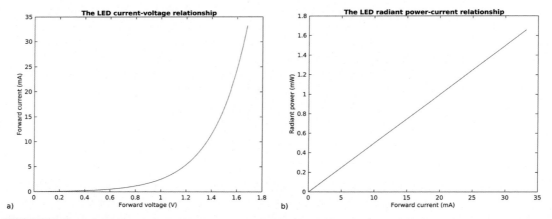

FIGURE 2.5   (A) LED measured using ammeter and voltmeter shown in Fig. 2.4. Input voltage V_{in} was varied from 0 V to 1.85 V. (B) Radiant power of LED versus the current through the LED when V_{in} is varied in the same way as in A.

In datasheets for LEDs, the radiant power is not explicitly presented. Instead, luminous intensity is normally given versus forward current, presented in mcd versus mA. Here, mcd mean milicandela. For example, for 20 mA of the forward current, the luminous intensity of a low-power LED is in the order of 100 mcd. For a beam angle of 90 degrees, the solid angle is $\Omega = 1.84$ sr. This corresponds to a luminous flux of 0.184 lm. Depending on the luminous efficacy of the LED, this luminous flux would correspond to several mW of the radiant power. For example, for the luminous efficacy η of 90 lm/W, the radiant power is 2 mW.

A typical LED is up to 10% efficient, meaning that most of the power dissipated by the LED becomes heat. In addition, the optical power absorbed by the tissue also becomes heat. Therefore, it is important to select LEDs carefully to prevent tissue warming in PPG and pulse oximeter devices.

Switching time is also important because LEDs get switched on and off from several hundred (for example, 480 Hz) up to several thousand times per second in pulse oximetry. However, LEDs' typical switching time is usually much faster than that.

Forward current and power dissipation are temperature-dependent. However, they are usually relatively stable up to 40 °C (Webster, 1997). Temperature coefficients that quantify changes of variables of interest with the temperature are commonly given for LED. For example, the temperature coefficient of the forward voltage describes changes in the forward voltage in mV per changes in temperature in K. For example, -2 mV/K can be considered normal. The Food and Drug Administration requires that the contact region between the skin and the oximeter probe does not exceed the temperature of 41°C. This is important in preventing skin burns. Typically, LED working in a pulse mode, that is explained in Chapter 3, with a maximum current of about 50 mA will not cause a significant increase in the tissue's temperature.

The beam angle is defined as the angle between two endpoints at which the radiated power is reduced by half (Webster, 1997). It is a measure of how focused the emitted light is, and it can vary from several degrees up to 180 degrees. Pulse oximeters in transmissive mode might require LEDs with smaller beam angles than those operating in reflectance mode (the operating modes of pulse oximeters will be detailed in Chapter 6). Radiation characteristics is a graph of relative luminous intensity versus the angle.

Peak wavelength is defined as the wavelength at which the radiated power of the device is maximum. LEDs should have a small peak wavelength shift. The temperature coefficient for the peak wavelength is also reported, and it is typically around 0.1 nm/K.

Spectral Bandwidth $\Delta\lambda$ is usually given at 20 mA and is defined as a half-power bandwidth of the light emitted from the LED. It can be between 15 and 60 nm, and it is typically wider for infrared than for red LEDs. Of course, LEDs need to have narrow spectral bandwidth.

Several parameters of LEDs are summarized in Table 2.4.

2.5.3 Photodiodes

Photodiodes are semiconductor-based *photodetectors* that absorb light and generate photocurrent or light current due to the photoelectric effect. Photodiodes are made from silicon, germanium, or indium gallium and arsenide phosphide alloy. Silicon-based photodiodes have the highest sensitivity in the infrared range, while they can detect light for wavelengths between 400 and 1000 nm.

TABLE 2.4 Typical characteristics of LEDs used in pulse oximetry.

Feature	Parameter	Typical values for the low-power red LED
Time	Switching time	Normally not given
Frequency	Peak wavelength	630 nm
	Spectral bandwidth	15 nm
Space	Beam angle 2 θ	140 degrees
Power and intensity	Power dissipation	48 mW
	Radiated power	Normally not given
	Luminous intensity at 10 mA	100 mcd
Temperature	Temperature coefficients of forward voltage	−2 mV/K
	Temperature coefficients peak wavelength	0.1 nm/K
Current	Peak forward current	60 mA
	Reverse current	10 μA
Voltage	Forward voltage	Typical 2 V, max 2.4 V
	Reverse voltage	5 V

The photocurrent is proportional to the incident light intensity over a wide range of intensities. The photodiode is connected in a reverse current mode so that the current is generated through the diode only when it is exposed to the light source. Compared with the circuit shown in Fig. 2.4, the reverse voltage is applied, and a much larger resistor should be used (for example, 100 kΩ) because the current is in the order of μA. Device sensitivity is the ratio of the current produced to the incident radiant flux density or irradiance. Reverse light current versus the irradiance is shown in Fig. 2.6. It is usually linear.

Reverse light current versus reverse voltage is shown in Fig. 2.7 for irradiances of 0.1 W/m^2 and 5 W/m^2. As the irradiance increases, the inverse current increases as well (becomes more negative), and the reverse current-voltage curve shifts down. As seen in Fig. 2.7, the reverse light current is in the range of μA. The forward voltage is smaller than the forward voltage of LEDs. If we keep increasing the reverse voltage on the photodiode, the photodiode might start malfunctioning. The maximum reverse voltage is called the reverse breakdown voltage.

An ideal photodiode can be modeled as a controlled current source and a diode connected in parallel. Therefore, the current-voltage relationship of an ideal photodiode is given as $I = I_p - I_d \left[e^{qV/kT} - 1 \right]$ where I_p is the generated photocurrent and the second term is a diode current where I_d is the dark saturation current, q is the electron charge ($1.602 \cdot 10^{-19}$ C), V is the voltage across the terminals, k is the Boltzmann constant ($1.381 \cdot 10^{-23}$ J/K), T is the absolute temperature in Kelvin (Gupta and Ballato, 2006). The non-ideal photodiode can be presented as an equivalent circuit that includes the current source connected in parallel with a diode, junction capacitance, shunt, and load resistances—see Problem 2.15 and Fig. 2.15. Junction

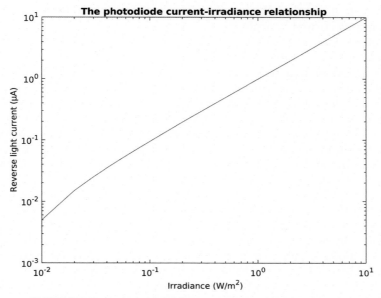

FIGURE 2.6 Simulated irradiance versus the photodiode current.

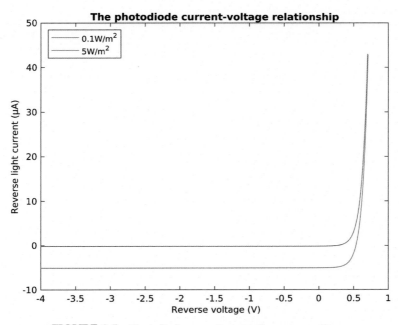

FIGURE 2.7 Photodiode current versus the reverse voltage.

capacitance directly determines the speed of response of the photodiode. The shunt resistance is very high, and its current is negligible.

The parameters of interest for the photodiode are electrical, temperature, space, frequency, and time parameters—see Table 2.5.

Timing parameters include rise and fall time. Photodiodes are quite fast for biomedical applications and their rise and fall time are in the order of 100 ns. Frequency parameters include peak wavelength—the wavelength with the largest relative spectral sensitivity. In datasheets, relative spectral sensitivity versus wavelengths is usually given, and it is normalized to the range between zero and one. Zero spectral sensitivity at a particular wavelength means that the photodiode is not detecting any incident light at that wavelength. Spectral bandwidth represents the spectral range where relative spectral sensitivity is high enough so that the photodiode can be used as a photodetector.

Space parameters include the angle of half sensitivity, defined as an angular displacement at which the relative radiant sensitivity of the diode is 0.5. Relative radiant sensitivity is maximum (one) when the light source is in front of the photodetector so that the angle is zero. The typical angle is wide, and it is, for example, ±65 degrees. The radiant-sensitive area is the photodiode-sensitive area for detecting visible or infrared light.

Reverse dark current represents leakage current when there is no light.

The photodiode's responsivity [A/W] is computed as the photocurrent divided by radiant power. Its maximum value is 0.5 A/W for silicon-based photodiodes.

TABLE 2.5 Characteristics of photodiodes using as an example TEMD5020X01 from Vishay semiconductors.

Feature	Parameter	Typical values for the photodiode
Time	Rise time and fall time	100 ns
Frequency	Peak wavelength	900 nm
	Spectral bandwidth	430–1100 nm
Space	The angle of half sensitivity	±65 degrees
	Radiant sensitive area	4 mm^2
Power	Power dissipation	200 mW
Temperature	Temperature coefficients of voltage	−2.6 mV/K
	Temperature coefficients current	0.1%/K
Current	Reverse light current	35 μA
	Reverse dark current	2 nA
Voltage	Forward voltage	Typical 1 V, max 1.3 V
	Reverse breakdown voltage	60 V
Other parameters	Diode capacitance	40 pF

2.6 Electrodes

Biopotential electrodes are used to measure bioelectric signals such as the ECG, bioimpedance measurements of the body tissues, application of the current to achieve therapeutic events, and so on. In this section, we will consider electrodes used to measure bioelectric signals. Common bioelectric signals include the following.

- Electrocardiogram (ECG), which records electrical heart activity,
- Electromyogram (EMG), which records muscle activity,
- Electrooculogram (EOG), which records eye movement,
- Electroencephalogram (EEG), which is the recording of the spontaneous electrical activity of the brain, and so on.

Among bioelectric signals, we are mainly interested in ECG measurements in this book.

When measuring bioelectric signals, electrodes are always used in pairs to measure the potential between two sites on the patient's body. Biopotential electrodes can be classified based on the contact with the skin as *contact* and *noncontact* electrodes, as shown in Fig. 2.8. Contact electrodes can be further divided into *wet* and *dry*. Wet electrodes are traditional electrodes that use conductive gel between the skin and the electrode. They are often adhesive, which, together with gel, makes them less susceptible to motion artifacts than other electrode types.

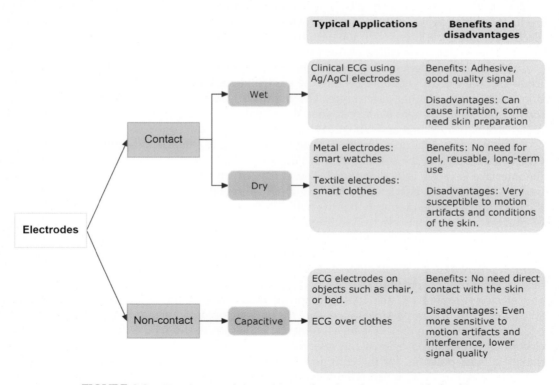

FIGURE 2.8 Classification of electrode types based on the contact with the skin.

Dry contact electrodes come into contact with the skin surface directly. Contact-based electrodes are commonly *passive*, meaning there are no amplifiers and other electronics next to the electrode.

Non-contact electrodes are dry electrodes (do not use gel or similar materials) that can measure a bio-signal signal through clothes with a typical separation from the body of several millimeters. They are implemented as one capacitor plate while the other plate corresponds to the human skin. Since these electrodes are susceptible to noise and interference, they need to be shielded, and the electronics need to be placed next to the electrode, making them *active electrodes*. The benefits and disadvantages of the electrode types are presented in Fig. 2.8.

This section will discuss and model electrode–electrolyte potential, electrode–electrolyte interface, and electrical properties of the skin. We will then compare wet electrodes with dry and capacitive electrodes. Dry electrodes are suitable for long-term ECG measurements, and some of them have already been implemented in smartwatches and other pervasive devices. We will start first with the models of traditional wet electrodes, and then we will show models of other electrode types.

2.6.1 Wet electrodes

2.6.1.1 Electrode–electrolyte interface

The electric current is carried by *ions* inside an electrolyte within the human body or electrode gel and *electrons* in the electrode, wires, and an electrical circuit. At the metal and the electrolyte interface, there is an ion–electron exchange resulting from an electrochemical reaction. This exchange is called *charge transfer*.

Oxidation is a tendency for metal atoms to lose electrons at the interface between the electrode and the electrolyte and pass into the electrolyte as metal ions. Therefore, during oxidation, metal ions enter the electrolyte. The opposite reaction is called *reduction*, where some ions in the solution take electrons from the metal and deposit them onto the electrode as metal atoms (McAdams and Yoo, 2011).

When we put metal into the electrolyte at the interface, the metal becomes more positive (loses electrons), and the solution is more negative. The potential difference that is formed at the interface is called *half-cell potential*. The rate at which metal atoms lose electrons and pass into the electrolyte is balanced by the rate at which metal ions in the electrolyte deposit onto the electrode as metal atoms so that the half-cell potential has a constant value. The separation of charges at the metal–electrolyte interface results in an electric *double layer*. In the double layer, one type of charge is dominant on the surface of the metal, and the opposite charge is in higher concentration in the layer of the electrolyte at the interface with the electrode. The half-cell potential is the result of the distribution of the charge at the electrode–electrolyte interface. Several factors contribute to the change in half-cell potential, including the type of metal and electrolyte involved, the concentration of metal ions in the solution, and temperature (Webster, 2010). It is impossible to measure the half-cell potential of one electrode as we always need two electrodes to make the measurements. Therefore, half-cell potential for major types of electrodes is measured against the reference hydrogen electrode under standardized conditions. Half-cell potential for the common silver/silver chloride (Ag/AgCl) electrode used in ECG measured at 25 °C when referenced to the hydrogen electrode is +0.223 V.

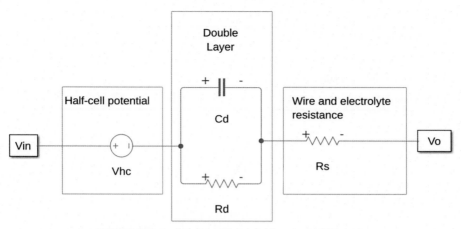

FIGURE 2.9 Overall circuit model of a wet electrode.

2.6.1.2 Electrode–electrolyte impedance

The electrode and electrolyte interface can be modeled as a resistance and capacitance in parallel. The equivalent model of the biopotential electrode is shown in Fig. 2.9. V_{hc} represents half-cell potential. The double layer at the electrode–electrolyte interface behaves like a parallel plate capacitor, and therefore, it is modeled as the capacitance C_d. There is some current flowing through the double layer, and to model that current, the resistor R_d is placed in parallel with the capacitor C_d. In wet ECG electrodes, current should flow through the electrodes, and therefore the resistance R_d should be small. R_s is the sum of the wire, electrode and electrolyte resistances (Box 2.3).

The impedance of the electrode is real and approximately equal to $R_d + R_s$ at very low frequencies, and it is also real and approaches the value of R_s at high frequencies. At frequencies between these extremes, the electrode impedance is frequency-dependent.

The electrode's potential changes when the current flows through the electrode. This change in the electrode interface's potential from its equilibrium value is called *polarization*. When there is no DC current flowing through the electrode, the electrode is called *ideally*

BOX 2.3

Circuit models for electrode –electrolyte interface

There has been a number of circuit models for the electrode–electrolyte interface proposed by the leaders in the field starting with Helmholtz in 1879, Fricke 1932, Geddes in 1968, and Sluters–Rehbach model in 1970. We are using a simplified model of a wet electrode in Fig. 2.9.

TABLE 2.6 Measured skin—electrode impedances for typical electrodes.

Electrode type	R_d	C_d	R_s
Wet Ag/AgCl (Chi et al., 2010)	5 kΩ	25 nF	500 Ω
Dry electrode (Chlaihawi et al., 2018)	1 MΩ	1 nF	500 Ω
Non-contact (Chi et al., 2010)	300 MΩ	35 pF	500 Ω

polarizable. For this electrode, $R_d = \infty$ in Fig. 2.9. When $R_d = 0$, DC current passes freely through the interface, and the electrode is called *ideally nonpolarizable*. In realistic non-polarizable electrodes, the value of resistance R_d in the order of hundreds Ω to several kΩ.

For wet electrodes, zero potential and zero impedance are impossible to achieve. Realistic goals include a low, stable potential and low and similar impedance at interfaces of both electrodes. Table 2.6 shows the capacitance and resistance of typical electrode types, including wet Ag/AgCl electrodes, and dry and non-contact electrodes. A wet electrode has very low resistance, and therefore it behaves as a non-polarizable electrode. A non-contact electrode is generally attached to the clothes and not directly to the skin. Hence, its resistivity is very high, and it behaves as a polarizable electrode.

2.6.1.3 Silver/silver chloride electrode

The Ag/AgCl electrode is a non-polarizable electrode with characteristics very close to ideally non-polarizable electrodes. Ag/AgCl electrodes are inexpensive and pervasively used in ECG applications.

The oxidation-reduction reaction that occurs at the surfaces of the Ag/AgCl electrodes is $Ag \leftrightarrows Ag^+ + e^-$.

Between the silver ions and the electrolyte, which is commonly gel that contains chloride ions, the reaction is

$$Ag^+ + Cl^- \leftrightarrows AgCl$$

where Ag is a silver atom (solid), Ag^+ and Cl^- are the silver and chloride ions (electrolyte), AgCl is the solid silver chloride molecule and e^- is the electron. The deposition of silver chloride on the silver electrode makes it suitable for any biological tissue because chloride ions exist in the tissues. The significant features of silver/silver chloride electrodes are as follows:

- A low, stable electrode potential
- Low level of noise
- A small value of resistance R_d.

The Ag/AgCl electrodes are also used as a benchmark to obtain a signal of high quality that can be compared against the signal obtained using other types of electrodes.

2.6.1.4 *Electrode–skin model*

When the wet electrode is applied to the skin, there is usually a gel layer between the electrode and the skin. This is shown on the left side of Fig. 2.10. To understand how biosignals are affected by electrode placement, sweating, the amount of gel, and other factors, we show a model (Fig. 2.10 on the right) that includes the impedances of the electrodes, gel, and skin layers.

The stratum corneum is the layer of dead cells at the skin's surface, which is in direct contact with the gel and has high impedance. It is shown as a yellow layer in Fig. 2.10 and it is a layer of the epidermis. Electrode–skin (epidermis) interface can be approximated by a capacitor C_{ep}. As pointed out in McAdams and Yoo (2011), the value of the capacitance of the skin is in the range 20–60 nF/cm^2 when the wet electrodes are applied, and the measurement is done several minutes after the application. The capacitance depends on the thickness of the stratum corneum. For a 1 cm^2 Ag/AgCl electrode, we can assume $C_{ep} =$ 20 nF.

The resistance of the epidermis R_{ep} depends on the placement of the electrode on the skin, the density of hair follicles under the electrodes, sweating, etc. In the case of sweating, both the resistance and the capacitance of the epidermis change. Sweating (perspiration) is modeled by adding the resistor R_p and the capacitor C_p in parallel with the epidermis resistor and capacitor. The dermis is modeled using the resistance R_{de}.

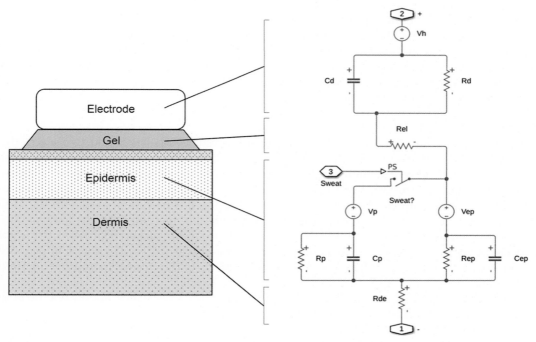

FIGURE 2.10 Electrode, gel, and skin layers on the left. Overall circuit model of a wet electrode and skin on the right.

A skin potential V_{ep} exists between the epidermis and the gel and is a result of the concentration difference of their ions. V_{ep} has a value of 15–30 mV negative with respect to inside the body and changes with sweating (McAdams and Yoo, 2011).

Motion artifacts affect all biomedical measurements. The artifacts can be included in the model as a controlled voltage source in series with the biopotential voltage source. If the thickness of the epidermis is changed by stretching or pressing down on the skin, the skin potential can vary by 5–10 mV (McAdams and Yoo, 2011).

Electrode *gel* has multiple purposes. First, it is placed to ensure good electrical contact between the electrode and the patient's skin. Also, the gel contains chloride ions, facilitating the transfer of charge at the electrode–electrolyte interface. In addition, after the gel is applied to the skin, the resistance of the skin can be reduced more than 10 times after several minutes. Other ways to reduce the resistance of the skin include cleaning the skin with alcohol wipes and applying abrasive pads. The gel is modeled as a resistor R_{el} whose resistance is assumed to be around 500 Ω.

The equivalent impedance of the wet skin–electrode interface Z_{eq} is given below:

$$Z_{eq} = R_{de} + \left(\frac{1}{j\omega C_p} + \frac{1}{R_p} + \frac{1}{j\omega C_{ep}} + \frac{1}{R_{ep}} \right)^{-1} + R_{el} + \left(\frac{1}{j\omega C_d} + \frac{1}{R_d} \right)^{-1}$$

where $\omega = 2\pi f$.

This model allows us to simulate the electrode, electrode–electrolyte, and skin interface. However, the model makes a number of assumptions such that the model's parameters do not change over time. Other issues with the layered skin–electrode model (Lu et al., 2018) are that RC circuits in parallel cannot characterize the biological characteristics of the skin, such as the dispersion effect of cell membranes. The skin layers are not uniform, nor is the electrode surface leading to non-uniform current density. However, modeling these issues is outside of the scope of this book.

2.6.1.5 Connecting wet electrodes

We present in Fig. 2.11 the model that includes a bioelectric signal and two electrodes attached to the skin. The model of the electrode–electrolyte and skin interface is the same as the one shown in Fig. 2.10 on the right. The input bioelectric signal is simulated as a square pulse wave with an amplitude of 1 mV, frequency of 1 Hz, and pulse width of 10%. The parameters are set to emulate the ECG signal. The pulse wave is used here instead of the ECG signal to better show the effect the electrodes and the interface have on the signal. The parameters of the model are given in Table 2.7. The impedance at the output represents the input impedance of the amplifier R_{in} that the electrodes are connected to. The input impedance is quite low in Table 2.7 (20 MΩ) to better show the effects of the electrodes and the interface on the signal.

We simulated several scenarios by varying the epidermis resistance R_{ep} on one or both electrodes and showed the potential difference over the resistor R_{in} in Fig. 2.12.

The scenarios include the following.

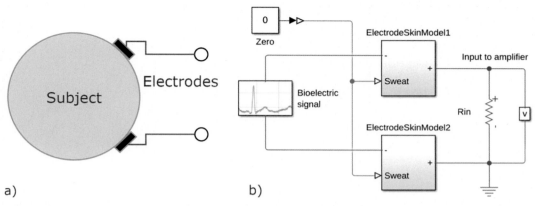

a) b)

FIGURE 2.11 (A) A setup showing two electrodes attached to the subject's body. The same setup is simulated using the equivalent circuit model in (B). It models the bioelectric signal, the interface between the skin and the electrodes, and the electrodes themselves. The electrodes are connected to the circuit, which is not modeled here. Only the input impedance of the amplifier R_{in} is included. The voltmeter is used at the output to measure the output voltage. The electrode models are based on the electrode–electrolyte and skin interface models shown in Fig. 2.10 on the right.

TABLE 2.7 Parameters of the circuit shown in Fig. 2.11B and simulation scenarios shown in Fig. 2.12.

Interface layer	Parameter	Value
Electrode	V_h	230 mV
	C_d	20 nF
	R_d	5 kΩ
Electrolyte gel	R_{el}	500 Ω
Sweat glands and ducts	V_p	5 mV
	C_p	5 nF
	R_p	100 Ω
	Sweating	Simulation scenario: (a) No, (b) No, (c) No, (d) Yes
Epidermis	V_{ep}	20 mV
	C_{ep}	20 nF
	R_{ep}	Simulation scenario: (a) $R_{ep1} = 500$ kΩ, $R_{ep2} = 500$ kΩ (b) $R_{ep1} = 500$ kΩ, $R_{ep2} = 250$ kΩ (c) $R_{ep1} = 5$ kΩ, $R_{ep2} = 10$ kΩ (d) $R_{ep1} = 500$ kΩ, $R_{ep2} = 500$ kΩ
Dermis and subcutaneous layer	R_{de}	1 kΩ
Impedance of the amplifier	R_{in}	20MΩ

2. Transducers

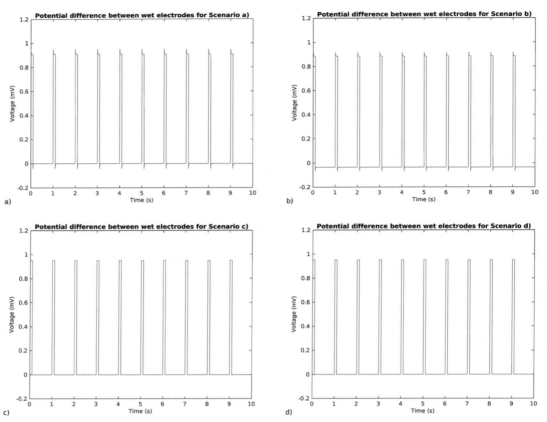

FIGURE 2.12 Signals at the output of the circuit model shown in Fig. 2.11B for input pulse signal with an amplitude of 1 mV and a frequency of 1 Hz. (A) All the parameters are the same between the electrode–skin interface, (B) there is a mismatch in epidermis resistances, but their resistances are still high. (C) Epidermis resistances are significantly reduced. (D) Epidermis resistances in the model are very high, and the sweating component is included.

(a) All parameters of the two electrodes are the same, and the sweating component is not included. This scenario simulates the situation in which there is no skin preparation so that the epidermis resistance R_{ep} under both electrodes is high $R_{ep} = 500$ kΩ. Since the interface parameters are the same, there is no baseline shift, as shown in Fig. 2.12A. However, the signal is distorted, and it has a bit lower amplitude than the input signal. The reason for the distortion is large $R_{ep}C_{ep}$ which results in a cut-off frequency of 15.9 Hz.

(b) In this scenario, there is a mismatch between the epidermis resistances under the two electrodes: $R_{ep1} = 500$ kΩ, $R_{ep2} = 250$ kΩ. Skin preparation is not done so that the epidermis resistances are still high. The sweating component is not included. Compared with the signal shown in Fig. 2.12A, the minimum value of the signal in

Fig. 2.12B is negative. So, the signal at the output is still distorted, and there is now a baseline drift.

(c) We simulated a scenario where skin preparation is done, and the epidermis resistance is significantly reduced: $R_{ep1} = 5$ kΩ, $R_{ep2} = 10$ kΩ. The sweating component is not included. The signal at the output is not distorted, and there is no baseline shift, as shown in Fig. 2.12C.

(d) This scenario simulates the situation in which there is no skin preparation so that epidermis resistance R_{ep} under both electrodes is high $R_{ep} = 500$ kΩ, but the sweating component is included. The signal at the output is not distorted, and there is no baseline shift, as shown in Fig. 2.12D.

As can be seen, sweating helps reduce the distortion of the signal because it reduces the resistance of the epidermis. However, usually sweating will cause baseline drift over time as the sweating is a relatively lengthy process.

2.6.2 Dry electrodes

Dry electrodes are more prone to noise due to high skin—electrode impedance, motion artifacts, and friction between electrodes and skin. Dry electrodes can be made of metals or integrated into flexible materials such as textiles. For example, the Apple watch incorporates dry metal electrodes for measuring ECG in their smartwatch. The Apple watch series 4, 5, and 6 use a titanium electrode in the digital crown and a special alloy on the back of the watch that has low electrical conductivity (Using Apple Watch, 2020).

Dry electrodes made of textiles potentially improve comfort and long-term usage. A textile electrode is a piece of textile with a conductive yarn knitted into the textile or with conductive ink printed on the textile. The research on textile electrodes shows that the electrodes have good durability in stretching, bending, and washing since there are no significant changes in electrical resistance of the electrode during the period of stretching and bending tests and there are minimal changes after washing. This encouraging result shows the viability of developing textile electrodes for wearable ECG monitoring (Li, 2018).

In order to collect bioelectric signals from hairy skin, a spiked electrode offers better performance than other dry electrodes. This is because the metal spikes can easily penetrate the hair and make direct contact with the skin (Xie, 2013).

A model of the dry electrode is shown in Fig. 2.13A. Electrodes made of conductive metal have the same circuit model as the circuit model of the wet electrodes, except that there is no electrolyte gel in between the electrode and the electrolyte, and therefore, C_d, R_d, and R_{el} are not included in the model (Chi et al., 2010). The resistance of the epidermis is higher than when the gel is applied. The resistance might decrease in the case of metal electrodes due to sweating. We modeled the same circuit as shown in Fig. 2.11B but with two dry electrode interface models and with the resistor $R_{in} = 1$ GΩ. The only difference between the two dry electrodes was in the epidermis resistance $R_{ep1} = R_{ep2} = 2$ MΩ all the other parameters are the same as in Fig. 2.13A. The voltage difference at the output of the circuit is shown in Fig. 2.13B. There is no distortion of the signal due to the large input impedance of the amplifier.

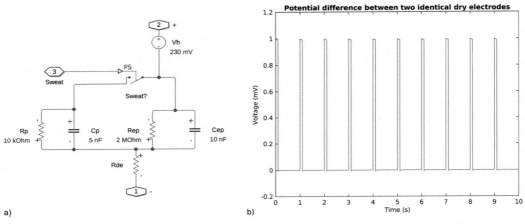

a)   b)

FIGURE 2.13   (A) The model of the dry electrode based on Chi et al. (2010). (B) The potential difference at the resistor R_{in} in Fig. 2.11B for the case when $R_{ep1} = R_{ep2} = 2\,M\Omega$ and the other parameters between the two dry electrodes are the same.

2.6.3 Non-contact electrodes

Non-contact electrodes aim to measure the ECG signal through clothing. They use an amplifier next to the electrode to provide a large impedance to overcome the problems of high resistance from clothes or hair and reduce the effect of stray capacitance.

In the non-contact electrode, there is an air gap between the electrode and the body. This gap is modeled using a capacitor C_{gap} in Fig. 2.14A (Chi et al., 2010). The effect of the clothes is modeled using an equivalent circuit with the capacitor C_{cotton} and and resistor R_{cotton} in parallel. Very large resistance of the clothes is used. Sweating would not have a significant effect here, and it is not modeled. When the two non-contact electrodes are attached, as in Fig. 2.11B, the input amplifier impedance needs to be very high. Fig. 2.14B and C show the potential difference between the electrodes in two cases where the output impedance is $R_{in} = 1\,G\Omega$ and $R_{in} = 10\,G\Omega$, respectively. It is clear that when the resistance R_{in} is larger, there is less distortion in the output signal. It is also obvious that the output signal is much more distorted than in the case of wet and dry electrodes.

2.7 Summary

In this chapter, we described transducers used to sense physical phenomena and convert one energy source into another more suitable for further processing. First, we covered displacement transducers, including strain gauges, piezoelectric and capacitive transducers. We then discussed LEDs and photodiodes. After optical sensors, we provided details about the modeling of biopotential electrodes and their interfaces with the skin. We showed simplified models of all transducers described in this chapter. The models of the transducers are

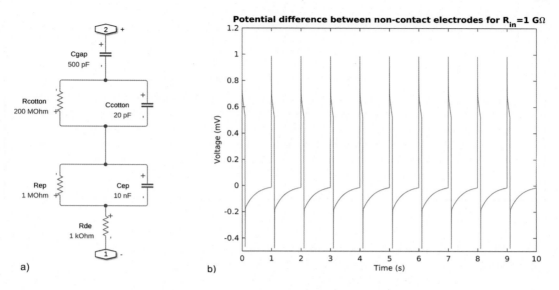

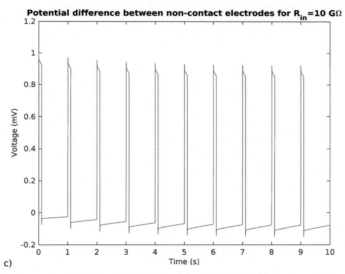

FIGURE 2.14 (A) The model of the noncontact electrode based on Chi et al. (2010). The potential difference over the resistor R_{in} in Fig. 2.11B for the following cases: (B) $R_{in} = 1\ G\Omega$ and (C) $R_{in} = 10\ G\Omega$.

circuit-based, and we mainly did not provide explanations of physical or chemical laws behind the operation of these transducers.

Research in transducers and sensors has been going in multiple directions, including miniaturizing transducers and sensors, making them ultra-low power, developing new technologies for the transducers, coming up with new applications, automated data processing, and

---

BOX 2.4

S m a r t p h o n e   s e n s o r s

Smartphone sensors can measure motion, orientation, sound, environmental conditions, and other parameters. Smartphone microphones detect sound and can be used to detect body sounds due to breathing or coughing. Accelerometers measure the movement of an object relative to the smartphone in x, y, and z directions. Gyroscopes are used to measure rotational displacement. Gyroscopes and accelerometers can be used to obtain a cardiac signal (seismocardiogram) when placed on the chest.

---

so on. Many of these research directions are related to complete sensors and not only transducers and, therefore, will be mentioned later in the book.

Regarding form factor, the transducers and sensors appear in many forms, including smartphones, shirts, smartwatches, jewelry, armbands, headphones, hearing aids, shoes, dental appliances, eyeglasses, skin-like sensors, and so on. The ability to design such small transducers and implement them in everyday items will make monitoring cardiac and respiratory signals truly pervasive.

Research on power consumption includes battery-less sensors that can operate only when in the proximity of an energy source, such as near field communication (NFC), rechargeable and flexible batteries, energy harvesting, and low power design that includes reduction of power consumption of the electronic components and communication.

The chapter did not discuss sensors because they commonly require conditioning logic or amplifiers, which are introduced in the next chapter. We briefly mention sensors used in smartphones pervasively nowadays that have the potential to be used in wellness and eventually in medical applications in Box 2.4.

2.8 Appendix

2.8.1 Derivation of the sensitivity of the piezoelectric equivalent circuit

Derivation of sensitivity related to Fig. 2.2 is given below based on the material presented in Webster (2010).

- K is a proportionality constant [C/m]
- dL represents changes in the depth of the transducer due to the applied force
- C represents the parallel combination of the capacitance of the sensor, cable, and amplifier input
- R is the transducer's leakage resistance and the amplifier input resistance

The piezoelectric device is represented as a charge generator. For a deflection L of the piezoelectric material, a charge q is generated as follows:

$$q = K \cdot L$$

or a current of the sensor i_s generated as

$$i_s = dq/dt = K \cdot dL/dt$$

From Kirchhoff's Current Law, we get $i_s = i_c + i_r$, where i_c is the current through the capacitor and i_r is the current through the resistor.

$$i_c = C \cdot dv_o/dt$$

$$i_r = v_o/R$$

After combing the equations above, we get

$$C \cdot dv_o / dt = K \cdot dL/dt - v_o/R$$

This first-order ordinary differential equation can be solved in the frequency domain.

$$C \cdot j\omega \, V_o \, (j\omega) = K \cdot j\omega \, L(j\omega) - V_o \, (j\omega) \, /R$$

After rearranging:

$$Cj\omega V_o \, (j\omega) = Kj\omega L(j\omega) - V_o(j\omega)/R$$

$$j\omega V_o \, (j\omega) = (K \, / \, C)j\omega L(j\omega) - V_o(j\omega)/RC$$

$$V_o \, (j\omega) = (K \, / \, C) \, L(j\omega) - V_o(j\omega)/ \, j\omega RC$$

$$V_o \, (j\omega) + V_o(j\omega)/ \, j\omega RC = (K \, / \, C) \, L(j\omega)$$

$$V_o(j\omega) \, / \, L(j\omega) = (K \, / \, C)/(1 + 1/ \, j\omega RC)$$

$$V_o(j\omega) \, / \, L(j\omega) = (K \, / \, C)j\omega RC/(1 + j\omega RC)$$

After defining K/C as the transducer sensitivity K_s and RC as the time constant τ, we get

$$V_o(j\omega) \, / \, L(j\omega) = K_s j\omega \tau/(1 + j\omega \tau)$$

2.9 Problems

Please note that questions that start with * are not covered in the book. Instead, they are added for readers interested in exploring different topics further.

2.9.1 Questions

2.1. Resistive displacement transducers
   (a) Define the sensitivity of a transducer in general. Derive formulas for sensitivity for resistive displacement.
   (b) What are the main problems related to metal strain gauges and to semiconductor-based strain gauges (consider gauge factor, temperature stability, length, etc.)?
   (c) *Describe the physics behind the piezoresistive effect.
   (d) *Find off-the-shelf resistive transducers, extract information from their datasheets and compare their sensitivity and linearity.
   (e) *Describe how these transducers are used to measure pressure, flow, and flow rate. Find three medical devices that rely on resistive transducers for measuring pressure, flow and flow rate and describe the characteristics of their transducers and their functionality and operating modes in devices.
2.2. Capacitive transducers
   (a) Classify capacitive transducers. List capacitive transducers that we use in everyday life and relate these transducers to the classification.
   (b) Derive expressions for the sensitivity of capacitive transducers that are based on changes in the (a) plate gap, (b) area, or (c) amount of dielectric material.
   (c) *What capacitive transducers are best for the following applications: non-contact measurements, very small displacements, and angular displacement measurements? Relate these applications to practical biomedical applications.
   (d) Do we increase the capacitance by increasing the number of plates from n to $n +$ 1 in the interdigitated capacitive transducer? What is the mechanism behind that?
2.3. Piezoelectric transducers
   (a) Derive formulas for sensitivity for piezoelectric displacement transducers and for the relationship between voltage at the output and the displacement.
   (b) *Describe the physics/chemistry behind the piezoelectric effect.
   (c) *Find off-the-shelf transducers, extract information from their datasheets and compare them.
   (d) *Describe how these transducers are used to measure blood pressure, pulse rate, acceleration, and foot sole pressure. Find medical devices that do that and describe their functionality.
   (e) *What are other biomedical applications of piezoelectric transducers besides measuring displacement? Describe them briefly.

2.4. Compare displacement transducers
   (a) Compare different displacement transducers, including capacitive, resistive, and piezoelectric, regarding their size, sensitivity, linearity, and temperature stability.
   (b) Compare displacement transducers regarding how complex it is to connect them with the rest of the electronics.
   (c) Describe how resistive, capacitive, and piezoelectric displacement sensors are used in biomedical applications where pressure and flow are converted into displacement.
   (d) *What displacement transducer is used in a breathing belt to measure respiration rate? What are the requirements for sensors for this application?
   (e) *How are displacement transducers used as components of pressure sensors in arterial tonometry? What are the requirements for sensors for this application?
2.5. LEDs and photodiodes
   (a) Derive the sensitivity of a photodiode.
   (b) *What is the photoelectric effect? Find out more about it.
   (c) *How does a photodiode work?
   (d) Explain why different wavelengths of LEDs are used in biomedical applications. What are these wavelengths?
   (e) *Describe one commercial photodetector—you can search for photodetector used in pulse oximeters or photoplethismographs.
2.6. Electrode–electrolyte potential
   (a) Where and why does an ion–electron exchange occur?
   (b) Explain the following terms/concepts: oxidation, reduction, half-cell potential, and reference electrode.
   (c) How is the electrode potential computed?
   (d) How can one make a battery if she/he uses two identical electrodes?
   (e) Do we want to have zero potential between the electrodes to be able to detect weak biosignals? When don't we have zero potential between two identical electrodes?
   (f) *Describe some commercial electrodes used for EEG in terms of their properties.
2.7. The electrode–electrolyte impedance
   (a) What happens when the electrode is in contact with a solution of its ions? What is the difference when it is in contact with a solution of ions of some other substance? What would happen if a metal electrode is used directly in touch with the human body?
   (b) How is the electrical charge carried in electrodes and wires, and how in the gel and the human body? Where does the transition happen?
   (c) What is the double layer? How is the double layer modeled using an electric circuit? What are the limitations of that model?
   (d) What is the frequency response of this model circuit?
   (e) What is the phase response? Why is the phase response important?
2.8. The electrode–skin impedance and Ag/AgCl electrodes
   (a) What are polarizable and non-polarizable electrodes? Where are they used? What types of electrodes do we need for measuring biopotential?
   (b) What are the features of silver/silver chloride electrodes? Describe the reduction and oxidation process of these electrodes.

(c) Why do silver ions not contaminate the skin when silver/silver chloride electrodes are used?

(d) Draw the model of the skin layers and the electrode on the skin and list the typical values of resistors and capacitors? Derive equivalent impedance, phase, and frequency response.

(e) Describe why high and variable impedance of electrode and skin is a problem. Why do the potential over the electrode and skin vary with time? Why is that a problem?

2.9. Problems with connecting the electrodes

(a) Do we want to have zero potential between the electrodes on the skin in order to be able to detect weak biosignals? When do we not have zero potential between two same electrodes?

(b) What can be done regarding electrode design and skin preparation when the difference in potential between the electrodes is not zero and varies in time?

(c) *What was done with Ag/AgCl electrodes to stabilize the voltage on the electrodes and reduce electrochemical noise? Does improving the effective area of the electrode help reduce electrochemical noise?

(d) *What design decisions need to be made when designing/choosing instrumentation amplifiers to handle the non-ideal characteristics of the electrodes?

(e) *Analyze several medical devices regarding other issues in connecting electrodes and instrumentation amplifiers.

2.10. Research directions and current trends

(a) *Describe skin-like pressure and strain sensors. What are the applications? Describe the types of sensors.

(b) *Describe MEMS sensor technology applied to the medical field. Describe medical devices that use MEMS sensors.

(c) *Describe state-of-the-art of 10 degrees of freedom (10DOF) sensors. Describe characteristics of sensors used as components of 10DOF sensors (3D accelerometer, 3D gyroscope, 3D magnetometer, a barometer). Describe biomedical applications of 10DOF sensors.

(d) *Describe state-of-the-art organic LEDs and photodiodes.

(e) *Describe sensors used in smartphones. How are these sensors used in medical applications?

(f) *What is Labs-on-chip, and how can they be used in biomedical applications?

(g) *What are nanochips, and how can they be used in biomedical applications?

2.9.2 Problems

2.11. Show that the capacitance of the multi-plate capacitor is $C = (n-1)\varepsilon A/L$. Find examples of biomedical applications where multi-plate capacitors are used.

2.12. For the electrode–electrolyte model in Fig. 2.9, explain why the resistance at DC frequency is $R_d + R_s$ and the resistance at high frequencies R_s. What can you tell about the phase of the impedance?

2.13. When explaining LEDs, the following values of LED parameters were given: the luminous intensity is $I = 100$ mcd, the solid angle is $\Omega = 1.84$ sr, and the luminous efficacy $\eta = 90$ lm/W. Compute the radiant power P and show your steps.

2.9.3 Simulations

2.14. A model of a piezoelectric transducer is shown on the book's webpage.
   (a) Plot the frequency response for Equation (2.2).
   (b) Add a capacitor in parallel to increase the time constant. Run the simulation and plot a figure similar to Fig. 2.3B.
2.15. This is an example of an equivalent circuit model of a photodiode based on (Webster, 2010). Consider that the photodiode current is proportional to the incident light. The model is shown in Fig. 2.15. The parameters of the model components are
   • photodiode current: $I_p = 520$ nA
   • forward voltage: $V_f = 1.3$ V
   • diode current has dark current: $I_d = 0.2$ nA
   • junction capacitance: $C = 1.8$ pF
   • shunt resistance: $R_{shunt} = 1000$ MΩ
   • series resistance: $R_s = 10$ Ω
   • load resistance: $R_{load} = 50$ Ω
   (a) Compare this model with the photodiode model that already exists in Simscape.
   (b) Draw output current versus light power.
   (c) Perform sensitivity analysis of the components of the model.
   (d) Use the datasheet of a photodiode of your choice and include the parameters of that photodiode into the model. Then, repeat steps (a) to (c).
   (e) Write the equation for the current through the load resistor.
2.16. A model of a system with two dry electrodes is shown on the book's webpage.
   (a) Modify the input impedance of the amplifier and comment on the distortion of the signal for the smaller values of the input impedance.

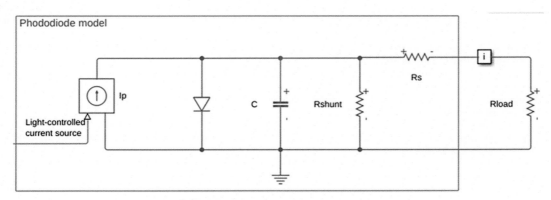

FIGURE 2.15 A model of a photodiode.

(b) Use the following values of the epidermis resistance $R_{ep1} = 1\,M\Omega$ and $R_{ep1} = 2\,M\Omega$ and observe and comment on the signal at the output.

(c) *Model the motion artifacts and determine if dry or wet electrodes are more affected by the motion artifacts. Discuss your results.

2.10 Further reading

In this section, we refer to several relevant resources for further reading. However, please note that this list is not comprehensive and that many excellent and relevant resources might have been omitted unintentionally.

A detailed description of different sensors can be found in Ripka and Tipek (2007). Transducers, sensors, and electronics for biomedical devices are covered in Webster (2010). A textbook introduction to biomedical transducers is given in Chan (2016).

Very detailed coverage of piezoelectric materials and devices is given in Vijay (2013). Excellent material on biomedical optics can be found at https://omlc.org/classroom/index.html. Similar material was very well presented in the slides by Jacques (2020). LEDs are described in detail in Chapter 10 of Dakin and Brown (2017). A comprehensive review paper that covers equivalent circuit models for dry electrodes is Chi et al. (2010).

References

Aezinia, F., 2014. Design of Interface Circuits for Capacitive Sensing Applications. Ph.D. dissertation. Simon Frazer University.

Chan, A.K., 2016. Biomedical Device Technology: Principles and Design, second ed. Charles C Thomas Publisher.

Chi, Y.M., Jung, T.P., Cauwenberghs, G., 2010. Dry-contact and non-contact biopotential electrodes: methodological review. IEEE Reviews in Biomedical Engineering 3, 106–119.

Chlaihawi, A., et al., May 2018. Development of printed and flexible dry ECG electrodes. Sensing and Bio-Sensing Research 20, 9–15. Elsevier.

Dakin, J.P., Brown, R., 2017. Handbook of optoelectronics. In: Concepts, Devices, and Techniques, vol 1. CRC Press.

Dong, J., Xiong, D., 2017. Applications of light emitting diodes in health care. Annals of Biomedical Engineering 45 (11), 2509–2523.

Gupta, M., Ballato, J., 2006. The Handbook of Photonics, second ed. CRC Press.

Jacques, S., 2020. Tissue Optics. SPIE Photonics West Short Course SC029, San Francisco.

Li, Z., 2018. Textile Electrode Design and Simulation for Bio-Signal Recording. Ph.D. thesis. University of Southampton.

Lu, F., et al., 2018. Review of stratum corneum impedance measurement in non-invasive penetration application. Biosensors, MDPI 8 (2), 31, 26 Mart.

Macy, A. "The Handbook of Human Physiological Recording," https://alanmacy.com/book/the-handbook-of-human-physiological-recording/, Last accessed on April 29, 2021.

McAdams, E., 2011. Biomedical electrodes for biopotential monitoring and electrostimulation. In: Yoo, H.-J., et al. (Eds.), Bio-Medical CMOS ICs. Springer.

Ripka, P., Tipek, A. (Eds.), 2007. Modern Sensors Handbook. ISTE USA, Newport Beach, CA.

Staworko, M., Uhl, T., 2008. Modeling and simulation of piezoelectric elements, comparison of available methods and tools. Mechanics 27 (4).

Using Apple Watch for Arrhythmia Detection, 2020. https://www.apple.com/ca/healthcare/docs/site/Apple_Watch_Arrhythmia_Detection.pdf. (Accessed 28 April 2021).

Vijay, M.S., 2013. Piezoelectric Materials and Devices: Applications in Engineering and Medical Sciences. CRC Press.

Wang, T.-W., Lin, S.-F., February 2020. Wearable piezoelectric-based system for continuous beat-to-beat blood pressure measurement. MDPI Sensors 20 (851).

Webster, J.G. (Ed.), 1997. Design of Pulse Oximeters. IOP Publishing Ltd.

Webster, J.G. (Ed.), 2010. Medical Instrumentation: Application and Design, fourth ed. John Wiley & Sons.

Xie, L., et al., 2013. Characterization of dry bio-potential electrodes. In: 35th Annual International Conference of the IEEE EMBS, Osaka, Japan, 3 - 7 July.

Xie, H., Fedder, G.K., Sulouff, R.E., 2008. 2.05 - accelerometers. In: Gianchandani, Y.B., et al. (Eds.), Comprehensive Microsystems. Elsevier, pp. 135—180.

3

Electronics

Acronyms and explanations

ADC   analog-to-digital converter
CDC   capacitance-to-digital converter
CMG   common mode gain
CMRR   common mode rejection ratio
DAC   digital-to-analog converter
DDS   direct digital synthesis
DMA   direct memory access
GBWP   gain-bandwidth product
I²C   interintegrated circuit
LSB   least significant bit
MAC   multiple and add/accumulate
MCU   microcontroller
PGA   programmable gain amplifiers
PGA   programmable gain amplifier
PWM   pulse width modulation
RAM   random access memory
SPI   serial peripheral interface

Variables used in this chapter include:

N number of bits at the output of an A/D converter or the input of
         a D/A converter
V_o analog output voltage
V_{in} analog input voltage
V_o sensitivity
V_o gain of an operational amplifier
A_d differential gain
A_c common mode gain
τ time constant
f_c cutoff frequency
T signal period and PWM period
T_s sampling period
f_s sampling frequency
f_h Nyquist frequency
V_{FS} full-scale voltage of A/D and D/A converters
W binary code word that represents the digital input to a D/A
         converter
D pulse width offset for PWM

3.1 Introduction

This chapter focuses on biomedical system components for signal conditioning and conversion to digital signals. The majority of biomedical devices nowadays are designed similarly to the system shown in Fig. 2. Transducers have been introduced in the previous chapter. This chapter presents analog processing components, including conditioning circuits, amplifiers, and filters. They are presented in blue in Fig. 2. Filters remove undesired frequency components of the signal. In a system where processing is done in software, the signal is filtered using an antialiasing analog filter and then converted into a digital signal using an analog-to-digital (A/D) converter. After that, the signal is processed in real time or stored and processed offline by a processor. The digital component in Fig. 2 is the signal processing and control block.

Table 3.1 shows the conditioning circuits commonly connected to the transducers that we covered in Chapter 2. For example, the resistive displacement transducers are typically connected to a Wheatstone bridge, which is further connected to a differential or instrumentation amplifier. In addition to traditional conditioning circuits, we also show some modern industrial integrated solutions that connect transducers directly to the A/D converter or the microcontroller.

In this chapter, we first start with bridge circuits. Next, we introduce operational amplifiers and show their different configurations. Then, we present different conditioning circuits based on operational amplifiers for connecting electrodes (instrumentation amplifiers), piezoelectric transducers (charge amplifiers), and capacitive transducers (synchronous demodulation). Also, we show some operational amplifier solutions for linearizing the bridge circuits. Next, we present passive and active filters. After that, we introduced A/D converters. We introduce concepts of sampling, aliasing, errors during the conversion, and other concepts needed to understand the functioning of A/D converters. We then show several integrated special-purpose configurations of converters, including capacitance to digital converters, time to digital converters, and others. Components used for generating signals are covered

TABLE 3.1　Transducers and their conditioning circuits.

Transducer	Conditioning circuit	Integrated circuit with the conditioning circuit and other components
Strain gauges	Wheatstone bridge and a differential amplifier	Bridge transducer A/D converter
Piezoelectric	Charge amplifier	Integrated solution that includes charge amplifier
Capacitive	AC bridge, synchronous demodulation	Capacitance to digital converter
Photodiodes	Transimpedance amplifier	Current to digital converters, current to frequency converters
Electrodes	Instrumentation amplifier	ECG integrated circuits

next, including digital-to-analog (D/A) converters and pulse width modulation (PWM) circuits. Integrated special-purpose D/A converters and PWM circuits are discussed as well.

3.2 Wheatstone bridge

Several transducers, such as temperature transducers and strain gauges, convert a physical signal into resistance. There are technical limitations that need to be addressed when using these transducers, including the accurate measurement of small resistance changes, commonly in the order of a few percent of the nominal resistance.

The simplest way to convert a resistance change into a voltage change would be to place the transducer in a *voltage divider* circuit. However, in this case, the output voltage change is a nonlinear function of the resistance change and its baseline voltage depends on the resistance of the selected resistor and the resistance of the transducer. We will show below that the baseline voltage of a Wheatstone bridge configuration has a baseline voltage of zero.

The Wheatstone bridge is commonly used for measuring small changes in resistance. The circuit is shown in Fig. 3.1A. It consists of four resistors connected as two voltage dividers, a voltage or current source connected across one diagonal. In our case, the bridge is excited using a voltage source V_E. The output voltage is measured across the other diagonal. Namely, the voltage difference V_o between the outputs of the voltage divider on the left (resistors $R1$ and $R2$) and the voltage divider on the right (resistors $R3$ and $R4$) is measured. The output voltage V_o is a of the change in the resistance:

$$V_O = V_E \left[\frac{R_1}{R_1 + R_2} - \frac{R_3}{R_3 + R_4} \right]$$

(3.1)

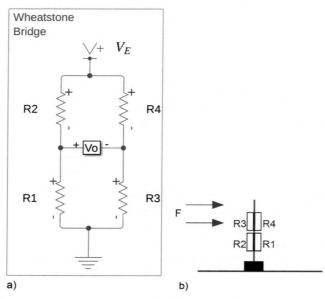

a)   b)

FIGURE 3.1   (A) Wheatstone bridge and (B) possible full-bridge configuration of strain gauges on a bending vane.

When $V_o = 0$ the bridge is balanced, meaning that $R_1/R_2 = R_3/R_4$. The voltage V_o is then amplified using an amplifier that we will introduce later in the chapter. The bridge with only one transducer is called the quarter bridge or the single-element varying bridge. The quarter bridge is suited for temperature sensing and for applications with a single resistive strain gauge. Let us assume that the transducer is placed in the bridge in the right leg and that its resistance is R_3. Further, let us assume that $R_1 = R_2 = R_4 = R$ and that the transducer resistance can be represented as $R_3 = R + \Delta R$ where ΔR represents changes in the resistance which can be, in general, both positive and negative. R is the nominal resistance of the transducer.

The *sensitivity* of the bridge is defined as the change in the output V_o over the change of the resistance of the transducer at the input ΔR. Sometimes, the sensitivity is defined as the ratio of the maximum expected change in the output voltage to the excitation voltage, but we will use the first definition here. Based on Eq. (3.1) and by replacing the values of all resistors with R and the transducer resistance with $R_3 = R + \Delta R$, we get for V_O:

$$V_o = \frac{-\Delta R}{2(2R + \Delta R)} V_E$$

If ΔR is very small, then the sensitivity of the bridge can be approximated as

$$S = \frac{V_o}{\Delta R} = -\frac{V_E}{4R}$$

If it is possible to use two or four transducers in the bridge configured so that the resistances of half of them reduce while the resistances of another half of the transducers increase after applying the force, then the sensitivity can be improved. In addition, for four transducer configuration, the relationship between V_O and ΔR is linear:

$$V_O = -\frac{\Delta R}{R_o} V_E$$

Four transducer configuration is called a full bridge or all-element varying bridge. The full bridge is an industry-standard configuration for load cells based on four identical strain gauges. An example of placing four transducers on a bending vane or a beam and their organization in the bridge is shown in Fig. 3.1B. The vane or the beam will bend due to the applied force F so that the resistance R_2 and R_3 will increase while the resistances R_1 and R_4 will decrease.

If it is impossible to use four transducers, we can linearize the bridge using an operational amplifier shown in Section 3.3.6.1. AC or capacitive bridges will also be introduced in Section 3.3.6.2.

Important aspects in designing the bridge circuit include selecting a configuration with 1, 2, or 4 transducers, techniques for linearizing the bridge if necessary, selecting the excitation (current or voltage, AC or DC), and the amplifier at the output of the bridge. Other design decisions include the tolerance of resistors, on-chip or discrete bridge solution, stability of the excitation, and so on.

3.3 Amplifiers and their configurations

3.3.1 Operational amplifier

An operational amplifier amplifies the voltage difference between its inputs. They are characterized using several parameters, such as.

- *Open-loop gain* V_o can be in the order of a million.
- The relationship between the input (V_{in}^+, V_{in}^-) and the output V_o voltages is $V_o = A(V_{in}^+ - V_{in}^-)$.
- *Input resistance*, V_o that should be very high, for example, in the order of GΩ.
- *Output resistance*, V_o that is low in the order of 100 Ω.
- Minimum and maximum output voltages that depend on the applied supply voltage.
- Maximum *slew rate*, which is defined as the maximum positive or negative rate of change of the magnitude of output voltage. A typical value is, for example, 2 V/μs.
- *Open-loop bandwidth* is the frequency at which the magnitude of the frequency response drops by 3 dB compared to the magnitude at low frequencies.

An equivalent circuit of a band-limited operational amplifier is shown in Fig. 3.2A. The model describes the effects of the input and output impedance and the limited bandwidth but does not include nonlinear effects such as the slew rate. The input resistance V_o connected to the voltage-controlled voltage source that acts as an ideal amplifier and amplifies the input by the gain V_o. In this example, the gain is 100,000. A voltage-controlled voltage source maintains the output voltage proportional to the input voltage. An ideal operational amplifier is a voltage source dependent on the voltage between its input terminals. Components V_o and V_o emulate the first order lowpass filter with a cutoff frequency of $1/(2\pi R_p C_p)$ that determines the open-loop bandwidth. The second voltage-controlled voltage source is set to a gain of one, and it is there to separate the input from the output. The magnitude and phase response of the equivalent circuit model of the operational amplifier is shown in Fig. 3.2A. The magnitude is 100 dB (100,000 gain) at low frequencies, and the cutoff frequency is 1 kHz $(1/2\pi R_p C_p)$. The phase response is zero degrees at low frequencies.

3.3.2 Operational amplifier parameters

We will introduce several additional parameters of the operational amplifiers.

The *offset voltage* is a differential voltage that needs to be applied at the input to obtain $V_o = 0\ V$. Ideally, it should be zero. The low offset voltage that is common nowadays in precision low noise amplifiers is, for example, 10 μV. An offset trim pin is commonly provided to help remove the offset externally.

An ideal operational amplifier will have zero current at its input terminals. An *input bias current* is an input current to the realistic operational amplifier and can be modeled as two current sources connected directly to the inputs V_{in}^+ and V_{in}^- in Fig. 3.2A. For example, 20 nA bias current is common. Bias currents on both input terminals should be equal. However, in realistic operational amplifiers, there is a difference in the bias currents between the input terminals, and the absolute value of the current difference is called the *input offset current*.

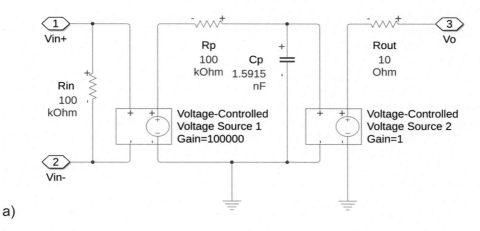

a)

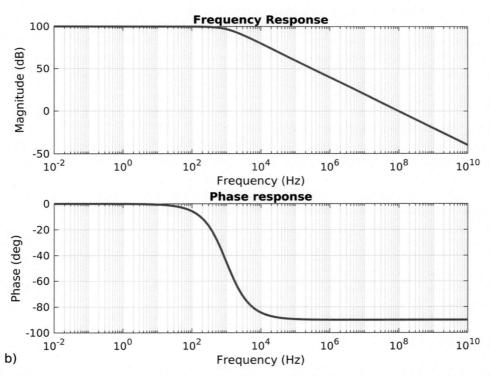

b)

FIGURE 3.2 (A) The equivalent circuit of a band-limited operational amplifier and (B) its magnitude and phase response.

The noise of the amplifier can be modeled as a voltage source connected in series to one of the inputs of the operational amplifier. Therefore, it can be considered as an offset that changes over time. In low noise amplifiers, noise can be as low as $3\,nV/\sqrt{Hz}$ (read as 3 nV per root Hertz).

Operational amplifiers are normally used in *negative feedback configuration* in which the output is connected to one input through a *feedback circuit*. By doing that, the ideal operational amplifier will adjust the output until the difference of the voltages at the input is close to zero. When used as amplifiers, several more performance metrics play important roles. One is *Gain BandWidth Product* (GBWP) which represents a product of the open-loop gain and the open-loop bandwidth. However, GBWP remains constant as the gain of the operational amplifier in the closed loop decreases, and the cutoff frequency of the operational amplifier increases. For example, from the magnitude response in Fig. 3.2B, we can see that the frequency is 100 MHz when the gain is 1 (magnitude is 0 dB), and therefore, GBWP = 100 MHz.

In operational amplifiers, if the voltage at both inputs is equal, the voltage at the output should be zero. This voltage is called *common mode voltage* and usually appears as interference at the inputs of the operational amplifiers. In realistic operational amplifiers, the output is not zero where the is a common mode voltage. The finite attenuation of the common mode voltage is called *common mode gain* (CMG)—we will use a symbol A_c for CMG. A *common mode rejection ratio* (CMRR) is an important characteristic of operational amplifiers that is defined as the ratio of the differential gain versus the common mode gain. It is commonly given in decibels. CMRR can be quite large, commonly more than 100 dB. It drops with increasing frequency.

3.3.2.1 *Operational amplifier supply voltage*

Double supply or dual voltage operational amplifiers have positive (V_{cc}) and negative ($-V_{cc}$) voltage supplies. The rail refers to the supply voltage levels. *Rail-to-rail* double supply operational amplifier is the amplifier whose output voltage can swing from $-V_{cc}$ to V_{cc}.

A *single supply* operational amplifier requires only one voltage supply (V_{cc}) and therefore, the upper rail is V_{cc} and the lower rail is the ground. Rail-to-rail single supply operational amplifier is the amplifier whose output voltage can swing from 0 to V_{cc}.

Many amplifiers are not rail-to-rail because the output voltage of the amplifier cannot reach the supply voltage. The difference between the supply voltage and the maximum voltage of the operational amplifier is sometimes called the headroom.

Single supply operational amplifiers are prevalent nowadays in battery-operated biomedical devices.

3.3.3 Processing circuits

Many analog processing circuits can be built based on operational amplifiers. They include inverting and noninverting amplifiers, adders, multipliers, differentiators, integrators, and so on. Special-purpose circuits, such as amplifiers with differential inputs and active filters, will be explained in Sections 3.3.4 and 3.4. Fig. 3.3 shows the circuits and the signals of inverting, noninverting, and differentiator operational amplifier circuits. The input signal for all three circuits is sine wave with an amplitude of 1 V and at frequency of V_o. The noninverting

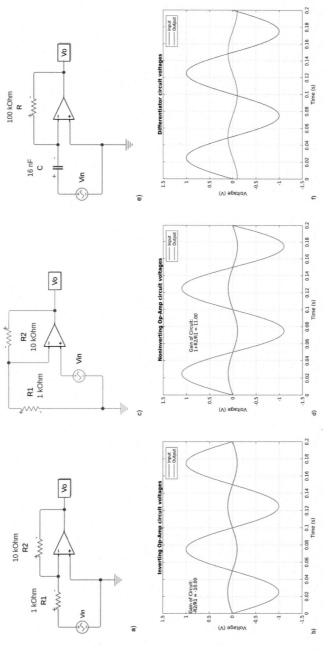

FIGURE 3.3   Three amplifier configurations and their signals at the input and the output: (A, B) inverting amplifier, (C, D) noninverting amplifier, and (E, F) differentiator.

amplifier does not change the phase of the input signal. The inverting amplifier changes the phase by 180 degrees while the differentiator changes the phase by 90°.

The gain of the *inverting amplifier* is $V_o/V_{in} = -R_2/R_1$. Therefore, in the circuit in Fig. 3.3B, the gain is -10.

The gain of the *noninverting amplifier* is $V_o/V_{in} = 1 + R_2/R_1$. Therefore, in the circuit in Fig. 3.3C, the gain is 11.

Let us derive the relationship between the input and the output for the *differentiator*. The current i through the resistor R is equal to the current through the capacitor if the operational amplifier is ideal:

$$i = -\frac{V_o}{R} = C \cdot \frac{dV_{in}}{dt}$$

The output voltage is then:

$$V_o = -RC \cdot dV_{in}/dt$$

The amplitude response can be computed in the frequency domain by applying Fourier transform. Therefore,

$$|V_o(j\omega) / V_{in}(j\omega)| = \omega RC$$

where ω is the angular frequency $\omega = 2\pi f$.

The differentiator is actually an active highpass filter. The problem with the differentiator implemented in this way is that it might become unstable at high frequencies and is sensitive to noise. Practical implementation of the differentiator usually involves the addition of a small capacitance in parallel with the resistor R as well as the addition of a resistor connected in series with the capacitor C. The additional capacitor prevents the circuit from becoming unstable at high frequencies, while the additional resistor will limit the increase in gain with frequency (please observe that the gain keeps increasing with frequency in the original differentiator implementation). For more details, please consult (Northrop, 2014; Webster, 2010).

The integrator circuit is shown in Problem 3.19, and the computation of its gain is given as an exercise.

3.3.4 Differential amplifier

The differential amplifier is shown in Fig. 3.4. The differential amplifier amplifies the difference between the input signals. Therefore, the output is computed as $V_O = A_d(V_2 - V_1)$ where A_d is the differential gain of the differential amplifier.

The amplifier is extremely sensitive to the imbalance of the resistor values, and therefore, it is critical to satisfy the following condition: $R_3/R_1 = R_4/R_2$. Consider the differential amplifier in Fig. 3.4, and let us assume that $R_1 = R_2$ and $R_3 = R_4$. The output of the amplifier is

$$V_O = \frac{R_3}{R_1}\left(\frac{1 + R_1/R_3}{1 + R_2/R_4}V_2 - V_1\right) \tag{3.2}$$

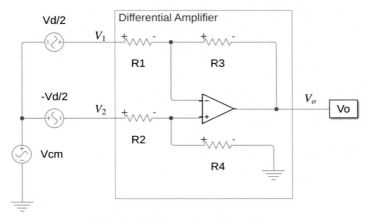

FIGURE 3.4 Differential amplifier.

$$V_O = \frac{R_3}{R_1}(V_2 - V_1) \tag{3.3}$$

This means that the amplifier works as a differential amplifier and that the differential gain is determined by the selection of resistors R_3 and R_1 and its value is $A_d = R_3/R_1$. CMG is zero in this case, and therefore, CMRR is infinite. In practice, differential amplifiers have high CMRR, but not infinite, due to mismatch of the resistors and other imperfections of the circuit.

3.3.4.1 CMRR of a differential amplifier

We defined CMRR in Section 3.3.2. In this section, we will show the effect of common mode rejection for differential amplifiers and derive CMRR for the case when one resistor has a higher tolerance than the others.

The output voltage of the differential amplifier V_o can be written as a sum of contributions from differential and common mode inputs:

$$V_o = A_d V_d + A_c V_{cm}$$

Next, differential voltage and common mode voltage can be written as follows:

$$V_d = V_2 - V_1$$

$$V_{cm} = (V_2 + V_1)/2$$

CMRR is defined as $CMRR = A_d/A_c$

Example 3.1. The amplifier has a differential gain $A_d = 1000$ and common mode gain $A_c = 0.003$. Let us assume that the signal of interest is a cardiac signal modeled as a sine wave with a frequency of 1.2 Hz and that the interference is a 60 Hz sine-wave signal. The amplitude of the interference signal is assumed to be much larger than the amplitude of the cardiac signal. Also, assume that the signal of interest is differential and that the interference is the common-mode voltage. The signal at the input of the differential amplifier is:

$$V_2 = 0.02 \sin(1.2 \cdot 2\pi t) + \sin(60 \cdot 2\pi t)$$

$$V_1 = -0.02 \sin(1.2 \cdot 2\pi t) + \sin(60 \cdot 2\pi t)$$

(a) What is the *CMRR* of this amplifier?
Find the signal-to-noise ratio:
(b) at the input, and
(c) at the output.

Solution.

(a) $CMRR = 1000/0.003 = 333{,}333$. It is common to present *CMRR* in decibels. Therefore, $CMRR = 20 \cdot \log(A_d/A_c) = 110.5$ dB, where log is the base 10 logarithm.

Signal to *noise* ratio is defined as $SNR = 20 \log(V_{signal}t/V_{noise})$, where V_{signal} represents the amplitude of the *signal*, and V_{noise} is the amplitude of the *noise*.

(b) The amplitude of the *signal* of interest is 0.04 V, while the amplitude of the interference is 1 V. Therefore, $SNR_{in} = 20 \log(0.04/1) = -28$ dB.

(c) At the output, the signal and the noise are multiplied by the differential and common mode gain and, therefore $SNR_{out} = 20 \log((0.04 \cdot 1000)t/(1 \cdot 0.003)) = 82.5$ dB
Therefore, we can see that the signal-to-noise ratio from the input to the output is improved by 82.5 dB − (−28 dB) = 110.5 dB, which is exactly the *CMRR* of this differential amplifier.

Now, let us assume that $R_1 = R_2$ and $R_3 = R_4$ and that the tolerances of R_1, R_2 and R_4 resistors are negligible, while the R_3 resistor's tolerance is $er = 0.5\%$ (mismatch of 1%) and it is presented as $R_3(1 - er)$. The common-mode voltage is no longer zero for this differential amplifier. To calculate the CMG, we will assume that $V_1 = V_2 = V_{cm}$ and plug these values into Eq. (3.2).

$$V_O = \frac{R_3 er}{R_1 + R_3} V_{cm}$$

However, the 0.5% error barely influences the differential gain. Therefore, the *CMRR* will be very much dependent on the CMG:

$$CMRR = \frac{A_d}{\dfrac{R_3 er}{R_1 + R_3}}$$

Table 3.2 shows the *CMRR* when $A_d = 1$ ($R_1 = R_3$). *CMRR* is low if the error is 1%. For the error of 0.01%, the CMRR is 86 dB, which is acceptable for most applications.

Several problems have been identified when using the differential amplifier, including.

- The input impedance is limited and depends on selected resistors.
- The differential gain is dependent on a good match of the resistors $R_3/R_1 = R_4/R_2$.
- CMRR drops significantly with increasing the tolerance in the resistor R_3.

Next, we will consider the signal at the input and the output of the differential amplifier in cases (1) when all resistors are perfectly matched and have zero percent tolerance, and (2) when the resistor R_3 has 1% tolerance. We use the differential amplifier shown in Fig. 3.4 and the resistors are selected so that $A_d = 10$. The signals V_1 and V_2 are the same as in Example 3.1. In Fig. 3.5A, we show the input signal V_1 and the output signal V_o. We can see that, at the input, the common mode signal is dominant. However, since the differential amplifier is ideal, the common-mode signal is completely removed, and the differential input is amplified 10 times. In Fig. 3.5B, the common-mode signal is not completely removed but is significantly attenuated. The CMG is -46 dB, based on Table 3.2. Therefore, the common mode signal still appears as noise over the amplified differential signal.

3.3.5 Instrumentation amplifier

Some of these issues recognized by the differential amplifier are addressed by using a more advanced configuration of the differential amplifier, which is called an instrumentation amplifier. In biomedical instrumentation, it is very important that the amplifier has a large input impedance. Therefore, two voltage followers are used, and they are connected to the differential inputs. The voltage follower is a configuration with a gain of 1 and the input impedance that corresponds to the input impedance of the selected operational amplifier, which is normally very high.

The instrumentation amplifier is usually based on three operational amplifiers, as shown in Fig. 3.6. We can see that the instrumentation amplifier is composed of two voltage followers followed by the differential amplifier. We have already computed the gain of the differential amplifier in Eq. (3.2) for the case when $R_1 = R_2$ and $R_3 = R_4$: $V_O = -R_3/R_1 (V_3 - V_4)$. Ideally the CMG = 0 for the differential amplifier.

TABLE 3.2 Effect of resistor accuracy on CMRR and differential gain. Please note CMRR = $-$CMG in decibels since $A_d = 1$.

Resistor error (R_3)	CMRR
1%	46 dB
0.1%	66 dB
0.01%	86 dB

FIGURE 3.5 (A) The input and the output signal of an amplifier shown in Fig. 3.4. The input signal is given in Example 3.1. (B) The output signal of an amplifier is shown in Fig. 3.4 when the value of the resistor R_3 is changed by $er = 0.5\%$.

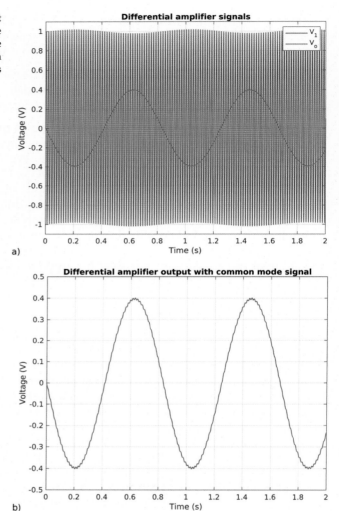

a)

b)

Next, we will consider the circuit on the left in Fig. 3.6. This circuit has high input impedance because the inputs are connected directly to the voltage followers. However, this configuration still has a problem: we need to address the problem of amplifying the common-mode voltage. This is achieved by adding resistors between the junctions, as shown on the left side of Fig. 3.6. R_6 is often called a gain resistor. The current i through the resistors R_5, R_6, and R_7 can be computed as

$$i = \frac{1}{R_5 + R_6 + R_7}\,(V_3 - V_4) = \frac{1}{R_6}\,(V_1 - V_2)$$

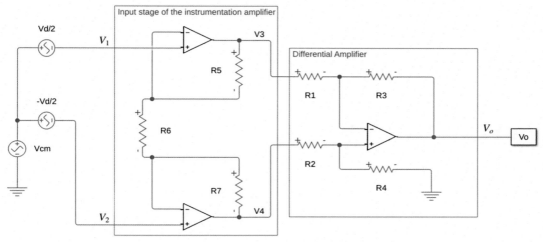

FIGURE 3.6 An instrumentation amplifier.

The differential gain of the left part of the circuit is then:

$$\frac{(V_3 - V_4)}{(V_1 - V_2)} = \frac{R_5 + R_6 + R_7}{R_6}$$

When $V_1 = V_2$, no current flows through R_6, and therefore $V_1 = V_3, V_2 = V_4$. So the common-mode gain is CMG $= 1$.

Therefore, the differential gain of the instrumentation amplifier is

$$A_D = -\frac{R_5 + R_6 + R_7}{R_6} \frac{R_3}{R_1}$$

and the CMRR is still theoretically infinite. Please note that in Fig. 3.6, the resistor R_6 is placed on the left. When the instrumentation amplifier is implemented as an integrated circuit, it is often possible to add the resistor R_6 externally and therefore to partially control the differential gain of the instrumentation amplifier. The symbol of the instrumentation amplifier with the gain resistor is shown in Fig. 3.7.

3.3.5.1 Practical considerations

Practical instrumentation amplifiers have an input impedance greater than 1 GΩ, and an input bias current in the order of several nA. Gain is normally determined by attaching the gain resistor R_6 externally or by selecting or combining the internal gain resistors. A gain-programmable instrumentation amplifier can have four or more internal gains that a microcontroller can set. An example of a programmable gain amplifier with a gain of 1, 2, 4, or 8 is AD8251 by Analog Devices. CMRR is normally 90–120 dB or more in modern instrumentation amplifiers (Kitchin and Counts, 2006).

The power supply current of typical instrumentation amplifiers is of the order of 1 mA. However, micro- and ultra-low-power instrumentation amplifiers are used in wearable

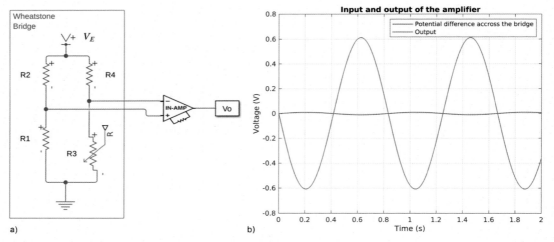

FIGURE 3.7 (A) instrumentation amplifier connected directly to the Wheatstone bridge. (B) The signal at the output of the bridge (blue line) and the signal at the output of the instrumentation amplifier (red line).

devices. An example of a low-power instrumentation amplifier is AD621 by Analog Devices, for which the power supply current is in the order of tens of μA.

Commonly, in practical applications, R_4 is not connected to the ground, but the reference voltage V_{REF} (Kitchin and Counts, 2006). The output voltage is measured with the reference to V_{REF}. This allows us to connect the differential input from a Wheatstone bridge that can have both positive and negative values to the instrumentation amplifier with a single supply voltage. To prevent the signal at the output of the single supply instrumentation amplifier from saturating at 0 V (having an output at 0 V when the input is negative), the voltage V_{REF} has to be set properly. An example of a single rail instrumentation amplifier is AD623 by Analog Devices.

The signal output of the bridge is the differential voltage, which can be connected directly to the instrumentation amplifier's inputs. This is shown in Fig. 3.7A, where the quarter bridge is used with all resistors having a nominal resistance of 120 Ω. The resistor R_3 is a variable resistor used to simulate a transducer. The resistance of the variable resistor changes as follows: $R_3 = 120\ \Omega + 1\ \Omega \cdot \sin(2\pi f t)$, where $f = 1.2$ Hz. Fig. 3.7B shows the signal at the output of the bridge and the signal at the output of the instrumentation amplifier that is amplified 60 times. The signal at the output of the amplifier is in the opposite phase compared to the resistance R_3. The advantage of using the instrumentation amplifier here is that the input resistance is very high and therefore does not reduce or modify the differential voltage at the input of the amplifier.

3.3.6 Using amplifiers as conditioning circuits

3.3.6.1 Linearizing Wheatstone bridge

Some transducers, such as resistive temperature transducers, cannot be connected in a full bridge. However, it is still possible to have a linear relationship between V_o and the resistance change ΔR. An example of such a system is shown in Fig. 3.8 (Kester, 2005b). The configuration looks like a differential amplifier. The operational amplifier used here is commonly a double supply voltage operational amplifier. In this circuit $R_1 = R_2 = R_4 = R$ and R_3 is the transducer with the resistance $R_3 = R + \Delta R$. The current i is the same through the resistors R_2 and R_4 if the operational amplifier is ideal. The relationship between the excitation V_E and the output voltage V_o can be written as follows:

$$V_E = V_o + (2R + \Delta R)i$$

$$V_E = 2Ri$$

By combining the two equations above, we get

$$V_E = V_o + \frac{(2R + \Delta R)V_E}{2R}$$

Therefore, the output voltage is

$$V_O = \frac{-\Delta R}{2R}V_E$$

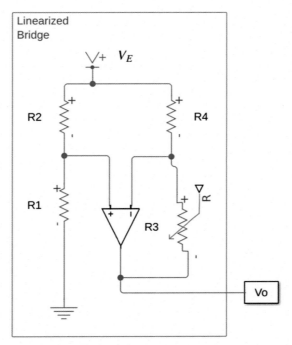

FIGURE 3.8 Linear quarter bridge.

We can notice that this configuration doubled the sensitivity of the quarter bridge while allowing a linear output even for large values of ΔR.

Fig. 3.9A shows the signal at the output of the bridge when the resistance of the variable resistor changes as follows: $R_3 = 120\Omega + 1\Omega \cdot \sin(2\pi ft)$, where $f = 1.2$ Hz. Fig. 3.9B shows the relationship between the changes in the resistor value and the output voltage. We see that the change of V_o is linear with regard to the change in resistance. The curve in Fig. 3.9B was obtained by running simulations in Simscape while linearly changing the resistance of the resistor R_3 in the range $[116.4\ \Omega, 123.6\ \Omega]$.

FIGURE 3.9 (A) The output signal of a linearized bridge. (B) The linear curve shows the dependence between the change in the resistance to the output voltage change.

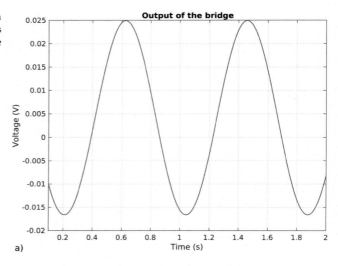

a)

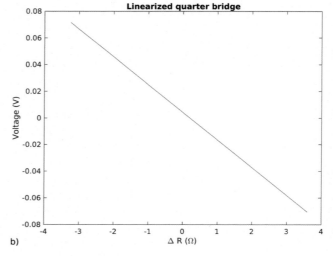

b)

3.3.6.2 *Interfacing capacitive sensors*

Several types of interface electronic circuits were developed to connect capacitive transducers. They include ones based on resonant oscillators, AC bridges, synchronous demodulation techniques, and capacitance to digital converters (Islam, 2017). Capacitance-to-digital converters are covered in Section 3.5.4.2.

In resonant oscillators, the capacitive sensor is the part of the oscillator circuits whose frequency is inversely proportional to the capacitance. Change in the capacitance will correspond to the changes in the resonant frequency. This technique is not used much nowadays because its accuracy is affected by stray capacitance.

An AC bridge normally has one capacitive transducer C_S and a reference capacitor C on one side of the bridge and two resistors R_3, R_4 on the other side—see Fig. 3.10. It is used to detect small capacitive changes. If two capacitances are not balanced, a current will flow through the bridge. The circuit can further be linearized by using the bridge in Fig. 3.8. The output voltage is calculated as:

$$V_o = V_E \left(\frac{\frac{1}{j\omega C}}{\frac{1}{j\omega C} + \frac{1}{j\omega C_S}} - \frac{R_3}{R_3 + R_4} \right) = V_E \left(\frac{1}{1 + \frac{C}{C_S}} - \frac{1}{1 + \frac{R_4}{R_3}} \right) \tag{3.4}$$

From Eq. (3.4), by having $R_4 = R_3$ and $C_s = C + \Delta C$, it is possible to obtain a similar relationship between changes in the capacitance and output voltage as for the standard quarter bridge. However, as we will explain later, synchronous demodulation is often used to obtain the output voltage.

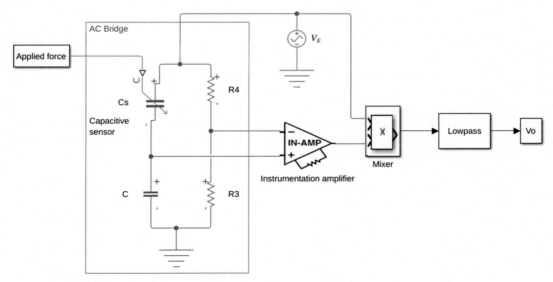

FIGURE 3.10 Synchronous demodulation circuit for interfacing a capacitive transducer.

A potential problem with the traditional nonlinearized AC bridge is the stray capacitance between the bridge output points and the ground, and this stray capacitance can imbalance the bridge.

Next, let us consider the demodulation technique for interfacing the capacitive sensor. The capacitive sensor is coupled with a reference capacitor through a voltage divider, as shown in Fig. 3.10. Actually, an AC bridge is used with two resistors in another branch of the bridge. The capacitive sensor is modeled as a variable capacitor whose capacitance is controlled by a sine-wave generator presented as the applied force block in Fig. 3.10. This bridge is connected to the input of the instrumentation amplifier, which amplifies the signal. The signal is then demodulated using a multiplier, also called a mixer. The mixer multiplies the input reference signal with the signal at the output of the instrumentation amplifier, which is of the same frequency as the reference signal. The product of two sine-wave signals with identical frequencies is the sum of two components: (1) a scaled applied force signal and (2) a sine-wave signal at twice the original frequency. This sine-wave signal at twice the original frequency is removed using a lowpass filter. Reference signal frequency is commonly high compared to the frequency of physiological signals and is in the order of kHz or tens of kHz. Demodulation-based architecture is one of the most popular techniques for measuring capacitance due to its high accuracy. Synchronous demodulators enable accurate measurement of small AC signals in the presence of noise that is several orders of magnitude greater than the signal's amplitude. An example of a synchronous demodulator circuit used for sensor interfaces is the ADA2200 chip by Analog Devices.

3.3.6.3 Charge amplifier

A charge amplifier is a charge-to-voltage converter. It is mainly used to connect piezoelectric transducers, including a number of sensors based on piezoelectric transducers such as microphones, accelerometers, ultrasonic receivers, and dynamic pressure sensors.

The charge amplifier is shown in Fig. 3.11. It includes the operational amplifiers with the capacitor C_f and resistor R_f in the feedback loop. The capacitance C_f is inversely proportional to the amplitude of the output voltage. The resistor R_f is selected to prevent the amplifier from drifting into saturation. The value of R_f and C_f set the low cutoff frequency of the amplifier: $f_L = 1/(2\pi R_f C_f)$. This charge amplifier acts as an active lowpass filter. Resistor R_i provides electrostatic protection.

The input pin is normally connected to a piezoelectric sensor with an equivalent capacitance V_o shown in Fig. 3.12. Resistor R_i and capacitor V_o provides an upper cutoff frequency $f_H = 1/(2\pi R_i C_p)$. Since we are using single-rail operational amplifier, the positive input of the operational amplifier is connected to $V_{CC}/2$. Therefore, the output voltage $V_o = V_{CC}/2$ with no input.

The circuit that includes a piezoelectric transducer, charge amplifier and additional amplifier at the output is shown in Fig. 3.12 (12). The piezoelectric transducer is modeled as a current source $i_{IN} = dq/dt$ with the capacitance C_f in parallel. Both amplifiers in the circuit are single rail amplifiers, and their voltage range is between $0\ V$ and $V_{CC} = 3.3\ V$. Therefore, $V_{CC}/2$ is used as an offset for both operational amplifiers. The output V_{O1} of the charge amplifier can be calculated as

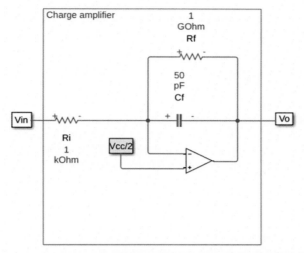

FIGURE 3.11 Charge amplifier with a single rail operational amplifier.

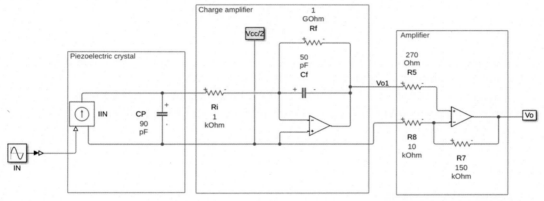

FIGURE 3.12 A circuit that includes a model of a piezoelectric sensor, a charge amplifier and a noninverting amplifier.

$$V_{O1} = \frac{V_{CC}}{2} - V_{Cf} = \frac{V_{CC}}{2} - \frac{q}{C_f} \qquad (3.5)$$

Therefore the change in the voltage dV_{O1} is proportional to the change in the charge of the piezoelectric sensor dq/C_f. The output voltage of the circuit is amplified using the noninverting amplifier and it can be calculated as:

$$V_O = \frac{V_{CC}}{2} - \left(1 + \frac{R_7}{R_8}\right)\frac{q}{C_f} \qquad (3.6)$$

Example 3.2. Assume that we connect a piezoelectric quartz-based accelerometer to the charge amplifier (12) shown in Fig. 3.12. The parameters of the accelerometer are as follows:

- Range: ± 5 g.
- Sensitivity: $S_a = 1\ pC/g$. The sensitivity is defined as the change of the charge over the change of acceleration $S_a = \Delta q/\Delta a$.
- Capacitance: $C_p = 90\ pF$.

(a) Select the values of the components in the circuit and determine low and high cutoff frequencies. Let us assume that the output of the charge amplifier can have a voltage swing of $\pm 0.1\ V$, meaning that it can change by 0.1 V around the reference value.
(b) Assume that the signal at the piezoelectric transducer is a sine wave at 20 Hz simulating charge in the range between ± 5 pC. Simulate the circuit and show signals V_{Cf}, V_{o1}, and V_o.

Solution.

(a) The voltage developed across C_f due to a charge Δq is $V_{Cf} = \Delta q/C_f$. By multiplying and dividing the right side of the expression for V_{Cf} with Δa, we get

$$V_{cf} = \frac{\Delta a}{C_f}\frac{\Delta q}{\Delta a} = \frac{\Delta a \cdot S_a}{C_f}$$

For the voltage swing of $\pm 0.1\ V$ at the output of the charge amplifier V_{o1}, we have the same voltage swing at V_{Cf} (see Eqs. 3.5). Therefore, we can compute C_f as $C_f = S_a \Delta a/V_{Cf} = (1\ pC/g \times 5\ g)/0.1\ V = 50\ pF$. Let us choose R_f to be 1 GΩ. We can compute the lower and upper cutoff frequencies of the circuit as:

$$f_L = \frac{1}{2\pi R_f C_f} = 3.18\ Hz$$

If we select $R_i = 1$ kΩ then

$$f_H = \frac{1}{2\pi R_i C_p} = 1.77\ MHz$$

To observe changes in the output V_o in the range from 0 to 3.3 V, we need to amplify the signal V_{o1} 16.5 times ((3.3 V/2)/0.1 V). Therefore, based on this computation and also based on Eq. (3.6), selected values for the resistors are $R_7 = 150$ kΩ and $R_8 = 10$ kΩ.

(b) Since the sensitivity of the accelerometer is 1 pC/g, a charge change of ± 5 pC corresponds to a change in acceleration of ± 5 g. In the model, the sine-wave signal is differentiated to obtain $i = dq/dt$ and used as an input of the current-controlled source I_{IN}, as shown in Fig. 3.12. As can be seen in Fig. 3.13A, the amplitude of the signal V_{Cf} over the capacitor C_f is exactly 0.1 V. The signals are shown in the following time interval [1 s, 1.5 s]. It can also be seen in Fig. 3.14B that signal at the output of the charge amplifier V_{o1} is amplified around 16 times so that the amplitude of the signal at the output V_o spans almost the whole range between 0 and 3.3 V.

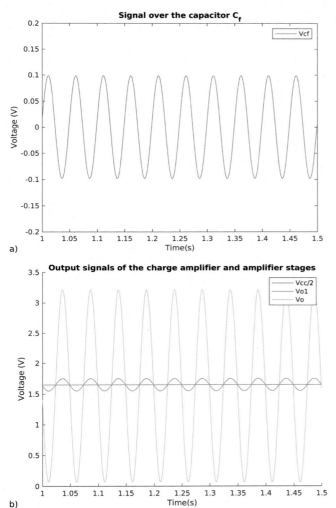

a)

FIGURE 3.13 Signals in the circuit shown in Fig. 3.12 in the interval between 1 and 1.5 s for 20 Hz input sine-wave signal, including (A) signal V_{Cf} and (B) signals V_{o1} and V_o.

3.3.6.4 Transimpedance amplifier

A transimpedance amplifier amplifies a photodiode's reverse-biased current and generates a voltage at the output. It is used in many applications outside of the biomedical area, including compact disc players, infrared remote controls, ambient light sensors, and laser range finding. In the biomedical area, they are used in plethysmography and all imaging applications based on photodetectors. The transimpedance amplifier is shown in Fig. 3.14A. Conceptually, it looks similar to the charge amplifier. The voltage V_o at the output is proportional to the resistance R_f and the photodetector current i_p. Therefore, V_o. The capacitor C_f determines the highest frequency of interest and prevents the oscillation of the

FIGURE 3.14 (A) Transimpedance amplifier and (B) the signal at the output V_o for the input current $i_p = (-1 - \sin(2\pi t \cdot 20Hz))\,\mu A$.

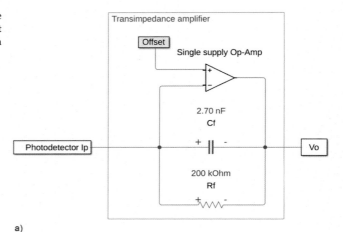

a)

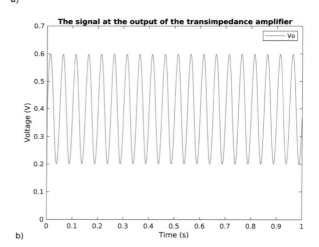

b)

transimpedance amplifier at high frequencies. The oscillations occur since the photodetector can be modeled as a current source, shunt resistor, and capacitor connected in parallel. These shunt resistance and capacitance are the reason for potential oscillations at high frequencies.

In Fig. 3.14A, the photodetector current is simulated as $i_p = (-1 - \sin(2\pi t \cdot 20\,Hz))\,\mu A$. Therefore, the expected peak-to-peak voltage at the output is 0.4 V, as shown in Fig. 3.14B. The offset voltage is chosen to be 0.2 V so that it can handle the dark current of the photodiode while still using the bulk of the output voltage range. The cuff-off frequency of the transimpedance amplifier is $f_c = 1/(2\pi R_f C_f) = 295$ Hz.

An example of a commercial transimpedance amplifier is OPA857 by Texas Instruments. The circuit has two transimpedance gains that can be selected by choosing one of two available resistor–capacitor (R_f, C_f) pairs. The amplifier uses a single supply voltage and has a very large bandwidth.

The output of the transimpedance amplifier is often connected to the input of the programmable gain amplifier that will be introduced next. In this case, the programmable amplifier can boost the signal in low-light conditions.

3.3.6.5 *Programmable gain amplifier*

Programmable gain amplifiers (PGA) commonly have a set of gains values that can be digitally programmed or set by a microcontroller or a microprocessor. PGAs are used in biomedical instrumentation in general, as a photodiode amplifier circuit, as an ultrasound preamplifier, and so on.

The amplifiers commonly come with four or eight different values for the gain. The gains are normally multiples of 2, 5, or 10. These gains can be set directly using two or three additional pins on the PGA chip. Alternatively, they can be programmed using a serial bus such as the Serial Peripheral Interface (SPI) bus (Serial Peripheral Interface, 2021). SPI is a serial bus used in embedded systems in which a microprocessor communicates in a full-duplex mode with SPI-supported peripheral devices. Multiple peripheral devices are normally supported through selection with individual chip select (CS) lines.

Programmable gain amplifiers are very important in applications where the signal level can vary significantly and cause the signal at the output of the amplifier to be either too small or saturated. They are often used as a signal conditioning circuit before the A/D converter to ensure that the full range at the input of the A/D converter is utilized. An example of a PGA with gains 1 or 2 is shown in Fig. 3.15. The PGA has only one control input that can be set to either 0 V or 5 V. When it is set to 0 V, the switch is up, as shown in Fig. 3.15, and the gain is 1. When the control input is set to 5 V, the amplifier acts as a noninverting amplifier, and the gain is $1 + R_3/R_1 = 2$. This setup of the PGA is used in order to remove the effect of the finite resistance of the switch when the switch is closed (Precision Variable Gain, 2008).

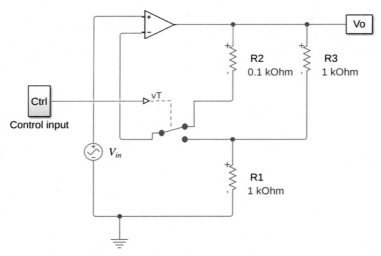

FIGURE 3.15 PGA with two selectable gains.

Of course, the bandwidth and the gain are inversely related, as mentioned before. Therefore, increasing the gain will decrease the bandwidth of the amplifier. An example of gain and bandwidth for the Burr-Brown PGA103 amplifier is shown in Table 3.3.

Besides programmable amplifiers, programmable instrumentation amplifiers also exist. An example is the AD8250 instrumentation amplifier from Analog Devices, which has two additional control pins for gain control and allows for setting the gain to 1, 2, 5, or 10.

3.4 Analog filters

In this section, we will briefly introduce analog filters. Filters that also amplify the signals are called *active filters*. Some of the active filters have already been mentioned before. For example, a differentiator corresponds to the first-order active highpass filter. We start this section with a simple RC filter followed by the second-order active lowpass filter. More details about filters can be found in (Wanhammar, 2009).

3.4.1 Passive filter

A simple RC lowpass filter is shown in Fig. 3.16A. Let us define the impedance $Z_c = 1/jC\omega$ as well as *the time constant* $\tau = RC$. The time constant is important since it determines the cutoff frequency of the filter as well as the rise time as a response to a step function. The *cutoff frequency* f_c of the lowpass filter is the frequency selected in a way that all the frequency components of the signal lower than f_c are passed through the circuit unchanged or amplified while the frequency components of the signal higher than f_c are attenuated. The definition related to the magnitude response of the filter will be given later. The cutoff frequency of the filter shown in Fig. 3.16A is about 2 Hz. The input of the circuit in Fig. 3.16A is the sine-wave signal of the amplitude 1 V and frequency 60 Hz. This signal is attenuated about 16 times and shifted in phase by $\pi/2$ at the output as shown in Fig. 3.16B.

The filter's response to the step function is shown in Fig. 3.17. The step function has a value of 0 V until the instant of 0.2 s, and then its value is 5 V. The capacitor is charged fully after about five time constants. After one time constant, the voltage across the capacitor is about $0.63 \cdot V_{in}$.

TABLE 3.3   Gain and bandwidth of a commercial PGA: Burr-Brown PGA103.

Gain	Bandwidth
1	1.5 MHz
10	750 kHz
100	250 kHz

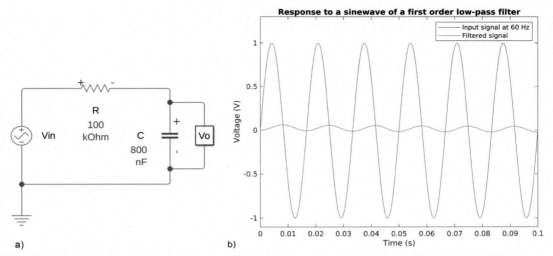

a)

b)

FIGURE 3.16 (A) The first-order lowpass filter. (B) Input signal at 60 Hz (blue line) and filtered signal at the output (red line).

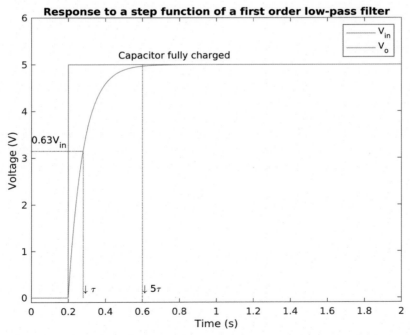

FIGURE 3.17 Filter's response to a step function. The input voltage is 0 V until 0.2 s; after that, it remains at 5 V.

Now, let us compute the magnitude and phase response of the RC filter. The transfer function in the frequency domain can be written as

$$\frac{V_o(j\omega)}{V_{in}(j\omega)} = \frac{\frac{1}{j\omega c}}{R + \frac{1}{j\omega C}} = \frac{1}{1 + j\omega RC}$$

where $\omega = 2\pi f$ is the angular frequency. The amplitude and phase responses are then:

$$A(j\omega) = \frac{1}{\sqrt{1 + (2\pi f)^2 (RC)^2}} = \frac{1}{\sqrt{1 + (2\pi f \tau)^2}}$$

$$\varphi(j\omega) = \text{atan}(-2\pi f \tau)$$

From the amplitude response, we can determine the frequency where the amplitude drops by half (by 3 dB). This frequency is the *cutoff frequency*, and it can be shown that it is

$$f_c = \frac{1}{2\pi RC} = \frac{1}{2\pi \tau}$$

For this filter, we used capacitor $C = 800$ nF and resistor $R = 100$ kΩ. Then the cutoff frequency is $f_c = 2$ Hz.

3.4.2 Active filter

Active filters are used frequently because they amplify while also performing filtering. They can be implemented in different configurations. One of the most popular architectures is the Sallen-Key second-order lowpass active filter shown in Fig. 3.18. This is a second-order filter with an attenuation of 40 dB per decade. This filter is widely used because it can be implemented to have characteristics of the most popular filter types, such as Butterworth or Chebyshev (Webb, 2018). The filters are also easily cascaded to create higher-order filters. They can be implemented as well as highpass or bandpass by exchanging places of resistors and capacitors.

The cutoff frequency of the filter is determined by

$$f_c = 1/2\pi\sqrt{R_1 R_2 C_1 C_2}$$

To design the filter, we use the following approximate formula for determining the cutoff frequency based on the required attenuation A_v at the frequency of interest. Let us assume that we would like to attenuate $f = 60$ Hz frequency 10 times ($A_v = 0.1$).

$$f_c = ft / t\sqrt[4]{(1 - A_v^2)/A_v^2}$$

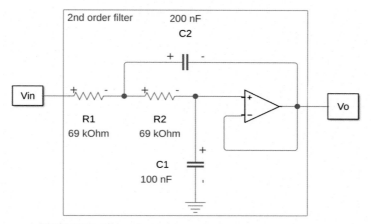

FIGURE 3.18   Sallenkey second-order lowpass active filter.

After plugging the numbers for the frequency of interest and attenuation, we obtain that $f_c = 20$ Hz. Then, let us assume that $C_2 = 2 \cdot C_1 = 200$ nF. Also, if we assume that $R_1 = R_2 = R$, we obtain from $f_c = 1/(2\pi\sqrt{2}\,RC_1)$ that the resistance is $R = 59$ kΩ. We can choose a standard resistor value instead, for example, $R = 69$ kΩ.

Its Bode diagram is shown in Fig. 3.19. For the selected resistors and capacitors, the cutoff frequency is 16.2 Hz and the attenuation at 60 Hz is 0.074 or -22.7 dB which is smaller than the required attenuation of 0.1.

Please note that this section does not introduce bandpass and highpass filters. However, they can be made by modifying the Sallen–Key topology (Wanhammar, 2009).

3.5 Analog-to-digital converter

3.5.1 Sampling

3.5.1.1 *Sampling theorem*

An analog signal is defined at every point t of the signal $x(t)$. Discrete signals, on the other hand, are defined only at the sampling points $x(nT_s)$, where n is the sampling number. Here, the signal $x(t)$ is sampled every T_s seconds, where T_s is called the *sampling period* and $f_s = 1/T_s$ is the *sampling frequency*.

To accurately reproduce the analog input data from its samples, the sampling rate f_s must be at least twice as high as the highest frequency expected in the input signal. This is known as a sampling theorem (Oppenheim and Schafer, 1998). Half of the sampling frequency is known as the *Nyquist frequency f_h*.

$$f_s = 2 \cdot f_h$$

3. Electronics

Bode Diagram

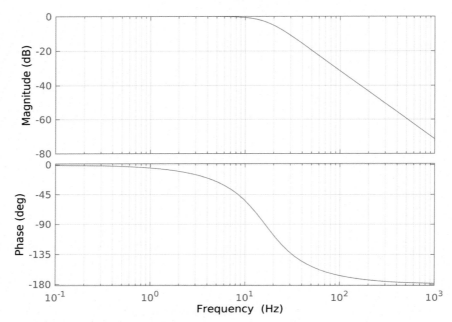

FIGURE 3.19   Bode plot of the amplitude and phase of the Sallenkey filter.

The discrete signal is commonly presented in a way that time t in the continuous signal is replaced with nT_s in the discrete signal representation. For example, let us consider a breathing signal at 15 breaths per minute (0.25 Hz) modeled as a sinusoid with a sinusoidal interferer at 60 Hz:

$$x(t) = 1.5 + \sin(2\pi \cdot 0.25t) + 0.25 * \sin(2\pi \cdot 60t) \tag{3.7}$$

After sampling, the discrete signal $x(nT_s) = x(n)$ is

$$x(n) = 1.5 + \sin(2\pi \cdot 0.25nT_s) + 0.25 * \sin(2\pi \cdot 60nT_s) \tag{3.8}$$

3.5.1.2 Antialiasing

Sampling analog signal below $2f_h$ generates *aliased components* at the frequencies determined by $f_{alias} = |f_l - kf_s|$ where f_l is the frequency component of the sampled signal with a higher frequency than the Nyquist frequency, and k is the harmonic number $k = 0, \pm1, \pm2, \ldots$. We are interested only in the aliased components in the range between $-f_s/2$ and $f_s/2$.

Example 3.3.   10 Hz sine-wave signal is sampled at 12 Hz.

(a) Show the signal at the output of the A/D converter.
(b) What is the fundamental frequency of the signal at the output?

Solution.

(a) The signals are shown in Fig. 3.20. The blue line represents a 10 Hz sine wave signal.

(b) After sampling, the fundamental frequency is $f_{alias} = |f_I - kf_s|$ which is 2 Hz for $k = 1$. Samples at the sampling period of $1/12$ s are shown using red dots. The red line is the interpolated signal that has a frequency of 2 Hz.

Example 3.4. The signal shown in Eq. (3.7) is sampled at $f_s = 80$ Hz.

(a) Determine if aliasing occurred.

(b) Simulate and draw the signal after sampling.

Solution.

(a) The sampled signal will then be:

$$x(n) = \sin(2\pi n \cdot 0.25/80) + 0.25 * \sin(2\pi n \cdot 60/80)$$

$$= \sin(2\pi n \cdot 0.25/80) - 0.25 * \sin(2\pi n \cdot 20/80)$$

As can be seen, $\sin(2\pi n \cdot 60/80) = -\sin(2\pi n \cdot 20/80) = -\sin(2\pi n/4)$. So, the frequency component from 60 Hz aliased to -20 Hz. The same result can be obtained using $f_{alias} = |f_I - kf_s|$ for $f_I = 60$Hz and $k = 1$.

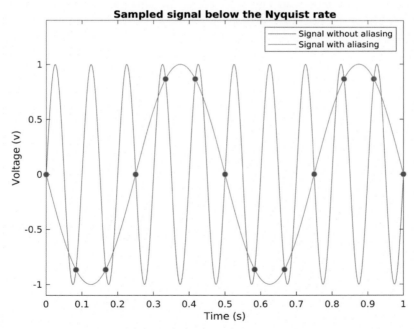

FIGURE 3.20 Blue line shows a 10 Hz sine-wave signal, while the red line with red dots shows samples from the 10 Hz sine wave sampled at 12 Hz. The red line is a sine wave that is interpolated between the red samples.

(b) The aliased signal is shown in Fig. 3.22A as a blue signal at the frequency of 0.25 Hz modulated with the frequency of 20 Hz.

Frequency components above the highest frequency of interest need to be removed. However, in practice, they cannot be removed but only attenuated, which is done using a lowpass filter called an *antialiasing filter*. Fig. 3.21 shows an antialiasing filter followed by an A/D converter ADC2. The filter has already been introduced in Fig. 3.18. In addition, we show the effect of sampling without using an antialiasing filter. This is done by directly connecting the A/D converter ADC1 to an input signal. We know that the filter presented in Fig. 3.18 attenuates the signal at 60 Hz by about 10 times (20 dB). Therefore, as shown in Fig. 3.22B, there is still aliasing even if the antialiasing filter is used; however, the aliased frequency component is attenuated by 22.7 dB, as explained in Section 3.4.2. In the time domain (Fig. 3.22A), the red line shows the sampling signal after aliasing. The 60 Hz interference signal is attenuated significantly, but it is not removed—it appears as a 20 Hz signal.

3.5.1.3 Quantization

Quantization is the conversion of a discretized analog signal into an integer number. The number can be positive or negative, or only positive depending on the range of the A/D converter. The transfer function of the A/D convertor is a staircase-like function with the input analog signal on the abscissa and the digital signal on the ordinate. The digital signal is presented using integers, often in a binary format. An example of the transfer function of an ideal 3-bit A/D converter is shown in Fig. 3.23. Here, the abscissa represents normalized input voltage where V_{FS} is the full-scale voltage of the A/D converter and V_{in} is the input signal at the A/D converter. In Fig. 3.23, the ordinate represents the binary value at the output of the ADC.

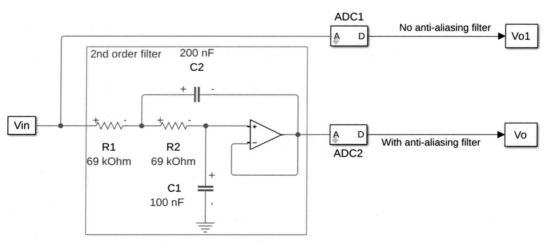

FIGURE 3.21   A circuit with 2 A/D convertors for analyzing the effects of aliasing with and without an antialiasing filter.

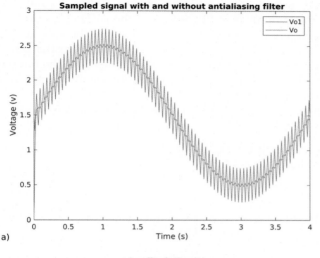

a)

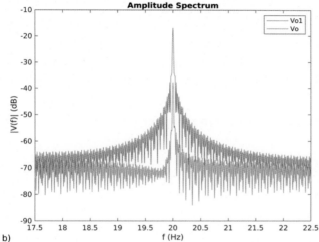

b)

FIGURE 3.22 Signal sampled at $f_s = 80\,Hz$ without (blue line) and with (red line) the antialiasing filter. (A) Time-domain signal. (B) Power spectrum. There is a frequency component at 20 Hz, which results from the aliasing of the 60 Hz sine-wave signal. The magnitude of the frequency component at 20 Hz is reduced by 22.7 dB using the antialiasing filter.

The minimum change in voltage required to change the output code level is called the *least significant bit voltage* V_{LSB}. The *voltage resolution* of the A/D converter is equal to V_{LSB} and it is computed as: $V_{LSB} = V_{FS}/2^N$, where N is the number of bits of the A/D converter. In Fig. 3.23, relative V_{LSB} corresponds to the change of $0.125 \cdot V_{FS}$. Also, the line of infinite resolution is shown as well—in the case of an A/D converter with a very large number of bits V_{LSB} would tend toward zero.

After quantization, the signal can be reconstructed as follows:

$$V_o = V_{FS} \sum_{i=0}^{N-1} b_i 2^{-(N-i)} \tag{3.9}$$

Here b_0 corresponds to the least significant bit (LSB) of the digital signal and $b_{N-1}b_{N-2}...b_1b_0$ represents the N-bit binary number at the output of the A/D converter.

Quantization error represents the difference between the input value and the value that is reconstructed at the output: $e_c = V_o - V_{in}$. Quantization error is zero only in the points $V_{LSB}\cdot n$, where V_o We will assume that the quantization error is equally likely in the interval $[-V_{LSB}/2, V_{LSB}/2]$, and therefore, it follows a uniform distribution. This will be important for uncertainty quantification in the following chapters.

Example 3.5.   A/D convertor operates between 0 and 5 V, and its transfer function is shown in Fig. 3.23.

(a) What is the output of the A/D converter if the input is $V_{in} = 2.25$ V
(b) What is the recovered output in volts?
(c) What is the quantization error?

Solution.
(a) Full-scale voltage is $V_{FS} = 5$ V.The relative voltage at the input is $2.25\ V/5\ V = 0.45$. The binary output for the A/D converter is $b_2 b_1 b_0 = 100$.
(b) The recovered output $V_o = 5\ V\cdot(b_2 2^{-1} + b_1 2^{-2} + b_0 2^{-3}) = 5\ V\cdot 2^{-1} = 2.5\ V$
(c) Therefore, the quantization error is $e_c = V_o - V_{in} = 0.25$ V

3.5.1.4 A/D conversion steps

A model of the A/D converter is shown in Fig. 3.24. The main blocks are the limiter, sample and hold circuit and quantizer. If the signal at the input is smaller than zero or larger than

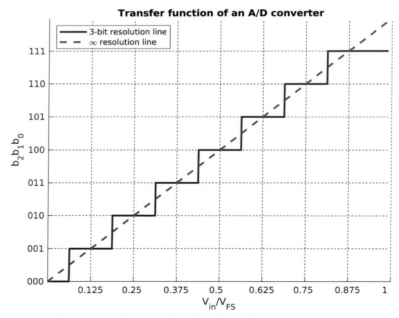

FIGURE 3.23   Transfer function of a 3-bit A/D converter.

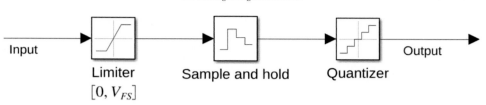

FIGURE 3.24 Functional block diagram of an A/D converter.

the maximum voltage applied to the A/D converter, the output of the limiter will be saturated to zero or maximum voltage, respectively. Of course, the minimum voltage does not need to be zero and can also be negative. The *sample and hold* circuit converts the analog signal into the discrete signal that keeps the same value for the duration of the sampling period. This value is then quantized and converted into a digital signal. Finally, the *quantizer* performs the following operation:

$$V_o = round(V_{in} / V_{LSB})$$

The difference between the discrete and the digital signal is that the discrete signal can take any value in the range of the A/D converter that the analog signal can take, while the digital signal can take only 2^N values where N is the *bit resolution* of the A/D converter. In this A/D converter, the output is zero between 0 and V_o, binary one between V_o and $3V_{LSB}/2$, and so on.

Fig. 3.25 shows the results of the simulation of the A/D converter shown in Fig. 3.24 with the following characteristics: sampling rate $f_s = 12\,Hz$, number of bits $N = 4$, minimum voltage level 0 V, maximum voltage level 5 V. The input signal shown using the blue line is a 0.25 Hz sine-wave signal whose amplitude is in the range [0.5 V, 2.5 V]. The signal at the output of the sample and hold block is shown as a red line. The dots on the red line are the voltage values sampled at the specific time instant. After the signal is sampled, its value is kept constant for the duration of the sampling period. The quantizer converts the value of the discrete sample after the sample and hold block into one out of 2^N integers. These values are normally given in binary form. However, here we modeled the output of the A/D converter as the reconstructed signal shown in Eq. (3.9). Of course, with increasing the resolution of A/D converter to, for example, $N = 8$ bits, the quantization error would become smaller, and it would be difficult to observe the difference between the output of the sample and hold circuit and the reconstructed output of the quantizer.

3.5.2 Errors during A/D conversion

Quantization error has already been introduced. This error contributes to the distortion of the signal at the output of the A/D converter. To understand this section, we will introduce several terms related to *periodic waveforms*. The *fundamental frequency* is the lowest frequency of a periodic waveform. Multiples of the fundamental frequency are called *harmonics*.

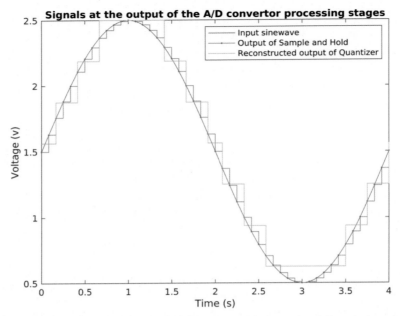

FIGURE 3.25 Blue line represents a sine-wave signal at the input of the 4-bit A/D converter, the red line is the signal obtained after the sample and hold block, while the yellow line is the scaled signal obtained at the output of the quantizer. Scaling is done by dividing the signal with V_{FS}.

Signal-to-noise ratio (SNR) of an A/D converter is the measured ratio of the power of the signal and the noise at the output of the A/D converter, where the noise can come from different sources and can include the quantization noise:

$$SNR_{DB} = 10\log_{10}\left(P_{signal} \,/\, P_{noise}\right) \tag{3.10}$$

The theoretical *signal-to-noise* ratio for an ideal N-bit converter with a sine-wave input is given by

$$SNR = (6.02N + 1.76) \text{ dB} \tag{3.11}$$

where N is the bit resolution of the A/D converter.

Let us consider a sine signal at the frequency of 1.1 Hz centered between 0 and 1 V corresponding to a pulse waveform sampled at 80 Hz. We compute *SNR* based on the definition Eq. (3.10) while changing the resolution of the A/D converter from 4 to 16 bits (Fig. 3.26). Computed SNR matches well the approximation shown in Eq. (3.11).

Signal-to-Noise-and-Distortion Ratio (SINAD) is the measured ratio of the power of the signal to the power of noise and distortion at the output of the A/D converter. Here, the signal is normally the sine wave at the fundamental frequency. The noise and distortion include the power of all nonfundamental components of the signals up to the frequency

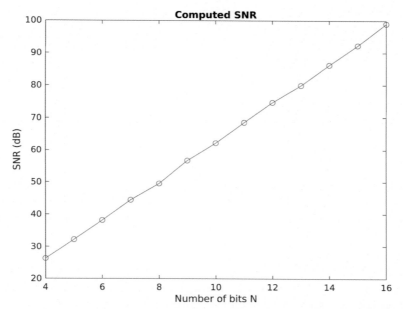

FIGURE 3.26 Computed SNR for the A/D converters based on Eq. (3.10) when the number of bits changes between 4 and 16 for the signal $V_{in}(t) = 0.5\,\text{V} + 0.5\,\text{V} \cdot \sin(2\pi f_i\, t)$.

$f_s/2$ excluding the DC value. Harmonics of the original signal are also considered distortion. Please note that different definitions of SINAD exist in other fields, such as communication.

The effective number of noise-free bits in an A/D converter is normally smaller than the resolution of A/D converter. Therefore, the *effective number of bits* is defined as ENOB=(SINAD-1.76)/6.02. Normally, the last 2 bits at the output of the A/D convertor vary when the input signal is constant, and we can effectively use only N-2 bits of the A/D converter.

Example 3.6. Let us consider a 16-bit A/D converter and the following signal: $V_{in}(n) = 0.5\,\text{V} + 0.5\,\text{V} \cdot \sin(2\pi f_i\, T_s\, n) + v \cdot \text{randn}(0, 1)$. Here we add to the signal random Gaussian noise whose standard deviation is v. Compute *SNR*, *SINAD*, and *ENOB* when:

(a) $v = 0$,
(b) $v = V_{LSB}$,
(c) $v = 2V_{LSB}$.

Solution. Matlab code is provided for computing these values on the book website. The results are summarized in Table 3.4. The results shown in the table are expected for the following reasons. SINOD and SNR have almost the same values since there are no additional harmonics. It is known that more than 95% of the normal random variable will be between two standard deviations $[-2v, 2v]$ in case (b). In the case (b), $[-2v, 2v] = [-2V_{LSB}, 2V_{LSB}]$, and therefore, 2 bits (representing $2^2 = 4$ LSB levels) are lost due to noise.

TABLE 3.4   Computed *SNR*, *SINAD*, and *ENOB* for Example 3.6 for 16-bit A/D converter.

Feature	SNR (dB)	SINAD (dB)	ENOB (bits)
(a)	99	98.8	16.14
(b)	87.7	87.6	14.26
(c)	81.6	81.5	13.2

For case (c), 95% of the generated noise will be in the range of $[-2v, 2v] = [-4V_{LSB}, 4V_{LSB}]$. Therefore, the range of 8 V_{LSB} can be presented using 3 bits ($2^3 = 8$). Therefore, we observe the loss of 3 bits in the effective number of bits.

3.5.3 A/D converter types

Four main types of A/D converter architectures include flash, pipelined, successive approximation register, and sigma–delta A/D converters. The choice of an A/D converter depends on the resolution and sampling rate, as well as the power consumption of the A/D converter. High resolution and high sampling rate cannot be achieved at the same time. The speed of the A/D converters drops, and their resolution increases from flash to sigma–delta A/D converters. Since speed is normally not a concern in biomedical applications, successive approximation register and sigma–delta A/D converters are the most common types. A successive approximation A/D converter converts a continuous analog waveform into a digital representation via a binary search through all possible quantization levels before finally converging upon a digital output for each conversion. Sigma–delta converters have high resolution, low power consumption, and low cost. A significant portion of the conversion is performed digitally. Details about types of A/D converters can be found in many textbooks, including Webb (2018), Pelgrom (2017). Instead of presenting them here, we will focus on A/D convertor types that are important in connecting different types of sensors/transducers.

3.5.3.1 *Multiplexed versus simultaneous sampling*

A/D converters commonly support multiple analog inputs; therefore, they are called *multichannel A/D converters*. Here, the channel means the input signal. The benefit of multiplexing is fewer numbers of single-channel A/D converter chips required, saving the area on a printed circuit board as well as power and cost. Multichannel A/D converters normally convert a signal from each input channel sequentially in a time-multiplexing fashion using an input multiplexer. An example of an integrated 3-channel data acquisition system that includes a multiplexer, programmable gain amplifier, and an A/D converter in the same chip is shown in Fig. 3.27. Most integrated multiplexed A/D converters provide automated channel switching, custom sequencing between channels, and configurations with different input ranges and error calibration options for each channel (Pachchigar, 2016).

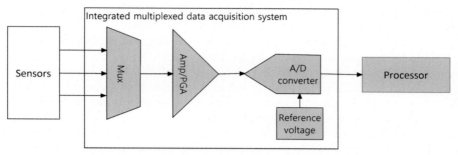

FIGURE 3.27   Multiplexed data acquisition system (Pachchigar, 2016).

Some common issues in multiplexed A/D converters are related to potential large amplitude changes between channels when switching from one channel to the next. To handle large amplitude changes, the amplifier (or PGA) must have a wide bandwidth and fast settling time. Therefore, dynamic performance is normally reported as output voltage settling time. An example settling time for Analog Devices AD74412R is 90 μs at a 10 V step.

In multiplexed A/D converters, each sample from a different channel is phase-shifted (delayed) relative to the sample in another channel. Simultaneous sampling means that the signals from different channels are all sampled at the same time. This is useful in applications that measure phase angle between signals, such as motor control or bioimpedance measurements. If it is necessary to sample signals simultaneously, then multiple A/D converters can be used. In the last 2 decades, main chip manufacturers, such as Analog Devices, Maxim Integrated, and others, developed integrated solutions allowing simultaneous sampling. The idea behind simultaneous sampling is based on using multiple parallel samples and holding circuits that sample signals from different channels simultaneously and then holding them until they are processed. Another integrated simultaneous sampling solution contains parallel A/D converters in a chip.

3.5.4   Integrated special-purpose A/D converters

A/D converters can be integrated with sensors, amplifiers, processors, and so on. In this way, they represent smart solutions that provide a digital interface to the rest of the system. We will cover below the following integrated A/D converters.

- Integrated *A/D converters with programmable gain amplifiers*
- *Capacitive-to-digital converters*
- *Time-to-digital converters.*

Some other examples of integrating A/D converters with other components include.

- *Audio A/D converters* are high-resolution A/D converters designed to sample several channels from microphones
- *Digital temperature sensors*, which include a temperature sensor and complete data acquisition system

- *Isolated A/D converters* that allow for galvanic isolation between the analog input and digital output, which is very useful for medical applications
- *Current sensing* A/D converters normally convert voltage on an external resistive load that is added externally to measure the current
- *Current-to-digital converters* convert the signals of photodiodes or other sensors with current signal output into a digital signal. An example includes a component AS89010, a 16-bit 4-channel A/D converter from AMS that has sensitivity up to 20 fA/LSB and allows for direct connections of four photodiodes.
- *Current-to-frequency converters* convert a current signal from the photodiode into a square wave with a frequency directly proportional to light intensity (irradiance) on the photodiode. If the photodiode is integrated into the chip together with the current-to-frequency converter, then they are called *light-to-frequency converters*.

3.5.4.1 A/D converters with PGAs

The magnitude of the input signal of the A/D converter should be fairly close to its full-scale voltage range. However, the output voltage of transducers can have a wide voltage range. In addition, the baseline voltage of the signal might change over time, causing saturation at the output of the amplifier. Therefore, a constant amplifier gain is not satisfactory in situations where the input of A/D converters can have a wide input range. So, PGAs are often integrated to increase the dynamic range of A/D converters, as shown in Fig. 3.27. This configuration allows for controlling the gain of the amplifier through the common A/D converter interface.

To understand the need for different gains, let us consider Eq. (3.11). If we are able to "add" one-bit resolution to the A/D converter, this will improve its SNR by about 6 dB. If the voltage resolution of the A/D converter is equivalent to $V_{LSB} = 2\,mV$, then the A/D converter cannot resolve signals smaller than $2\,mV$. However, if there is a built-in PGA that can amplify the input voltage by 2, for example, then the A/D converter will be able to resolve changes of 1 mV at the input. Therefore, amplification, in this case, would effectively increase the dynamic resolution of this A/D converter by 6 dB.

A/D converters with PGA allow for direct connection with sensors. In the case of strain gauges, the differential output of the bridge would be connected to the input of the A/D converter with PGAs. Therefore, inputs to the A/D converters with PGA chips are usually differential.

3.5.4.2 Capacitance to digital

There are several architectures of capacitance-to-digital converters (Jia, 2012). We will explain an architecture with a sigma–delta modulator followed by a digital filter. The capacitive transducer is connected externally. One plate is connected directly to the excitation signal. A fixed frequency is applied to the capacitor during the conversion. The other plate is connected directly to the input of the converter. As the capacitor is not connected to the ground, it is a floating capacitive transducer. It is possible to measure the capacitance of a single or differential floating capacitive transducer. The accuracy of the conversion is in the order of fF. The capacitance is measured as $C = Q/V$, where the V represents the excitation voltage, and Q represents the measured charge.

3.5.4.3 Time-to-digital converters

Time-to-digital convertors measure the time between the start and stop of an event. They are used wherever time-interval measurement is required, such as time-of-flight measurements, or measurement and instrumentation applications such as digital scopes (Henzler, 2010). As a large number of applications in biomedical instrumentation require time measurements, we see a potential for using the time to digital converters in this field.

The traditional approach to time-to-digital conversion is first to convert the time interval into a voltage using an integrator. Then, in a second step, this voltage is digitized by a conventional A/D converter. Another fully digital approach includes a counter that counts between the start and the stop event. The resolution of this time-to-digital converter depends on the counter's clock frequency—the higher the frequency, the higher the time resolution. In this case, the time resolution is limited to several hundreds of picoseconds since the clock frequency can be at most in the order of several GHz. However, this is more than enough resolution for the biomedical applications discussed in this book.

3.5.4.4 Integrating conditioning circuitry and A/D converters

Several different integrated solutions require only sensors to be added externally—all other conditioning and conversion electronics are in the chip. One example is the bridge transducer A/D converter. This circuit requires only a bridge to be connected externally. It produces a digital signal at the output. An example is the component AD7730 by Analog Devices. It consists of a high-impedance input buffer connected to a 4-gain PGA followed by a sigma–delta A/D converter and a digital lowpass filter. In addition, the component allows for serial communication with a microcontroller for parameter setting and A/D converter outputs, as well as offset control, placing the device in a standby mode for power reduction, and so on.

High integration of electronic components significantly reduces the number of components on the medical device's printed circuit board, allowing for the design of miniature wearable devices.

3.6 Generating signals

3.6.1 D/A converters

A *digital-to-analog converter* is a device that converts a digital code to an analog signal in the form of a current or voltage. The output of a D/A converter is used to drive various devices, including motors, mechanical servos, bioimpedance monitoring devices, and so on. D/A converters are also commonly used as a component in A/D converters, microcontrollers and digital systems. D/A converters produce discrete output in response to a digital input word; therefore, their transfer function is the inverted transfer function of A/D converters shown in Fig. 3.23. Let N be the number of bits at the input of the D/A converter. The analog output of the D/A converter is proportional to the ratio of the digital word (code) at the input and the maximum number of discrete output values 2^N.

3.6.1.1 The architecture of a D/A converter

As with A/D converters, we will not present different D/A converter architectures in detail. Several common architectures include voltage divider, segmented, ladder, and sigma-delta. To find out more about D/A architectures, please refer to Kester (2005a).

Fig. 3.28A shows a high-level block diagram of a D/A converter, including an input register, decoder, reference voltage source, D/A architecture itself and an amplifier. These components are commonly integrated on a chip. The input to the D/A is an N-bit word. The simplest D/A converter architecture is a register string network that has 2^N equal resistors and 2^N switches. Therefore, the input N-bit word must be decoded and converted into a 2^N-bit word that contains only one logic one at the position that corresponds to the digital input. The output is determined by closing one out of 2^N switches to point into a particular location on the resistor string, as shown in Fig. 3.28B. When there is a code transition, one switch will be turned off, and another turned on; therefore, we say this is a low glitch architecture. The major disadvantage of this architecture is a large number of resistors and the need for the resistors to have the same values. However, the resistor string D/A architecture can be used as a component (or segment) in a more complex D/A architecture. For example, D/A converters with resolutions from 8 to 16 bits are commonly designed as segmented architectures.

The output of the D/A converter is determined as

$$V_{OUT} = A \cdot V_{FS} \cdot W/2^N$$

where W is the decimal equivalent of the binary code word that represents the digital input to a D/A converter, N is the resolution of D/A converter, V_o is the gain of the output amplifier. V_{FS} is the full-scale reference voltage. In Fig. 3.28A, $V_{FS} = V_{cc}$.

3.6.1.2 Errors during D/A conversion

Specifications related to D/A converters can be grouped into static or DC specifications and time and frequency domain specifications (Mercer, 2021). Common static specifications include gain error, offset error, and differential and integrated nonlinearities. Common time domain specifications are output voltage settling time and digital-to-analog glitch impulse. Frequency domain specifications include spurious-free dynamic range (SFDR), total harmonic distortion (THD), and signal-to-noise ratio (SNR).

Let us start with static specifications. *Gain error* is the deviation of the slope of the converter's transfer function from that of the ideal transfer function over the full scale, and it is given as a percentage of the full-scale range. In the absence of a gain error, the *offset error* is the deviation of the output of the D/A converter from the ideal output, which is constant for all the input codes. It is expressed in mV normally. For example, a 10-bit D/A converter with 5 V reference voltage can have a 1.5 mV offset error. *Integral nonlinearity* (INL) is the deviation of the actual output voltage measured for a certain input code from the ideal output voltage for that code. INL is calculated for all input codes after offset and gain errors are removed. It is normally given as a figure showing INL versus the digital input codes and also as the maximum value of the deviation over the full scale in the number of LSB units (Johns and Martin, 1997; Sansen, 2006).

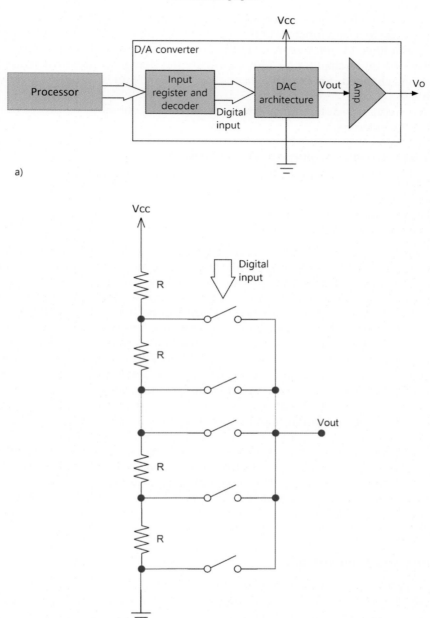

a)

b)

FIGURE 3.28 (A) a block diagram of a D/A converter and (B) simplified resistor string structure.

Settling time is the time needed for the output to settle to a value within its specified error limits in response to a step function at the input. It is normally given in μs for the output gain of 1 of the amplifier. *A glitch* represents the area of the unwanted signal generated at the output when the input changes by 1 bit. Normally the change is measured when the largest number of bits changed, for example, from 0x7FF to 0x800 for a 12-bit D/A converter. It is given in nV × s.

The same frequency domain characteristics are used here as those described in Section 3.5.2 for A/D converters.

3.6.1.3 *Integrated special-purpose D/A converters*

Special-purpose integrated circuits based on D/A converters include

- Waveform generators
- Bioimpedance measurement chips
- D/A converters specialized in audio or video signals

Waveform generators are used in many applications, including frequency stimulus, bioimpedance measurements, sensor applications for proximity and motion, clock generators, and so on. These waveform generators can produce sine, triangular, and square wave outputs. The output frequency and phase are commonly software programmable. An example of a circuit is AD9837 by Analog Devices, which can be tuned to 0.02 Hz resolution with up to 5 MHz frequency. Basic waveform generators are based on the sequential reading of waveform amplitudes from an internal lookup table and converting them into an analog signal using a D/A converter. For example, the AD9837 has a 10-bit D/A converter.

Waveform generation based on direct digital synthesis (DDS) allows for generating sine waves with precise phase and frequency control, including single-frequency sine-wave and sweeping sine-wave signals. Frequency resolution and maximum frequency are normally much better than the resolution and frequency achieved using other types of waveform generators.

Bioimpedance measurement chips are based on DDS techniques that allow for generating sine-wave signals. They are used for electrochemical analysis, bioelectrical impedance analysis, and impedance spectroscopy. The waveform generator provides a signal of known frequency at the output of the chip that is used to excite an external complex impedance. The response signal from the impedance is sampled by an A/D converter. An example of that is the AD5933 integrated circuit by Analog Devices.

3.6.2 Pulse width modulation

PWM generates square-wave pulses of different widths. The width of the pulse, sometimes called pulse width offset, is commonly proportional to the amplitude of the input signal. PWM is used to control motors or to power LEDs. The main reason for using PWM is that it allows for controlling the average amount of power delivered to a load or the output. They are also used for voltage regulation and modulation in communications.

A PWM system normally has several parameters for tuning, including the PWM period (reciprocal of the pulse frequency), the pulse width offset, and the pulse delay time. Pulse delay time represents the shift in time compared to the pulse starting at 0 s. The values can be given in percentages of PWM period or seconds. Pulse width offset D is given in percentage of the PWM period T. Sometimes, offset is also given in seconds. PWM output signal $y(t)$ is logic one (y_{max}) during the time DT and logic zero (y_{min}) during the period $(1 - D)T$. The average amplitude of the signal $\bar{y}$ at the output of PWM is

$$\bar{y} = \frac{1}{T} \int_0^T y(t)dt = \frac{1}{T} \int_0^{DT} y_{max}dt + \frac{1}{T} \int_{DT}^T y_{min}dt = Dy_{max} + (1 - D)y_{min}$$

When $y_{min} = 0\ V$ then $\bar{y} = Dy_{max}$. The simplest way to generate a PWM signal $y(t)$ is to use a triangular wave generator as a modulation waveform and a comparator. When the amplitude of the input signal is larger than the amplitude of the modulation waveform, the output is logic one—otherwise, the output is a logic zero. Fig. 3.29 shows a 10 Hz modulation waveform in blue. The input signal is the sine wave with a frequency of 1 Hz, shown as a black line in the upper figure. The lower figure shows the output PWM signal.

PWM to voltage conversion can be done by integrating the PWM signal by connecting the PWM signal to a lowpass filter. This emulation of D/A converter has traditionally been

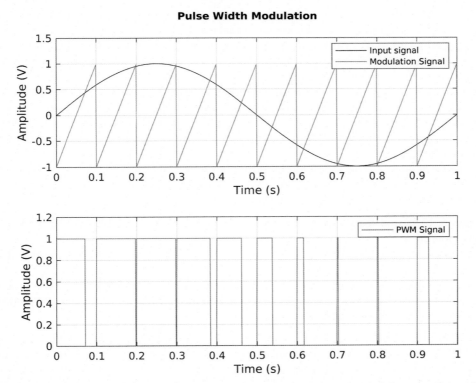

FIGURE 3.29 PWM signal generation. The upper figure shows the input signal and the triangular modulation waveform, while the lower figure shows the PWM signal output.

done for low-speed, low-resolution applications. The accuracy of this circuit depends on the properties of the lowpass filter. If the lowpass filter has a bandwidth that is too small, then it will take much time for the signal to change from one state to the next. On the other hand, if the bandwidth of the filter is high, then the signal at the output of the filter will oscillate around the mean value of the PWM signal, which means that the resolution of the emulated D/A converter will be low. Therefore, this solution is used in low-cost microcontroller-based applications where the PWM signal is generated at the output of the microcontroller (Alter, 2008)—please see Problem 3.23. There are also PWM to voltage integrated solutions mainly based on converting the PWM signal into a binary signal and then doing D/A conversion. In that case, good signal resolution at the output can be achieved.

3.7 Other components of the system

Until now, we introduced transducers, amplifiers, analog filters, and A/D and D/A converters. Very often, these components are combined to form *sensors*. We define a sensor as a device that converts the physical measurand into voltage or current output. Sensors will be introduced in Chapters 5—9 when we discuss different devices. The sensor's output can be analog; in this case, the sensor is connected to a processing unit through an A/D converter. The sensor-integrated circuit can also contain an A/D converter producing a digital signal at the output. The digital-output sensor is connected to a processing unit through a digital communication interface. Common interfaces that are used nowadays in wearable devices include SPI and I²C. I²C, or Inter-Integrated Circuit, is a chip-to-chip interface supporting two-wire communication. SPI, or Serial Peripheral Interface, is a synchronous serial data link standard that operates in a full-duplex mode. Both interfaces connect the microcontroller with relatively slow peripherals, A/D and D/A converters, and digital sensors.

Other system components include a processing unit, power management unit, battery and battery charger, and digital electronic components. Some of these components will be introduced in Chapter 10.

Wearable devices are designed as embedded systems. *Embedded system* is a special-purpose computer system optimized to execute several specialized functions while the optimizations normally include (Valvano and Yerraballi, 2022)

- Software designed for the particular hardware, including available sensors and other components
- Real-time constraints meaning that certain processing tasks need to be completed before the deadline
- Need to minimize memory requirements since the amount of memory in the embedded system is normally small in comparison with memory available on personal computers
- Need to minimize power consumption.

Microcontrollers (MCU) are used in embedded systems to perform a certain task, handle communication, and control other hardware components. A microcontroller normally contains a relatively simple central processing unit (CPU) combined with peripheral components such as memories, I/O devices, and timers (Valvano and Yerraballi, 2022). Some of the components commonly integrated into a chip are:

- Interrupt controller.
- On-chip memory.
- Power management control unit that is used for entering/leaving power-down modes and enabling or disabling the peripherals.
- Direct memory access (DMA) controller that allows for data transfer between the memory and peripherals without using the CPU.
- One or more A/D converters.
- One or more D/A converters.
- Comparators that compare the analog input with a preset threshold and provide a binary output that is logic one if the input signal is greater (or smaller depending on the implementation) than the threshold.
- Multiple timers/counters that can generate interrupts after some programmed time intervals.
- Real-time clock (RTC) that is used to track clock time and calendar date, and it is normally turned on all the time.
- I^2C and SPI buses for connecting MCU to other peripherals.

Some MCUs also have integrated Bluetooth or other communication modules. As an example, ARM Cortex-M0+ core can run up to 48 MHz and has on-chip 512 KB flash memory and on-chip 128 KB of random access memory (RAM).

Digital signal processors are processors designed for completing signal processing tasks such as filtering or computing Fast Fourier Transform (FFT). These tasks involve repetitive numeric calculations, such as multiple and add/accumulate (MAC) operations. Digital signal processors implement fixed-point or floating-point algorithms requiring high memory bandwidth for the streaming sensor data. Modern digital signal processors include special hardware for performing MAC operations, can execute multiple operations and memory accesses per clock cycle, and have multiple addressing modes to support the implementation of algorithms such as FFT.

3.8 Summary

This chapter focuses on the electronics required to interface the sensors with processors. We covered bridge circuits, operational amplifiers, analog filters, and A/D converters. In addition to the data acquisition path, we briefly analyzed circuits required to generate signals and drive the loads. These circuits include D/A converters and PWM circuits.

In this chapter, we introduced a number of newer solutions that are commonly used in design nowadays. These integrated single-chip solutions include multiple components we studied in this chapter placed in one chip. Examples include transimpedance amplifiers, different types of bridges, waveform generators in D/A converter section, capacitance to digital and time to digital converters in the A/D converter section, and so on. These should give the readers ideas about current trends in the industry.

There are several limitations in this chapter. Sometimes, examples of integrated solutions from the industry are chosen. These examples are chosen randomly without the intention to favor any particular company—nevertheless, it turned out most examples are from Analog Devices. Power consumption is not analyzed, which should have been done if these components are used in wearable biomedical devices. Digital electronics is not introduced here, and we assume that the majority of digital processing will be performed in software on a computer, processor, or microcontroller. Processor architectures are not discussed.

3.9 Problems

Please note that questions that start with * are not covered in the book. They are added for readers interested in exploring different topics further.

3.9.1 Simple questions

Questions: Wheatstone bridge

3.1. Describe the Wheatstone bridge and describe how it is used to connect different transducers.

3.2. Describe configurations and derive an input—output relationship for bridges with 1, 2, and 4 transducers in the bridge. Compare results.

3.3. What can be done when it is not possible to place transducers so that two of them increase resistances and two decrease resistances?

3.4. How are capacitive transducers connected to the amplifier?

3.5. *Describe temperature effects on the bridge. How is temperature compensation done?

Questions: Instrumentation and differential amplifiers.

3.6. What are the differential gain, common mode gain, and common mode rejection ratio of amplifiers? What are their typical values for amplifiers used in ECG and EEG?

3.7. Derive the differential gain and the common mode gain of an instrumentation amplifier.

3.8. Why is it better to use an integrated instrumentation amplifier in a chip and not build your own using operational amplifiers and resistors? Discuss the need to match resistors up to 0.02% accuracy.

3.9. Describe programmable gain instrumentation amplifiers.

3.10. Is there a relationship between the gain and the bandwidth of the instrumentation amplifier?

3.9.2 Problems

3.11. Derive the output voltage and the output current of a circuit model presented in Fig. 3.2A.

3.12. Derive formulas (3.5) and (3.6) for the output voltage of the charge amplifier.

3.13. In Example 3.3, the interpolated signal has a fundamental frequency of 2 Hz. Why is the interpolated sine wave at 2 Hz shifted by 180 degrees?

3.14. Consider a system for monitoring breathing using an accelerometer shown in Example 3.2. The accelerometer is placed on the subject's chest. Please assume that the breathing signal is periodic and that it is causing changes of 10 mg in one of the axes of the accelerometer (see, e.g., Schipper et al., 2021). What would be the amplitude of the voltage V_{o1} and V_o in Fig. 3.12.

3.15. Show expressions for the gain of the PGA shown in Fig. 3.15. Modify the PGA to support the gain of 4 as well.

3.9.3 Simulations

3.16. Compute linearity error $\%e_{L_{max}}$ for the quarter bridge using both the terminal and the best-fit straight-line methods described in Chapter 1. Please note that the nonlinearity error computed here is for the bridge itself without including the transducer.

3.17. Modify the equivalent circuit model shown in Fig. 3.2A to include the amplifier's offset voltage and input bias current noise. Find a datasheet of a precision amplifier of your choice to obtain realistic values of these parameters.

3.18. Draw V_0 vs. v_{in} when $\Delta R/R$ changes from -10% to 10% in the steps of 1% for the resistive bridge with:

a) One sensor

b) Four sensors

3.19. The ramp generator is shown in Fig. 3.30. The input signal is a pulse with a frequency is 1 kHz, 50% duty cycle, a minimum amplitude of -1 V and a maximum amplitude of 0 V. The output is a triangular shape pulse with a minimum amplitude of 0 V and a maximum amplitude V_{max}. Implement this ramp generator in Simscape and show how you selected components to get the desired output.

3.20. Consider Eq. (3.4) for interfacing a capacitive sensor that changes its capacitance with the distance between the electrodes. Then, compute the formula for the output voltage versus distance change between the electrodes.

3.21. Show the Bode plot for the charge amplifier in Fig. 3.12. Compare lower and upper cutoff frequencies with the ones computed in Example 3.2. To do this, make sure that you label input and output signals in the circuits by clicking on the line of

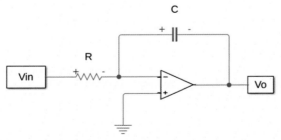

FIGURE 3.30 Integrator (ramp amplifier).

interest and then selecting Linear Analysis Points. After selecting the input and output signal, go into the Frequency Response app in Simulink and select Bode plot.

3.22. Modify the code, for Example 3.3 for the following sampling rates 10 Hz, 20 Hz, 27 Hz, and 30 Hz.

3.23. This question is related to PWM signal generation.

    (a) Implement PWM signal in Simscape with the pulse with offset $D = 50\%$, $T = 10\mu s$ and $y_{max} = 3.3\,V$ and $y_{min} = 0V$ followed by a first-order passive lowpass filter. Analyze the output of the filter when the cutoff frequency is (1) 50 kHz, (2) 1 kHz, (3) 100 Hz.

    (b) Assume that this circuit is used as a D/A converter. Analyze the performance of this converter for cases (1), (2), and (3) above.

3.10 Further reading

In this section, we refer to several relevant resources for further reading. However, please note that this list is not comprehensive and that many excellent and relevant resources might have been omitted unintentionally.

A comprehensive overview of instrumentation amplifiers is given in (Kitchin and Counts, 2006). Active filters are detailed in the book Wanhammar (2009). A very comprehensive resource for A/D and D/A converters is (Kester, 2005a). An excellent book that focuses only on A/D converters is Pelgrom (2017).

Many excellent books are available for an introduction to the sampling theorem, antialiasing, and other topics related to signal processing, including Oppenheim and Schafer (1998).

A good practical resource for learning about embedded systems is Valvano and Yerraballi (2022).

References

12-Bit, 1 MSPSSingle-Supply, Two-Chip Data Acquisition System for Piezoelectric Sensors, Analog Devices, Circuit Note CN-0350, 2017.

Alter, D.M., 2008. Using PWM Output as a Digital-to-Analog Converter on a TMS320F280x Digital Signal Controller," Application Report. SPRAA88A, Texas Instruments.

Henzler, S., 2010. Time-to-Digital Converters. Springer.

Islam, T., 2017. Advanced interfacing techniques for the capacitive sensors. In: George, B. (Ed.), Advanced Interfacing Techniques for Sensors: Measurement Circuits and Systems for Intelligent Sensors. Springer.

Jia, N., 2012. ADI Capacitance-to-Digital Converter Technology in Healthcare Applications, Analog Dialogue. https://www.analog.com/en/analog-dialogue/articles/capacitance-to-digital-converter-technology-healthcare. html. (Accessed 14 March 2021).

Johns, D.A., Martin, K., 1997. *Analog Integrated Circuit Design*. Wiley.

Kester, W., 2005a. Data Conversion Handbook. Newnes.

Kester, W., 2005b. Bridge circuits. In: Jung, W. (Ed.), Op Amp Applications Handbook. Elsevier Inc, pp. 231–256.

Kitchin, C., Counts, L., 2006. A Designer's Guide to Instrumentation Amplifiers, third ed. Analog Devices, Inc.

Mercer, D.A., 2021. Current Steering Digital-to-Analog Converters, Analog Devices. https://wiki.analog.com/ university/courses/tutorials/cmos-dac-chapter. (Accessed 31 March 2020).

Northrop, R.B., 2014. Introduction to Instrumentation and Measurements, third ed. CRC Press.

Oppenheim, A.V., Schafer, R.W., 1998. Discrete-time Signal Processing, second ed. Prentice Hall.

Pachchigar, M., 2016. Design Trade-offs of Using Precision SAR and Sigma-Delta Converters for Multiplexed Data Acquisition Systems, Analog Devices, https://www.analog.com/en/technical-articles/precision-sar-sigma-delta-converters.html, Last accessed on March 11, 2021.

Pelgrom, M., 2017. Analog-to-Digital Conversion, third ed. Springer.

Precision Variable Gain Amplifiers (VGAs), Analog Devices, MT-072 TUTORIAL, Rev.0, 10/08, 2008.

Sansen, W., 2006. *Analog Design Essentials*. Springer.

Schipper, F., van Sloun, R.J.G., Grassi, A., Derkx, R., Overeem, S., Fonseca, P., 2021. Estimation of respiratory rate and effort from a chest-worn accelerometer using constrained and recursive principal component analysis. Physiol. Measure. 42 (4).

Serial Peripheral Interface, 2021. Wikipedia. https://en.wikipedia.org/wiki/Serial_Peripheral_Interface. (Accessed 24 February 2021).

Valvano, J., and Yerraballi, R., 2022. Embedded Systems - Shape The World, online book, http://users.ece.utexas.edu/~valvano/Volume1/E-Book/, last accessed on June 20, 2022.

Wanhammar, L., 2009. Analog Filters Using MATLAB. Springer.

Webb, A.G., 2018. Principles of Biomedical Instrumentation. Cambridge University Press.

Webster, J.G. (Ed.), 2010. Medical Instrumentation: Application and Design, fourth ed. John Wiley and Sons.

4

Modeling and simulation of biomedical systems

Pervasive Cardiovascular and Respiratory Monitoring Devices
https://doi.org/10.1016/B978-0-12-820947-9.00012-X

Acronyms and explanations

DFT　discrete Fourier transform
kSQI　kurtosis-based SQI
LSB　least significant bit
pSQI　spectral power ratio based SQI
PWV　pulse wave velocity
REB　Research Ethics Board
SDR　spectral distribution ratio
SQI　signal quality index
STFT　short-time Fourier transform
tSQI　template matching based SQI

Variables used in this chapter include:

B　amplitude of a narrowband signal
E　energy
f_{clk}　clock frequency
Kurt　kurtosis
L　the number of samples in a block or a window needed to produce one output sample
P　power

$S(f)$	spectral density function
T_K	the temperature in degrees Kelvin
tol	tolerance of a resistor
T_{proc}	time to process L samples
T_s	sampling period
f_s	sampling frequency
T_{sy}	sampling period of the signal at the output (if it is different from the sampling period of the signal at the input)
V_{dd}, V or V^+	supply voltage
$v(t)$	white Gaussian noise
$x(t)$	signal of interest
$y_j, y(t)$	the output of a function, model or a device
y_{REF}	the reference measurements
β	noise color parameter
μ	the mean
σ	the standard deviation

4.1 Introduction

In this chapter, we will discuss the modeling of a physiological system and circuits and provide examples of different models, performance analysis, parameter optimization, uncertainty, sensitivity analysis, etc. This chapter is needed to prepare for the next part of the book, where we will model particular devices, such as blood pressure monitoring devices. We have already seen in Chapters 2 and 3 several models of transducers and circuits; however, these transducers and circuits were not presented as parts of the biomedical devices, and additional analyses that include uncertainty propagation and sensitivity analysis were not performed.

This chapter starts with a description of the data collection process and some common physiological signal databases. Even if we include a model for signal generation, the parameters of the model need to be evaluated with experimental data. Data can be obtained in the following ways.

1. Performing data collection, which is necessary when developing a new device to fit the parameters of the device and to evaluate the performance of the device
2. Using already collected data from databases such as Physionet
3. Using in silico data which is generated using computational models
4. Generating data from developed models of the physiological system as well as models of interaction between the device and the subject.

The first three items will be covered in Section 4.2, while the models (item 4) will be described using an example in Section 4.3. Item 4 is important for us because it will allow us to evaluate the effect physiological and other parameters have on the accuracy of the device. In Section 4.3, we will also perform parameter estimation using provided experimental data.

A general model of a biomedical system is shown in Fig. 4.1. The model includes not only the device itself but also the model of an interaction of the physiological system and the

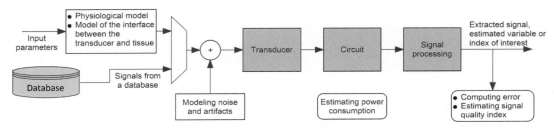

FIGURE 4.1   General model of a biomedical system. The block on the left of the plus sign represents a multiplexer that will allow us to use either signals from a database or signals generated from our models.

sensor, as well as the noise and artifacts model. This approach will allow us to understand the effect of all simulated parameters of the system on the output, including some physiological parameters. However, physiological models are normally very complex, and we will not focus on modeling the physiological system itself in detail because this is outside the scope of this book.

Noise and artifact models are covered in Section 4.4.2. Analysis of signal quality is provided in Section 4.4.3. In Fig. 4.1, the signal quality index (SQI) is computed at the end of the pipeline. However, we can compute the SQI after each block as well as at the input to the device. An error is normally evaluated at the end, as shown in Fig. 4.1.

A circuit-based model is presented in Section 4.5 to analyze uncertainty propagation.

After converting the signal from analog to digital, the signal is further processed in software (Section 4.6). We will briefly discuss modeling real-time signal processing in software followed by offline processing. Finally, as an example, we will consider the design of a system that generates an alarm in case of breathing cessation. We will simulate a signal obtained from a breathing belt. A breathing belt is a device that a subject wears around her or his chest used to extract the breathing signal based on the displacement of the chest during inspiration and expiration. This signal can be processed to provide breathing rates over time, analyze breathing patterns, and indicate whether the breathing rate is too high or too low. We will model and simulate a simplified system that can extract the breathing signal and detect that the subject stopped breathing for a longer period than a predefined threshold. This design will be based on two design paradigms. The first design paradigm is completely hardware-based, while the second design paradigm is based on the approach where the minimum required processing is done in hardware to obtain the digital signal, and further processing is performed in software.

Several example models are explained and simulated in this chapter. First, we introduce the Windkessel model to model the aortic pulse and show the process of estimating model parameters. The signal quality index was assessed on ECG data. Uncertainty is propagated through the circuit that converts the signal from the piezoelectric transducer into voltage to measure acceleration. The power consumption of the A/D converter is evaluated in the end. We have already mentioned a simulated system for the detection of breathing cessation.

4.2 Data collection and databases

When developing any biomedical device, we need data to set the parameters of the device. Data is normally obtained by performing pilot studies in which a new device is tested on human subjects against a reference device or a trained medical practitioner. To perform the pilot study, Research Ethics Board (REB) approval must be obtained first. Next, the study participants need to agree with the proposed protocol by signing an informed consent form. This section starts with a description of the ethics application process.

We often start data collection in the environment where the device is developed. However, colleagues in companies and students/researchers at universities and research centers might not be the appropriate population for testing the device. Nevertheless, some physiological changes can be caused by simple maneuvers, which might be good enough for the initial test of some devices. Hence, a subset of common maneuvers is also presented in this section.

We can use the data from existing databases or simulated data before data collection. A description of several online databases and simulators is presented at the end of this section.

4.2.1 Ethics application

Before any data collection, the first step is obtaining REB approval. The ethics board focuses on three major aspects of the study: foreseeable risks, potential benefits, and ethical implications of the project. Risks are related to protecting the subjects' privacy by not exposing their identity and ensuring they are safe. In Canada, the Tri-Council Policy Statement: Ethical Conduct for Research Involving Humans—TCPS 2 (2018) is commonly followed when submitting and reviewing Ethics applications (Tri-Council, 2018).

The ethics application normally includes the main form, recruitment letter, and informed consent form. In the main form, the researchers address the team of experts in the REB regarding the way the study will be conducted and the privacy will be preserved. In the consent form, the researchers address the subjects and explain, at a very high level, all the aspects of the study so that the subjects can understand the study. The experiments cannot begin before obtaining approval from the REB and the subject's signature on the informed consent form (Box 4.1).

The REB is looking for the following when reviewing the application:

- Description of the project,
- Description of the recruitment process,
- Description of what the subject needs to do during the study (participation),
- Risk for the subject,
- Benefits to the subject and the community,
- Privacy, confidentiality and data conservation, and
- Consent process and consent document.

The description of the project provides a rationale for the project and the background literature. The recruitment process deals with several important aspects, including the

---

BOX 4.1

Who does evaluate REB applications?

After submitting the application, the application gets evaluated first by the Ethics officer. The application can be considered a minimum risk; in this case, it is commonly evaluated by two members of the REB as well as the Ethics officer. If the Ethics officer assumes that the application has some issues, mainly regarding privacy or risks to participants, then the application goes to the full board review, where it is discussed and evaluated by all members of the REB during a face-to-face meeting. Being an REB member, I saw many engineering REB applications written in a hurry. These applications take most of the time for the REB to discuss. REB members are volunteers, and it is important to think of their time too. The informed consent form also needs to be carefully written too. The researcher and the subject sign it, and it represents the researcher's promise to the subject that the study will be conducted in a way it is claimed in the form which was approved by the REB.

---

participants' inclusion and exclusion criteria. Sometimes, including people with particular conditions is unnecessary or too dangerous. The inclusion criterion could be the age range (e.g., subjects should be at least 18 years old). An example of an exclusion criterion is excluding cardiac patients from studies that require significant physical effort. The recruitment process also needs to describe if there is a risk of coercion and how it is handled, and how potential participants will be notified about the incoming study. Next, the subjects' participation during the study needs to be detailed, including the type of activities, duration, location, physical effort required, and so on. In addition, it needs to include the description of a device or a method that is tested, and the description of reference devices. Potential harmful effects and risks to the participants need to be discussed in detail. Next, ways of protecting participants' private information need to be explained. In some cases, it will not be possible to guarantee the protection of the identity of participants—then limits of privacy and anticipated risks to participants need to be clearly explained. The last part is related to the consent process and consent document. The consent process needs to be clearly explained. The consent document should contain information needed for participants to understand the study, its goals, and risks.

REB's mandate is not to evaluate the scientific aspects of the project; however, a project description with references needs to be included so that the REB has the complete information required to evaluate the application. It is also very important for the researchers themselves to write the full application and put down on paper the study plan that is then evaluated by themselves as well as by the third party (REB), thereby improving the study.

4.2.2 Maneuvers

In this section, we will discuss some additional aspects of data collection. The first step in testing pervasive wearable devices is testing on healthy subjects. However, these subjects often do not show the same level of variations of physiological variables of interest as subjects with conditions would show. For example, the variations of some variables of interest can be smaller (blood pressure values) in a healthy group without maneuvers or larger (heart rate variability) than in the group with medical conditions. This is why several different maneuvers have been introduced to emulate variations of some of the physiological parameters of interest. By performing the maneuvers, subjects can rapidly change, for example, blood pressure which can help evaluate a newly developed blood pressure device. Maneuvers can also be performed on patients to evaluate the effects of some actions.

4.2.2.1 *Valsalva maneuver*

The Valsalva maneuver is an attempt at expiration made with the closed nose and mouth. It can result in a very rapid increase in blood pressure and heart rate. Therefore, it can be used in pilot studies where it is required to increase blood pressure intentionally.

4.2.2.2 *Muller maneuver*

After a forced expiration, an attempt at inspiration is made with a closed mouth and nose, representing the reverse of a Valsalva maneuver. It is normally performed when testing sleep apnea since it emulates the obstruction of breathing.

To understand the effect of change as well as the range of physiological parameters that are affected by the maneuvers, we show through simulations in Fig. 4.2 the effects of the Valsalva and Muller (sometimes written as Mueller) on breathing by observing the tidal volume, oxygen saturation, arterial blood pressure, and heart rate. A number of other parameters are affected by these two maneuvers but are not shown here. The simulated data is obtained using the software package PNEUMA 3.0 (Cheng et al., 2010). Valsalva maneuver starts about a time instant of 40 s with inspiration and breath-hold shown in Fig. 4.2A as a constant large volume of air in the lungs. Valsalva maneuver has four phases, where the first two phases are related to the period of forced expiration, and the other two phases occur after the occlusion is released. In the first phase, there is a sharp increase in arterial blood pressure and some increase in heart rate. In phase 2, the arterial blood pressure decreases while the heart rate increases significantly. In phase 3, breathing is resumed, and arterial blood pressure drops significantly while the heart rate is still high. Finally, in phase 4, there is an increase in blood pressure before returning to normal and a decrease in the heart rate.

We can also observe the changes in the measured parameters due to the Muller maneuver during the period between 130 and 150 s. During that time, the subject exhales, and therefore subject's tidal volume is low and relatively constant. There is a significant decrease in oxygen saturation (SaO_2) during the maneuver and immediately after. Please note that we will analyze and introduce these signals in Part II of the book. Here, we just want to show how significant the effect of different maneuvers can be.

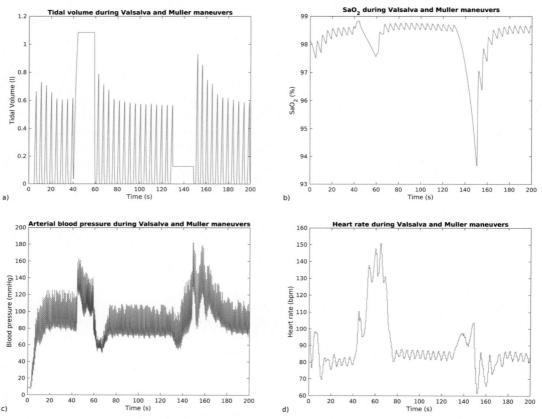

FIGURE 4.2   (A) Tidal volume, (B) oxygen saturation (SaO$_2$), (C) arterial blood pressure, and (D) heart rate during 200 s interval. The Valsalva maneuver is performed between 40 and 60 s, while the Muller maneuver is performed between 130 and 150 s. The figures are obtained using the simulation packet PNEUMA 3.0 (Cheng et al., 2010).

4.2.2.3 Deep breathing maneuvers

A deep breathing test requires the subject to breathe in and out at least 6—10 times. Respiratory sinus arrhythmia is usually measured using the deep breathing test (Löllgen et al., 2009). Respiratory sinus arrhythmia represents variability in the heart rate in synchrony with respiration by shortening the R—R interval in the ECG during inspiration and prolonging it during expiration.

4.2.2.4 Ewing's battery of tests

Ewing's battery of tests was intended to test cardiovascular autonomic function in diabetes patients. Ewing's battery of tests is divided into two groups, where the first group evaluates the sympathetic nervous system functions using, for example, the handgrip test. In contrast, the second group investigates the operation of the parasympathetic nervous system by analyzing, for example, heart rate response to deep inspiration and expiration and the

Valsalva maneuver (Ewing et al., 1985; Spallone et al., 2011). Some of these tests are used to induce changes in physiological parameters when developing blood pressure and other biomedical devices.

4.2.3 Databases

PhysioNet is a database of recorded physiologic signals (PhysioBank) freely available online. PhysioNet is supported by the National Institute of General Medical Sciences (NIGMS) and the National Institute of Biomedical Imaging and Bioengineering (NIBIB). One of the PhysioNet datasets is the MIMIC dataset. MIMIC III data set is described in (Aew et al., 2016) and contains thousands of recordings of multiple physiologic signals collected by patient monitoring devices in adult and neonatal intensive care units.

A large dataset of signals was collected in Great Britain from 4378 subjects. Among many other parameters, it contains finger photoplethysmogram (PPG) pulse waveforms alongside brachial-femoral pulse wave velocities (Elliott, 2014). Although this database is not available online, data can be accessed after obtaining approval from the data access committee and the Ethics committee.

Another large dataset on atherosclerosis, an important sub-type of arterial hardening and narrowing caused by plaque build-up, is available to researchers through the Multi-Ethnic Study of Atherosclerosis (MESA) (Burke et al., 2016). This dataset includes several biological, behavioral, and environmental markers obtained from over 6800 men and women living in different communities in the United States.

In the next chapters of the book, we will mention important databases for each device we describe.

4.2.4 Databases based on computational models

Computational (in silico) modeling is the direct use of computer simulation and modeling in biomedical research to simulate physiological processes. Computational modeling of cardiovascular and respiratory functions is important as it allows for:

- Evaluating the influence of individual cardiovascular properties and their variations over time on the signal of interest (e.g., the arterial pulse wave),
- Evaluating the signal of interest at different locations on the body,
- Analyzing the effect of aging and different cardiovascular diseases on the signal of interest,
- Providing exact measurement references, which might be difficult to obtain from in vivo studies,
- Developing methods for estimating the uncertainty of measurements,
- Generating the signal of interest of virtual patients and,
- Performing patient-specific modeling and potentially developing a digital twin.

Major disadvantages of this approach include reliance on modeling hypothesis and setting different parameters of the models where some parameters might be unknown or based on previous studies that might not be completely relevant. Nevertheless, computational modeling can provide additional understanding of the physiological processes of interest.

Modeling and simulation of arterial blood pressure, blood flow velocity, volume flow rate, and PPG pulse wave at common measurement sites was described recently in Charlton et al. (2019). The research resulted in a database of pulses of more than 4000 virtual patients. MATLAB (The MathWorks Inc., Natick, MA) code is provided, allowing for generating new virtual patients with varying parameters.

A database of simulated arterial waves of 3325 virtual subjects, each with distinctive arterial pulse waveforms, is described in Willemet et al. (2015). The following parameters are varied between subjects: arterial pulse wave velocity (PWV), the diameter of arteries, heart rate, stroke volume, and peripheral vascular resistance. It is applied for foot-to-foot PWV simulation and estimating arterial stiffness.

An example of a physiological simulator is PNEUMA which has already been mentioned in the previous section (Cheng et al., 2010). PNEUMA is a Matlab-based software that simulates the autoregulation of the cardiovascular and respiratory systems under normal conditions, during sleep, and under different interventions. It allows for obtaining many simultaneous respiratory and cardiovascular signals, including those presented in Fig. 4.2. It can simulate cardiorespiratory responses to different phases of sleep, the effects of mechanical ventilators, Valsalva and Mueller maneuvers, Cheyne-Stokes respiration during sleep, and so on. The simulator is very interesting for this book since it allows for generating a number of signals that can be used for testing different devices/systems with known preset physiological parameters. However, the simulator is designed for physiologists and not for engineers and is not easy to expand to simulate signals from different places on the body.

4.3 Models for signal generation

In the first part of this section, we discuss model types based on their interpretability and how parameter estimation is performed. Then, we show, as an example, the Windkessel model that describes the hemodynamics of the arterial system in terms of arterial resistance and compliance. This model is based on an electrical circuit analogy. Finally, the model parameters are estimated, and the sensitivity analysis is performed.

4.3.1 Types of models

In this book, we will often work with simulated physiological signals. We will review several modeling approaches that are helpful for understanding models used later in the book. They include:

- Physiology-based models are normally presented through a set of differential equations
- Physiology-based models based on the electrical circuit analogy
- Empirical data-based models

Physiological models are white or gray-box models, and as such, their parameters are often interpretable. They can model the way the signal is generated. The parameters used in the model should have physical or physiological meaning.

Models normally attempt to simulate different patient conditions; therefore, these conditions are inputs to the models. For example, the type of arrhythmia could be modeled by an ECG simulator.

As pointed out in Chapter 5 of Cerutti and Marchesi (2011), there are two main ways to assign values to the parameters of an interpretable model:

W1. Based on the previous knowledge (forward modeling)—we assign typical parameters to an entire class of subjects with a given pathology. We can use this model to generate signals that are then used as input to a simulated device. In this way, we can better understand a system's behavior for specific pathology or specific ranges of parameters of interest.

W2. Based on parameter estimation or model fitting (inverse modeling) that minimizes the error between the model output and the experimental data. The selection of the cost function to minimize the error and the selection of proper parameters for estimation are normally a challenge. It is pointed out in Cerutti and Marchesi (2011) that the unknowns regarding the choice of the parameters and the cost function are one of the bottlenecks in the widespread use of interpretative models in clinical practice.

In the next section, we will show these two ways of assigning the parameters to the model on an example of modeling the arterial pulse based on the Windkessel model. Table 4.1 gives an overview of the models used in the next chapters.

4.3.2 Modeling and parameter estimation example: Windkessel model

Three types of computational models are commonly applied as low-dimensional physics-based models of arteries: the Windkessel model, one-dimensional (1-D) models, and tube-load models (Zhou et al., 2019). The Windkessel model presents the pulse as a function of

TABLE 4.1 Different models used in the next chapters.

Signal or variables of interest	Physiological/data-driven model	Device-subject interaction
Arterial pulse	The arterial pressure-volume curve for different arterial stiffness/blood pressure Windkessel model	Effect of the cuff on the arterial volume
Photoplethysmography	Not used in the book. A potential data-driven model can be found in Tang et al. (2020)	Light propagation through the tissue
ECG	A potential mathematical model based on differential equations can be found in Chapter 3 of Clifford et al. (2006)	Modeling tissue-electrode contact and the effect of sweat
Pulse transit time for continuous blood pressure monitoring	Moens—Korteweg equation	—
Breathing signal	Sine-wave signal Breathing pattern simulator	—

time only and is a function of equivalent inertance, compliance, and resistance. 1-D and tube-load models represent distributed properties of the arterial system. The 1-D model is based on the Navier–Stokes equation and is used to simulate pressure and flow at any position in the entire arterial tree. Tube-load models are transmission line models, which are made up of multiple parallel tubes with loads representing arteries, and as such, they can model wave propagation and reflection with only a few parameters (Zhou et al., 2019). Multiple models connected commonly in series can represent multiple arteries. In this book, we will limit the analysis to the simplest Windkessel model.

4.3.2.1 Example: Windkessel model

The Windkessel model is one of the most used models in studying a systemic arterial system. Windkessel effect explains the phenomenon of the conversion of the pulsatile blood flow into the continuous blood flow in blood vessels.

The systemic arterial system is represented using a three or four-element circuit. In the circuit, the pressure $p(t)$ plays the role of the voltage at different points and the blood flow $Q(t)$ of the current. Pressure is given in mmHg. Flow represents volumetric flux and is defined in units of volume per unit time or ml/s. The four-element circuit representing the Windkessel model is shown in Fig. 4.3. The circuit parameters are:

- Z_0 is the characteristic impedance of the artery [mmHg·s/ml]
- R is the peripheral resistance [mmHg·s/ml]
- C_a is the arterial compliance [ml/ mmHg] which is the change in arterial blood volume (ΔV) due to a given change in arterial blood pressure (Δp), that is, $C_a = \Delta V/\Delta p$ [Spencer63]. Compliance is equivalent to the capacitance in electrical circuits in the following way: Capacitance accumulates the charge and then discharges it to the same circuit. Similarly, compliance represents the ability of arteries to expand during the

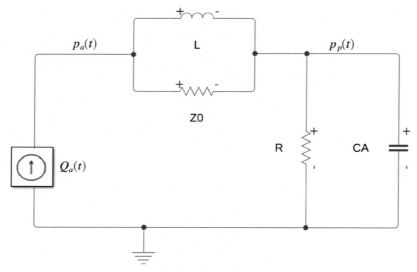

FIGURE 4.3   Four-element Windkessel model.

systole accumulating a volume of blood. During the diastole, this volume of blood is discharged along the arteries.

- L is the arterial inertance [mmHg·s^2/ml] and it is defined as the change in pressure in the arterial segment versus the acceleration of the blood (Spencer, 1987).

The controlled current source represents flow through the aorta $Q_a(t)$. Aortic pressure is $p_a(t)$ while $p_p(t)$ is the pressure over the peripheral resistance and arterial compliance. With the Windkessel model, we are interested in modeling $p_a(t)$.

The Windkessel model is an interpretative model that allows us to achieve the following:

1. Given the aortic flow and the parameters of the model, we can generate aortic and peripheral pressure signals. This corresponds to item W1 in the previous section and will be elaborated on in Section 4.3.2.2.
2. Given the aortic flow and the aortic pressure, we can estimate the values of parameters such as arterial compliance that have physiological meaning and can be interpreted. This corresponds to item W2 in the previous section and will be elaborated on in Section 4.3.2.2.

4.3.2.2 Windkessel model with parameters assigned based on previous knowledge

In this section, we will generate an aortic pressure signal based on the known parameters of the model shown in Table 4.2. As mentioned in Section 4.3.1, item W1, we assign typical parameters to an entire class of subjects with a given pathology. The parameters are obtained and modified from Kind et al. (2010). Pathologies considered are hypotension (low blood pressure) and hypertension (high blood pressure). Also, the parameters for a subject with normal blood pressure are included. The resistance of the aorta and the peripheral arteries is expected to increase with increasing blood pressure, and arterial compliance will decrease. The inertance also increases with the increase in blood pressure.

Fig. 4.4A represents the blood flow at the input of the model. Fig. 4.4B shows a pressure pulse generated from the Winkessel model based on parameters from Table 4.2 for hypotensive, normal, and hypertensive subjects. Normal diastolic blood pressure is 77 mmHg, and normal systolic blood pressure is 115 mmHg (minimum and maximum of the red pulse). Please note that the pulses are shown for the steady state. It takes about 5–10 s for the pulses to reach the desired amplitude (steady state) when simulating the circuit presented in Fig. 4.3.

TABLE 4.2 Parameters for the circuit elements of the Windkessel model shown in Fig 4.3 for hypotensive, normal and hypertensive subjects—modified from Kind et al. (2010).

Parameter	Z_0 [mmHg·s/ml]	R [mmHg·s/ml]	C_a [ml/mmHg]	L [mmHg·s^2/ml]
Hypotensive	0.039	0.71	1.95	0.011
Normal	0.043	0.95	1.5	0.015
Hypertensive	0.05	1.4	0.7	0.02

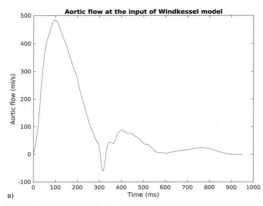
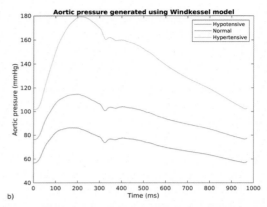

FIGURE 4.4 (A) Arterial flow $Q_a(t)$ measured from a healthy subject used as an input to the model. (B) Aortic pressure generated from the Windkessel model with the parameters for hypotensive, normal, and hypertensive subjects from Table 4.2. Waveforms for the flow and pressure are shown for the steady state.

Therefore, the pulses shown in Fig. 4.4B were extracted from the pressure signal at the time range between 15.9 and 16.9 s.

The arterial pulses can then be used as an input to the systems for oscillometric or continuous blood pressure measurements. We can evaluate how different physiological parameters affect blood pressure as well as the outcome of our device/system.

4.3.2.3 Windkessel model with sensitivity analysis and parameter estimation

In this section, we will assume that the values of the circuit parameters are unknown and need to be estimated. They will be estimated based on the known input, which is the flow $Q_a(t)$ given in Fig. 4.4A and the known pressure signal given in Fig. 4.4B, which is the blue signal for a hypotensive subject. The pressure signal corresponds to the hypotensive subject, and therefore, we should expect to obtain similar parameters as shown in the second row of Table 4.2. The following steps were followed:

1. Sensitivity analysis to determine the initial values of the parameters for the estimation
2. Selection of parameters that need to be optimized and their minimum and maximum values
3. Fitting the parameters (parameter estimation) of the model.

Parameter estimation through optimization could be performed without the sensitivity analysis. However, we expect that the optimization algorithm will perform better if good initial points are given and the range of the parameters over which the optimization is performed is reduced. Therefore, we will perform sensitivity analysis first to find the values of the parameters that will be used as initial values for the parameter estimation.

Sensitivity analysis was performed by generating 100 random values for each of the circuit parameters. The random numbers are generated from a uniform distribution. We will assume that we have a general idea about the range of parameters which is about 50% larger than the largest values of the parameters in Table 4.2 and 50% smaller than the smaller values of parameters in Table 4.2. We assume there is no correlation between parameters when generating random numbers, even though we know this is not the case. When determining what values to take out of 100 random values, we used mean square error (MSE) metrics between the samples of the measured pressure pulse and the generated pulse from the model during the steady-state (the segment of the pulse was taken between 15.9 and 16.9 s). MSE is computed as

$$\text{MSE} = \frac{1}{L} \sum_{k=1}^{L} \left[p_a(k) - \widehat{p}_a(k) \right]^2$$

where L is the number of samples in the segment, $p_a(k)$ is the true aortic pressure, and $\widehat{p}_a(k)$ is the estimated aortic pressure obtained at the output of the model. The MSE versus different parameter values is shown on a scatter plot in Fig. 4.5A. We can see in this figure that only values of the peripheral resistance R show correlation with the mean square error. Therefore, the peripheral resistance R is the most sensitive parameter in the model.

After performing sensitivity analysis, the *selection of parameters* for which the MSE was minimum was performed. These parameters were considered as initial values for the optimization step. The selected values for our implementation were : $Z_0 = 0.013$ mmHg·s/ml, $R = 0.73$ mmHg·s/ml, $C_A = 1.65$ ml/mmHg, $L = 0.0015$ mmHg·s²/ml. If we compare these values with the values of the parameters for the hypotensive subject, we will see that they are close but that there is still space for improvement. The minima and maxima of the optimization parameters are set to be 50% smaller and 50% larger than the parameters' values, respectively.

These parameters are used as input to the optimization algorithm to perform *parameter estimation*. The optimization was done using the Matlab function *fmincon*, which finds the minimum of a constrained nonlinear multivariable function. We used MSE as our cost function. The optimization reduced MSE from 2.35 mmHg² to 0.81 mmHg². The original pulse as well

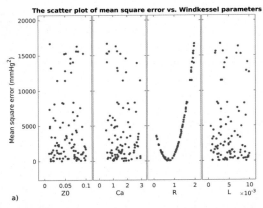
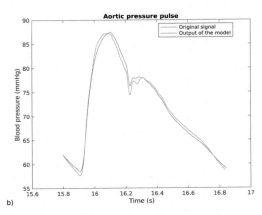

FIGURE 4.5 (A) Scatter plot of the 100 randomly generated values of the circuit parameters versus the mean square error (mmHg²) between the estimated and the measured pulse. (B) Reference or measured aortic pressure pulse in blue and the pulse obtained from the Windkessel model after parameter estimation in red.

as the pulse obtained from the model after optimization, is shown in Fig. 4.5B. The obtained parameters after optimization are $Z_0 = 0.039$ mmHg·s/ml, $R = 0.71$ mmHg·s/ml, $C_a = 1.95$ ml/mmHg, $L = 0.011$ mmHg·s^2/ml. We conclude that the optimization methods provided acceptable results with low MSE. For more advanced methods for estimating the parameters of the Windkessel model, please refer to Kind et al. (2010).

4.4 Modeling noise and assessing signal quality

In this section, we will describe the following topics:

1. Presenting uncertainties in the parameters of the electrical components
2. Modeling noise and motion artifacts
3. Quantifying the effects of noise and motion artifacts on signal quality.

Tolerance of the resistors and noise will be used later in the chapter. Therefore, in this section, we will cover the blocks of Fig. 4.1 that are not part of the processing pipeline: modeling noise and artifacts and estimating SQI.

4.4.1 Modeling uncertainties in parameters of electrical components

4.4.1.1 Sampling from distributions related to tolerances or errors of common components

Let us consider several electronic components. The parameters of resistors are their resistance and tolerance. Examples of tolerance values are 0.1%, 1%, 2%, and so on. It is commonly assumed that the resistor values can be modeled with a uniform distribution. If the tolerance is given as the variable *tol*, then the values of the resistor can be sampled from the uniform distribution in the range $[R \cdot (1 - tol), R \cdot (1 + tol)]$.

If the resistor values follow a normal distribution and tolerance is presented using a coverage factor k (for example, $k = 2$), then we would first obtain standard uncertainty $u = tol/k$ and then sample from the distribution $N(R, u^2)$.

Least significant bit (LSB) error of an A/D converter is commonly presented using a uniform distribution, and it takes values [0, LSB] in the case of A/D converters that perform truncation and [−LSB/2, LSB/2] for A/D converters that perform rounding. Both resistor tolerance and A/D converter resolution errors are Type B errors based on GUM (the Guide to the expression of uncertainty in measurement) (see Chapter 1).

4.4.1.2 Determining uncertainty of the components based on data from the datasheet

Let us consider a sensor whose data sheet includes several errors, such as linearity and hysteresis errors. The total uncertainty of the sensor is then computed as the combined uncertainty. Let us assume that the linearity and hysteresis errors of a fictitious force sensor are 0.2 N over the full-scale and 0.3 N over the range, respectively, and they are both given at 95% confidence. They are both type B measurements. In this case, combined uncertainty is computed as $u_c = \sqrt{(0.3\,\text{N})^2 + (0.2\,\text{N})^2} = 0.36\,\text{N}$ at 95% confidence (Figliola and Beasley, 2012).

4.4.2 Noise

4.4.2.1 Types of noise

Noise from sensors, electrodes, and amplifiers is often significant compared to the signal level because many bioelectric signals are in the μV range. In addition, interference from other devices and the environment, along with interference from other organs in the body, are also significant sources of disturbances.

Noise can be classified in different ways. Considering the portion of the overall signal bandwidth that the noise occupies, it can be classified as *narrowband* and *wideband/broadband*. *White noise* (such as thermal noise) occupies the whole bandwidth and is, therefore, broadband. Narrowband noise would be 60 Hz noise from the device itself.

Noise can also be classified as *low*, *middle*, and *high*-frequency noise. In biomedical instrumentation, low-frequency noise can cause baseline wander.

If the unwanted signal comes from external sources, it is called *interference*. For example, electrodes and electric cables can be capacitively coupled to a nearby electric field. When the impedance of the electrode and cable pairs differ, the capacitive coupling will result in a potential difference due to interference. For example, a 60 Hz signal picked up from the power lines is an interference signal.

We have already introduced and explained the quantization noise when discussing A/D converters. For more information about quantization noise, please refer to Chapter 3.

Noise can also be additive or multiplicative. *Additive noise* refers to the unwanted signal added to the signal of interest. *Multiplicative* noise modulates the amplitude of the signal. In biomedical instrumentation, we commonly consider additive noise.

We can model the noise by looking at its spectral content. This can be modeled by a single parameter representing the slope of a spectral density function $S(f)$ that decreases monotonically with frequency (Sameni et al., 2008):

$$S(f) \propto \frac{1}{f^{\beta}}$$

where β is *the noise color parameter*. Different types of noise can be modeled in this case, including:

- *White* noise ($\beta = 0$),
- *Pink* or *flicker* noise ($\beta = 1$),
- *Brown* noise or the *random walk process* ($\beta = 2$).

The simulation of pink or brown noise is normally done by generating white noise, transforming the noise signal into the frequency domain by applying Discrete Fourier Transform (DFT), altering frequency components of the DFT according to the defined slope of the spectrum and then performing the inverse DFT to transform the signal back into the time domain.

4.4.2.2 *Modeling noise added to a biomedical signal*

The signal with additive narrowband and white noises can be modeled as shown below:

$$y(t) = x(t) + B \sin(2\pi f t) + v(t),$$

where $y(t)$ is the noisy signal at the output, $x(t)$ is the clean signal of interest, B is the amplitude of the narrowband noise, and f is the interference or narrowband noise frequency. In addition, $v(t)$ is the Gaussian white noise that is usually zero mean and has a standard deviation σ that is either considered known or estimated from the data. Let us assume that the signal of interest is a cardiac signal with an amplitude of 1 V, simulated in Fig. 4.6 as a sine-wave signal of frequency 1 Hz sampled at 250 samples per second. An example of incorporating high-frequency noise is shown in Fig. 4.6A, where an interference sine-wave signal with an amplitude of 0.4 V and frequency of 60 Hz is added to the signal of interest. This is an example of a powerline interference. Another example of a narrowband signal that is added to the original sine-wave signal is shown in Fig. 4.6B. Here, the baseline interference is modeled as a sine-wave signal with an amplitude of 1.5 V, frequency of 0.1 Hz, and a mean of 0.5 V. This can be used to model the baseline drift of ECG and PPG signals.

4.4.2.3 *Modeling motion artifacts*

Motion artifacts are one of the most common disturbances for biomedical devices and they affect all the signals discussed in this book. The frequency range of motion artifacts overlaps with the frequency range of the biomedical signals, and therefore, it cannot be filtered out easily using traditional filtering methods. These artifacts can arise due to factors such as walking, the motion of the patient's arm on which the sensor is attached (such as moving

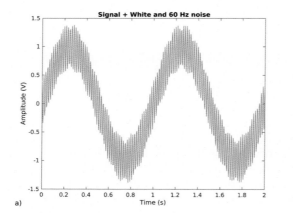
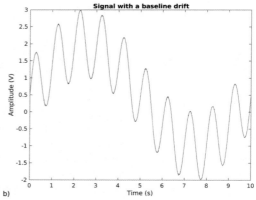

FIGURE 4.6   Input sine-wave signal with an amplitude of 1 V and a frequency of 1 Hz corrupted with Gaussian white noise with standard deviation 0.01 V and (A) 60 Hz noise corresponding to power line interference and (B) 0.1 Hz corresponding to baseline drift.

the arm up and down), and environmental conditions such as taking measurements in a moving vehicle. These artifacts tend to degrade the quality of the measured signal.

Motion artifacts resulting from patient and vehicle movements are inherently low-frequency signals. There are multiple ways to simulate continuous motion artifacts, such as ones due to walking. A simple way is to generate Gaussian white noise and then pass it through a lowpass filter, thus only keeping the low-frequency components of the signal, which resemble the interference introduced by motion artifacts (Thakkar, 2004). This approach is shown in Fig. 4.7, which shows the motion-corrupted signal in the time domain as well as the time-frequency domain representation of the noise. The signal of interest is the sine wave that simulates a cardiac signal with an amplitude of 1 V and frequency of 1 Hz sampled at 250 samples per second. The continuous motion artifact is modeled as a Gaussian noise that is zero mean and 0.4 V standard deviation and is filtered by a lowpass filter with a cutoff frequency of 10 Hz. We can see that this motion artifact is causing changes in the amplitude of the signal of interest. Time-frequency domain representation in Fig. 4.7B is obtained by applying the Short-time Fourier Transform (STFT) to the motion artifact signal with the following parameters: Discrete Fourier Transform (DFT) length of 8096 points windowed by a Hamming window of 1 s length with 50% overlap. STFT is a DFT computed over short and overlapping signal segments, and it provides information about both time and frequency for situations in which frequency components of the signal vary over time. On the other hand, the standard DFT provides the frequency information averaged over the entire time interval. The STFT spectrogram shows an intensity plot of the STFT magnitudes over time. As shown in Fig. 4.7B, the signal with motion artifacts occupies an almost fixed portion of the frequency spectrum over time. For more information about STFT, please refer to Mitra (2001).

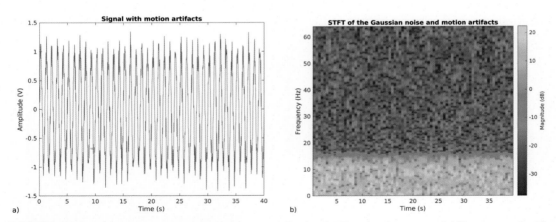

FIGURE 4.7 (A) Input sine-wave signal with an amplitude of 1 V and a frequency of 1 Hz corrupted with Gaussian white noise with standard deviation of 0.01 V and a lowpass filtered Gaussian noise with an amplitude of 0.4 V that simulates motion artifacts. (B) Time-frequency domain of the Gaussian white noise and the motion artifact signals, shown in the frequency range from 0 to 32 Hz.

Several more realistic motion artifacts models are shown in Farago and Chan (2021), including the hidden Markov model and recurrent neural networks. However, describing these models is outside of the scope of this book.

Very often, the motion artifacts are not continuous but occur over a short time intermittently. An example includes a quick movement of the arm while measuring blood pressure. That movement will result in a pulse on top of the signal of interest. Impulse artifact can be simulated by the *sinc* function (Li et al., 2009) in the discrete domain as $a_{imp}(n) = m \cdot \sin(\pi n T_s / \eta) / (\pi n T_s)$. The central lobe of the *sinc* function can be used as an impulse artifact and can be added to the signal of interest. The duration of the artifact is determined by the factor η. Fig. 4.8A shows the signal of interest (i.e., the same as the one described in Fig. 4.7) with the impulse motion artifact. The impulse is modeled to occur anywhere randomly. For the *sinc* function, we selected $\eta = 1.5$ resulting in about 5 s long movement artifacts. The amplitude of the movement artifact is $m = 1.2$ V. Fig. 4.8B shows the time-frequency domain representation of the motion artifact signal computed using the Short-time Fourier Transform with the same parameters as in Fig. 4.7. As seen from the figure, change in the frequency components over time occurs only for the duration of the motion artifact. A similar motion artifact model can be applied to biomedical signals such as the PPG with some modifications in amplitude and duration.

In Li et al. (2009), motion artifacts are modeled as Brown noise, filtered using a band-pass filter in the frequency range between 1.8 and 18 Hz. This corresponds to the motion artifacts obtained by, for example, dragging the clothes over the electrode or the transducer (Li et al., 2009). Fig. 4.9A shows the signal of interest that is the same as the one described in Fig. 4.7 corrupted with the described motion artifact in the interval between 8 and 17 s, while Fig. 4.9B shows the time-frequency representation of the motion artifact signal computed using the Short-time Fourier Transform with the same parameters as in Fig. 4.7. The time when the motion artifact occurs as well as the duration of the artifact, are modeled as uniform random variables. The amplitude of the motion artifact before being filtered was 5 V.

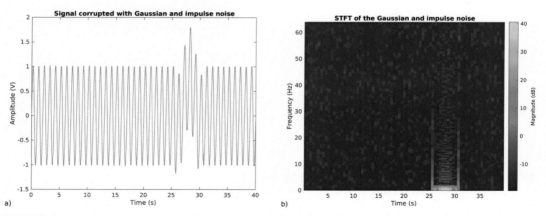

FIGURE 4.8 (A) Input sine-wave signal with an amplitude of 1 V and a frequency of 1 Hz corrupted with an impulse noise represented using the sinc function in the time domain. (B) Time-frequency domain of the Gaussian white noise and the motion artifact signals shown in the frequency range from 0 to 32 Hz.

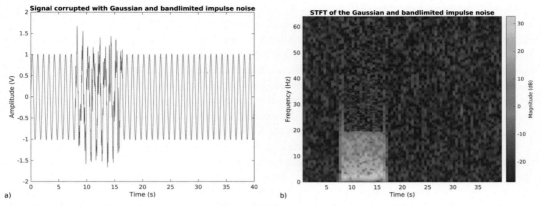

FIGURE 4.9 (A) Input sine-wave signal with an amplitude of 1 V and a frequency of 1 Hz corrupted with an impulse brown noise. (B) Time-frequency domain of the Gaussian noise and motion artifact signal signals shown in the frequency range from 0 to 32 Hz.

4.4.3 Signal quality

Signal quality is the degree to which a set of signal characteristics fulfill the requirements.

As presented in Orphanidou (2018), quality assessment aims to identify and quantify the instances of artifacts in a signal segment so that the information of interest extracted from that signal segment can be ignored or corrected. Acceptable signal quality depends on the application of interest. The output of the algorithm is a *Signal Quality Index* (SQI) that can be binary or have several degrees of acceptability.

The *quality factors* are factors that affect the quality of the data. They include quantization noise, noise introduced by filters and amplifiers, motion artifacts, environmental noises, and interference. The procedure for extracting the SQI normally involves several steps, including:

- Feature extraction, where these features are supposed to carry important information regarding signal quality. Some of the features include:
- Morphological features of the signals or pulses,
- Physiological feasibility,
- Time-domain and spectral features,
- Statistical features,
- Comparing outputs of multiple different algorithms that process the same data.
- Using features and decision methods based on empirical thresholds or machine learning to come up with *quality indices*.
- Combining the indices into a single SQI.

Algorithms for extracting some of the features are universal, while the thresholds or methods for comparison are different among biomedical signals. For example, to determine the signal-to-noise ratio (SNR) of any signal, the ratio between in-band spectral power and out-of-band spectral power is computed. In-band frequencies are different for different physiological signals. For example, the range between 5 Hz and 14 Hz could be used as an in-band frequency for the ECG signal.

Morphological features are different for every physiological signal. For example, for an ECG signal, the time and amplitudes of waves of interest, including P wave, T wave, and QRS complex, are the features. *Template matching* could be used to compare different pulses against a reference pulse. It is important to check if the extracted features are physiologically feasible. Features such as heart rate, the time difference between different points on the pulse, extracted amplitudes, and their ratios can all be checked for feasibility.

Time-domain features include, for example, the rise and fall time of the pulse. *Spectral features* include ratios of spectral power among different frequency regions of interest.

Statistics features include the mean, variance, skewness, kurtosis, and others.

Outputs of different algorithms processing the same signal segment should be the same when the signal is of good quality. However, if the signal is not of good quality, then different algorithms will be affected differently by noise or interference. Therefore, we should expect that the results of the algorithms would be significantly different. An example of this method is using two different algorithms for detecting the QRS complex in the ECG signal. The SQI is defined as the total number of matched QRS complex beats between the two algorithms divided by the total number of beats.

SQI can be computed for a *segment of interest,* such as a 10 s interval or *each detected pulse.* Below, we will show examples of SQIs computed based on statistical and spectral features where SQI is computed for the overall segment of interest. SQI based on template matching example is shown as well—it allows for evaluating SQI of every pulse.

4.4.3.1 Statistical feature: kurtosis

The reason behind using kurtosis of the biomedical signal is to distinguish between noise and the signal. The random noise samples often follow Gaussian distribution, whereas biomedical signals do not follow it. The kurtosis of any univariate Gaussian distribution is 3 and therefore the kurtosis of the empirical distribution obtained from the signal samples is compared to this value. The estimate of kurtosis of a signal can be represented as

$$Kurt(x) = E\{x - \mu\}^4 / \sigma^4$$

where $x(n)$ is the signal segment of L samples, μ and σ are the mean and the standard deviation of the signal $x(n)$, respectively, and $E\{x - \mu\}$ is the expected value for $x - \mu$. To compute the SQI for the ECG signal, the authors in Li et al. (2008) selected a threshold of 5 for kurtosis. Therefore, if the kurtosis of the distribution of a signal in a given sequence is greater or equal 5, the kurtosis-based SQI (kSQI) is $kSQI = 1$, otherwise $kSQI = 0$.

4.4.3.2 Signal-to-noise ratio based on the spectral power ratio

We have already mentioned that an estimate of the SNR can be computed as the ratio between in-band spectral power and out-of-band spectral power. In the case of an ECG signal, with a typical QRS, all of the power is concentrated below 30 Hz, and the peak for the power occurs in the range of 5–14 Hz (Li et al., 2008). The spectral distribution ratio is calculated as the sum of the power of the ECG in the range of 5–14 Hz to the power of the same signal in the range of 5–60 Hz. The spectral distribution ratio (SDR) can be represented as

$$SDR = \int\limits_{f=5}^{f=14} P(f)df \Big/ \int\limits_{f=5}^{f=60} P(f)df$$

The signal quality index (pSQI) is then computed by comparing SDR with a predefined threshold. SDR has a value between 0 and 1. If all the energy of the ECG signal is contained in the range between 5 Hz and 14 Hz, SDR would be 1, but that would not correspond to the spectrum of a real ECG signal that also has frequency components outside of this range. Therefore, it was empirically determined that signals with SDR above 0.8 and below 0.5 are unacceptable (Li et al., 2008). As we will show in Section 4.4.3.1, we selected 0.5 and 0.9 empirically as our thresholds.

We showed an example of computing pSQI for an ECG signal, but a similar analysis can be made for almost any biomedical signal. For other physiological signals, different frequency ranges and thresholds should be selected.

4.4.3.3 Template matching using average correlation coefficient

Template matching requires that all the pulses are first detected. Then, the Pearson correlation coefficient is computed between each pulse and the *waveform template*. We expect the correlation between the pulses and the waveform template would be high if there is no noise or distortions in the signal. The waveform template is normally unavailable but is computed as the average pulse waveform in a segment of the biomedical signal of interest by performing beat/pulse detection and then ensemble averaging over all pulse waveforms. Ensemble averaging all pulse waveforms in a given segment provides the pulse waveform template.

In Orphanidou (2018), the template waveform of PPG signals was obtained by an ensemble averaging 10 pulses. The correlation coefficient was computed between each of the 10 pulses and the template waveform and averaged. An average correlation coefficient greater or equal to 0.86 was considered acceptable, resulting in the template matching SQI or 1 (tSQI = 1). The average correlation threshold was determined empirically.

To summarize, the following steps are needed:

1. Some applications require that the signal is first filtered using a low-pass filter
2. Finding time instances and amplitude of the fiducial points of interest of each pulse (e.g., time instances and amplitude of R peaks of the ECG signal)
3. Extracting pulses and performing ensemble averaging to find the waveform template
4. Calculating the correlation coefficient between each pulse and the template
5. Determining tSQI for each pulse by comparing the correlation coefficient of each pulse with the threshold
6. Determining tSQI of the overall segment by averaging the correlation coefficients for the segment.

4.4.3.4 Example of signal quality assessment: ECG signal

In this section, we will compute previously described SQIs for an example ECG signal. A high-quality ECG signal was collected for about 27 s and sampled at 250 samples per second.

We show the SQIs of the ECG signal in three cases:

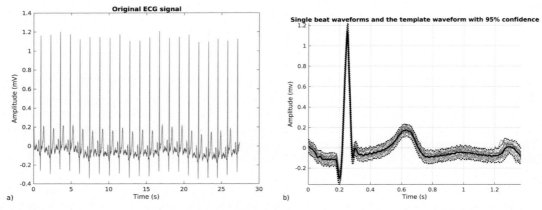

FIGURE 4.10   Case 1: Original ECG signal without added noise. (A) The waveform of the ECG signal and (B) overlapped waveforms for each ECG beat (different color lines) and their ensemble-averaged template waveform (black line) together with 95% confidence intervals (black dotted lines).

1. No noise is added to the original ECG signal
2. 60 Hz noise with an amplitude of 0.1 mV and random Gaussian noise with a mean of 0 mV and standard deviation of 0.1 mV is added to the original ECG signal resulting in a low SNR of 3.2 dB.
3. Random Gaussian noise with a mean of 0 mV and standard deviation of 0.01 mV was added together with the impulse noise represented using the brown noise described in Section 4.4.2.1 with the amplitude of 2 V followed by a bandpass filter with the frequency range of [1 Hz, 10 Hz] resulting in an SNR of 4.2 dB.

In Case 1, the original signal is considered without additional artificial interference and noise, as shown in Fig. 4.10A. Fig. 4.10B shows the template waveform obtained by performing an ensemble averaging of 19 beats. About 1.05 of the total averaged pulse duration is taken when performing averaging. The bold black line in Fig. 4.10B represents the mean of the averaged pulses, which is actually the template waveform. Dashed lines represent the averaged value of the signal ± two standard deviations. All other lines represent 19 beats overlapped one on top of the other. The same representation of the averaged pulses is used in Fig. 4.11 and Fig. 4.12. Segmentation of the ECG signal to find the R peaks and detect every beat is done based on the work (Sedghamiz and Santonocito, 2015). Ensemble averaging is performed based on the work presented in Parker, 2013.

In addition, even though R peaks of the ECG signal are detected, we start averaging 0.25 s before the R peak. Therefore, the R peaks of all beats are positioned exactly at 0.25 s. Table 4.3 shows that all SQIs are one. The correlation coefficient between the template waveform and every single beat is high, which means that the tSQI for every beat is 1.

In Case 2, the signal is noisy, as shown in Fig. 4.11A. Fig. 4.11B shows the template waveform obtained by averaging the 19 beats of the signal shown in Fig. 4.11A. It is quite clear that the noise is high; therefore, the correlation coefficient between each pulse and averaged template waveform is low. Therefore, tSQI for each pulse is tSQI = 0, as shown in Fig. 4.11A. The average correlation coefficient shown in Table 4.3 is $r = 0.82$.

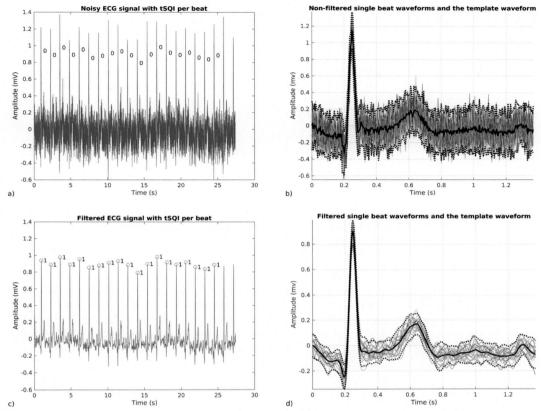

FIGURE 4.11   Case 2: ECG signal with Gaussian white noise and 60 Hz interference. (A) The waveform of non-filtered ECG signal with tSQIs per beat, and (B) overlapped waveforms of each beat of the ECG signal from (A) and the ensemble-averaged template waveform. (C) The waveform of the low-pass filtered signal with identified R peaks and tSQIs per beat, and (D) overlapped waveforms of each beat of the ECG signal from (C) and the ensemble-averaged template waveform.

A lowpass filter with a cutoff frequency of 15 Hz was applied to the signal. The filtered ECG waveform, as well as its template waveform, are shown in Fig. 4.11C and Fig. 4.11D, respectively. The average correlation coefficient, in this case, is increased to 0.97, and all tSQIs per pulse are 1, as shown in Fig. 4.11C. Kurtosis and power ratio are shown for the non-filtered signal in Table 4.3. Kurtosis dropped in comparison to Case 1, but it is still much higher than 5, which means that the kSQI = 1. The power ratio dropped significantly compared to the power ratio of Case 1, but it is still above 0.5 and, therefore, pSQI = 1. This power ratio drop is expected because of the 60 Hz interference since much of the noise spectral components are outside the 5–14 Hz range. Therefore, it is clear that multiple SQIs need to be considered when analyzing the signals.

In Case 3, the signal is noisy only in the range between 18 and 25 s, as shown in Fig. 4.12A. This situation corresponds to a short-term motion artifact, and it was simulated by adding

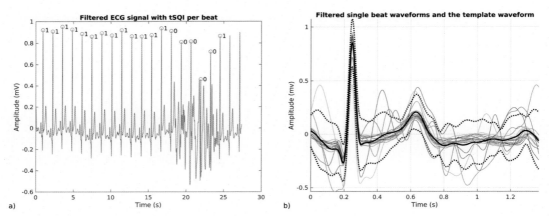

FIGURE 4.12 Case 3: ECG signal with brown noise. (A) The signal waveform with identified R peaks and tSQIs per beat—the signal is filtered with a low-pass filter with 15 Hz cuff-off frequency and (B) overlapped waveforms for each ECG beat (different color lines) and their ensemble-averaged template waveform (black line) together with 95% confidence intervals (black dotted lines).

bandlimited brown noise in the frequency range of 1–10 Hz, which overlaps with the frequency range of the ECG signal. Fig. 4.12A shows the signal after low-pass filtering with a 15 Hz cutoff frequency. Fig. 4.12B shows the template waveform obtained by averaging the 19 beats of the signal. It is clear that the pulses corrupted by noise have a different shape than the original ECG pulses. Fig. 4.12A also shows the tSQI computed for every pulse. The correlation coefficient of corrupted pulses is significantly below the threshold; therefore, the tSQI for these pulses is zero. The average correlation coefficient is around 0.87 for both filtered and unfiltered ECG signals, as shown in Table 4.3. As expected, filtering did not help much in this case to improve signal quality, and this type of noise will require a different kind of filter which is outside of the scope of this book. As expected, the power ratio did not change much compared with Case 1—it is very high. Kurtosis is reduced in comparison to Case 1 because of the added noise.

TABLE 4.3 Features and SQIs for the ECG signal with different types of noise.

	Kurtosis, kSQI	Power ratio, pSQI	Average r — no filtering, tSQI	Average r with filtering, tSQI
Case 1	Kurt = 22.2, kSQI = 1	ratio = 0.88, pSQI = 1	$r = 0.99$, tSQI = 1	$r = 0.99$, tSQI = 1
Case 2	Kurt = 11.8, kSQI = 1	Ratio = 0.63, pSQI = 1	$r = 0.82$, tSQI = 0	$r = 0.97$, tSQI = 1
Case 3	Kurt = 11.9, kSQI = 1	Ratio = 0.89, pSQI = 1	$r = 0.88$, tSQI = 0	$r = 0.87$, tSQI = 0

4.5 Uncertainty propagation in systems

The goal of uncertainty propagation is to show how uncertainties in the parameters of the model or the circuit affect the confidence intervals of the estimate at the output. Uncertainty propagation will be done here using Monte Carlo methods. We will show the propagation of uncertainties through a charge amplifier circuit described in Chapter 3.

4.5.1 Example: uncertainty propagation through a charge amplifier

In this example, we will show uncertainty propagation through the charge amplifier presented in Fig. 3.12 and Example 3.2. To remind the reader, this circuit uses the piezoelectric transducer as an accelerometer. The objective is to amplify the signal obtained from the piezoelectric sensor and estimate the uncertainty at the output if we assume that resistors and capacitors have a 1% tolerance.

The input to the circuit represents a sine wave with 20 Hz frequency and an amplitude of 5 pC (peak to peak is 10 pC). This signal is differentiated to obtain the current and then used as an input to the controlled current source $i_{IN} = dq/dt$. The only difference in comparison with Fig. 4.12 is that the resistor $R7$ has a value of 130 kΩ to reduce the likelihood of saturating the amplifier.

Uncertainty propagation is performed in the following way.

- $M = 200$ values of the resistors R_f, R_i, R_5, R_7 and R_8 and the capacitor C_f were sampled from a normal distribution with the mean value equal to their nominal value and the standard deviation equal to 0.5% of their nominal value. We assume that 1% tolerance of the component means that 95% values of that component will be in the $\pm 1\%$ of the nominal value (2σ).
- We computed the peak-to-peak amplitude of the signal at the output by subtracting the maximum and minimum values of the signal.
- The simulation was run M times, and M peak-to-peak amplitude values were obtained.
- From these M peak-to-peak amplitude values, we computed the standard error in percentage and the 95% confidence intervals. The histogram of M peak-to-peak amplitude values is also presented. Finally, 95% confidence is computed by sorting the output results and then taking the value for the low confidence interval, which corresponds to the 2.5% of the peak-to-peak amplitudes, and for the high confidence interval, which corresponds to the 97.5% of the peak-to-peak amplitudes. More accurate results would be obtained if we selected a larger value for M.

Derivation of the sensitivity analysis is given in (12-Bit 1 MSPS, 2017), and it is performed using a derivative approach based on the derived expression of the gain: $\left(1 + \frac{R_7}{R_8}\right)\frac{1}{C_f}$. Please note that in the calculations in (12-Bit 1 MSPS, 2017), it is assumed that the standard error is 1%.

We evaluated the uncertainty at the output using the Monte Carlo method presented above. The uncertainty mainly depends on the variations of the components R_7, R_8 and

C_f, as expected. The histogram of the magnitudes is shown in Fig. 4.13. The standard error at the output is estimated to be 0.8%, while the 95% confidence intervals are [2.73 V, 2.81 V]. The mean value is 2.76 V.

Confidence intervals computed in this way can be used in more complex circuits in which we can assume that the charge amplifier presented here is a black box with the uncertainty quantified using the confidence intervals.

4.6 Modeling software

4.6.1 Real-time signal processing

Real-time applications are unique in regard to how data is received and processed, including:

- The data comes as a continuous stream
- The data is processed continuously
- Processing time is important since the *latency* (the input-to-output delay of the system) introduced by the processing algorithm needs to be small.

Real-time processing can be classified based on how frequently the output is obtained (Ackenhusen, 1999) as:

- *Stream processing* where all computations related to one input sample are completed before the next input sample arrives
- *Block processing* where each input sample $x(n)$ is stored in a memory or a buffer before any processing is performed on it. After L input samples have arrived, the entire block of samples is processed at once.

Stream processing includes processing the last L samples, where $L \geq 1$, and producing the output signal at each sampling period T_s. If $L = 1$, then the output is produced for each input without taking into account any historical data. If $L > 1$, then processing is based on a sliding window that requires storing the last L input samples for processing. If the processing is completed before the next sample arrives, then the system is performing real-time processing. Therefore, if the time required for processing L samples is T_{proc}, the condition for real-time processing is that $T_{proc} \leq T_s$.

Block processing includes processing the last L samples, where $L \geq 1$, and producing the output signal after L samples, which means that the data at the output appears at a different rate from the data at the input. The output sampling period is $T_{sy} = (L - L_{over})T_s$. Blocks can be overlapped by the number of samples L_{over}, where $0 \leq L_{over} < L$ samples. $L_{over} = 0$ means that the blocks are not overlapped. If $L_{over} = L - 1$, then this is stream processing and $T_{sy} = T_s$. Therefore, we can see that stream processing can be presented as a special case of block processing. To perform real-time processing, the following condition needs to be met $T_{proc} \leq (L - L_{over})T_s$. Block processing is useful for implementing algorithms, such as discrete Fourier transform, that require processing vectors. The disadvantages of block processing are additional memory requirements and latency. Block processing with

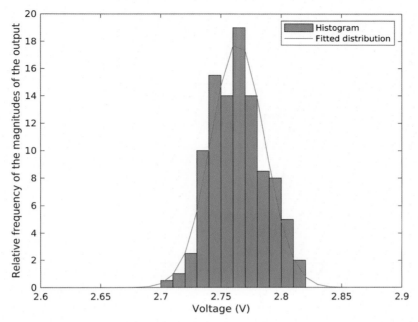

FIGURE 4.13 Histogram of the peak-to-peak amplitudes of the output voltage of the charge amplifier in case it is assumed that the tolerance of each component is 1%.

non-overlapping blocks is shown in Fig. 4.14, where $L = 4$ input samples are used to produce one output sample and $L_{over} = 0$. An output sample appears after $L \cdot T_s + T_{proc}$.

4.6.2 Real-time processing: hardware versus software

Signal processing can be done in hardware using *analog or digital electronics* or *in software*. Both approaches support stream and block processing; however analog processing is better suited for stream processing. We will show in the next section both types of processing on an example of a circuit for generating an alarm if breathing cessation takes longer than expected. Analog implementation in Section 4.6.3.2 is done using an integrator followed by a comparator. In Section 4.6.3.3, the analog signal is converted into the digital signal, which is then processed in software.

4.6.2.1 *Processing using analog hardware*

In Chapters 2 and 3, we covered several analog components that are commonly used in biomedical devices. We also described components used for conversion between the analog and digital domains: A/D and D/A converters. However, we did not present digital electronics. In this book, we will not use digital electronics such as flip-flops, registers, counters, accumulators, and so on—processing will be done either using analog electronics or using the software. Analog electronics allows us to perform basing arithmetic operations, filtering,

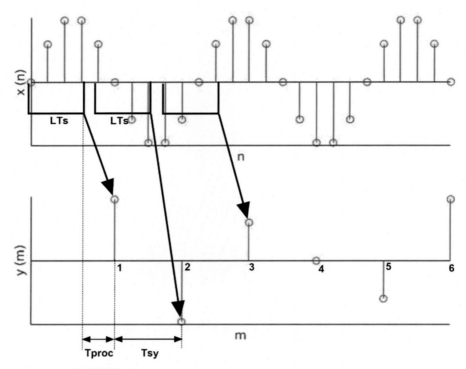

FIGURE 4.14   Block processing where blocks are not overlapped.

integration, differentiation, and similar. It provides very limited opportunities for processing in comparison to the processing opportunities in software.

4.6.2.2 *Processing in software*

Almost all algorithms nowadays are executed in software. Processing in software is done after the signal is first converted into a digital representation using A/D conversion. Then, the processing is done on a processor which is a part of a device or on a computer. The output of the A/D converter is normally an integer that is in the range defined by the resolution N of the A/D converter $[0, 2^N - 1]$. This integer is normally first scaled back into the range that depends on the application. For example, if the A/D converter is used for blood pressure measurement and the range of blood pressure that is measured is between 0 mmHg and 300 mmHg, then $2^N - 1$ is scaled to correspond to 300 mmHg. After that, the scaled signal is either immediately processed or buffered for block processing.

Please note that processing in software can be done in real-time as well as offline. *Offline* or *batch* processing requires that the data is collected and available after the experiment or the simulation. An example of a device that performs offline processing is an oscillometric blood pressure device. There, all the signals of interest are collected and then processed. There is no real deadline for processing except that the results should be presented to the user as soon as possible to improve the user experience and reduce the wait for the results. However, even if there is a delay in processing data, there will be no consequences like in real-time processing, where delays in processing might result in missing data at the output. As some medical devices rely on online real-time processing and some on offline processing, we will use both concepts in the book.

4.6.3 Example: stop-breathing detection

A breathing belt is a device a subject wears around her or his chest that can be used to extract the respiration signal based on the displacement of the chest during inspiration and expiration. The breathing belt can be used to estimate breathing rates over time, analyze breathing patterns and indicate that the breathing rate is too high or too low. Here, we will show a simplified system that can extract the breathing signal and detect that the subject is not breathing for some time that is longer than a predefined threshold.

4.6.3.1 *Description of the model*

In this section, we show an example of a circuit that collects the breathing signal and generates an alarm if the amplitude of the breathing signal is zero for some time. This system can be used as a central sleep apnea detector to detect the number of times the patient had apnea episodes or as a baby monitor system. The system is shown in Fig. 4.16. The model has six sub-models, including the breathing signal and noise generation model not shown in the figure, strain gauge with Wheatstone bridge, differential amplifier, low-pass filter, pulse generation, and alarm generation. An alarm is generated if the breathing ceases for about 10 s. Please note that $V^+ = 5$ V and $V^- = -5$ V in this circuit.

4.6.3.1.1 Signal and noise generation

The first block, called Input, represents the breathing simulator. We present breathing as a sinusoidal signal with the frequency f_B. Breathing cessation is modeled as a DC signal with an amplitude of zero. Let us assume that one breath involves inhaling air to full lung capacity, where full lung capacity corresponds to the maximum strain of the strain gauge, and then exhaling all air out of the lungs ending up with zero strain. Assuming a maximum strain of 0.02 and a strain change of 0.01, we represent breathing as signal $-0.01 \cos(2\pi f_B t) + 0.01$. In our example, the breathing frequency is 15 breaths per minute which means that the breathing frequency is $f_B = 0.25$ Hz. Gaussian white noise is also added. The signal together with the activity periods (high level means breathing and 0 level means no breathing) are presented in Fig. 4.15.

4.6.3.1.2 Transducer and amplifiers

For the strain gauge, the Gauge factor G is set to 2, and the nominal resistance of the strain gauge is 100 Ω. The change in resistance ΔR for the maximum strain of 0.02 is given as $\Delta R = GR\varepsilon = 4\ \Omega$. In this circuit, $R = R_1 = R_2 = R_3 = 100\ \Omega$. Please note that the Wheatstone bridge is balanced to $V^+/2$ and not to 0 V.

The voltage V_1 in Fig. 4.16 is the voltage at the output of the bridge, which is also the input to the differential amplifier. The voltage V_1 for a strain of 0.02 is

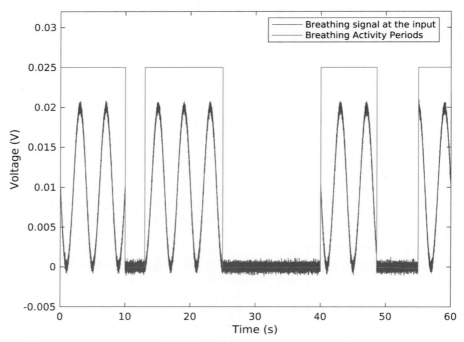

FIGURE 4.15 Waveforms at the input including blue: noisy breathing waveform, and red: a waveform that shows intervals of active breathing and breathing cessation.

$$V_1 = V^+ \left(\frac{R_2}{R + \Delta R + R_2} - \frac{R_3}{R_1 + R_3} \right) = V^+ \frac{\Delta R}{2(2R + \Delta R)} = 49 \, mV$$

A differential amplifier is used to amplify the signal V_1. We chose resistors so that the amplification is 100 ($R_5 = R_7 = 10 \, k\Omega$ and $R_4 = R_6 = 100 \, \Omega$). Therefore, the maximum voltage at the output of the differential amplifier is about 4.9 V.

4.6.3.2 Analog circuit for stop breathing detection

Detection of stop breathing is based on an integrator circuit. When the amplitude of the breathing signal is small, the capacitor C_1 is being charged. During charging, the amplitude of the output of the integrator increases. If the amplitude is greater than some predefined threshold, this means that the alarm needs to be generated. If the amplitude of the breathing signal is large, the integrator will be reset to zero and will not be charged. This is the basic idea behind the analog processing performed to detect stop breathing that takes more than 10 s. Please note that this is a very simplified implementation and is susceptible to motion artifacts. Implementation details are presented next.

The lowpass filter is used to filter out random noise. Components $C_f = 500 \, nF$ and $R_8 = 100 \, k\Omega$ are selected to achieve $f_c = 1/(2\pi R_8 C_f) = 3.2 \, Hz$, which is satisfactory for the breathing signal. The pulse detector circuit converts amplified and filtered breathing

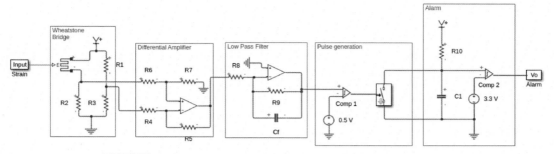

FIGURE 4.16 Stop-breathing detection circuit with analog processing.

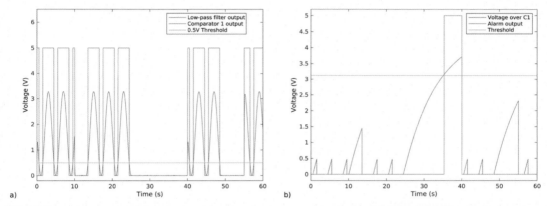

FIGURE 4.17 Waveforms for the circuit presented in Fig. 4.16 for the input signal presented in Fig. 4.15 (A) Blue: output of the low-pass filter, red: output of comparator 1, yellow: the threshold at 0.5 V. (B) Blue: the voltage over the capacitor C_1, red: output alarm signal, yellow: the threshold at 3.15 V.

signals into pulses shown in Fig. 4.17A. There, the signal at the output of the low-pass filter presented using a blue line in Fig. 4.17A is compared to a threshold that is set to 0.5 V using the comparator Comp 1. When the output of the comparator is one, the switch is closed, and the capacitor C_1 is connected to the ground through this voltage controlled switch and discharged. The capacitor is charged during the periods of logic zero at the output of the comparator Comp 1, which correspond to the low breathing amplitudes or no breathing signal. The time constant is selected to be 10 s ($C_1 = 1\ \mu F$ and $R_{10} = 10\ M\Omega$).

The output of the RC integrator in the block Alarm is compared against the fixed threshold at 0.63 * 5 V = 3.15 V using the comparator Comp 2. The alarm is generated when the signal is above the threshold. This threshold is selected like that because the voltage at the output of the RC circuit will reach 0.63 of its maximum at the time that corresponds to one time constant (see Section 4.4.1). In Fig. 4.17B, the blue line shows the voltage level at the capacitor C_1. When the signal at the output of the low-pass filter is smaller than 0.5 V, the capacitor C_1 will start charging. However, it will not reach the threshold at 3.15 V shown as a yellow line in Fig. 4.17B unless the signal amplitudes are lower than 0.5 V for more than 10 s. This occurs

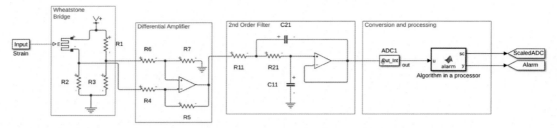

FIGURE 4.18   Stop breathing detection circuit where the algorithm for detecting stop breathing is implemented in software.

only about 35 s from the beginning. At that time, the output is one, and it remains one for the rest of the breathing cessation. This is presented as a red signal in Fig. 4.17B.

4.6.3.3 Software-based stop breathing detection circuit

The design is shown in Fig. 4.18. The major difference is that analog electronics for detecting stop breathing are replaced with an antialiasing filter, A/D converter, and processor. The processor is modeled in Matlab code. It processes data in real-time and outputs 1 V if the breathing signal is below a certain threshold for more than 10 s. The A/D converter is selected to have 12 bits and a sampling rate of 80 Hz.

The antialiasing filter is designed to perform low-pass filtering of the signal at the output of the differential amplifier. The components were selected as $C_{11} = C_{21} = 800\,nF$ and $R_{11} = R_{21} = 10\,k\Omega$ resulting in a cutoff frequency of 19.9 Hz.

The algorithm performs the following steps.

- Stores the samples that arrive from the A/D converter sequentially in a memory (modeled as an array)
- Scales data to the range from 0 V to 5 V because the data from the A/D converter is in the range between 0 and $2^{12} - 1$
- Compares the last 10 s of data against a fixed threshold of 0.5 V and, if the output is smaller than the threshold, generates the alarm.

Timing diagrams are shown in Fig. 4.19. The blue line represents the scaled output of the A/D converter. The red line represents the alarm generated 10 s after the signal was below the threshold of 0.5 V.

4.7 Modeling power consumption

Power is defined as $P = V \cdot I$ where V is the voltage, and I is the current through the component. The energy that the component or a device requires to operate is

$$E = \int_{t=0}^{T} Pdt$$

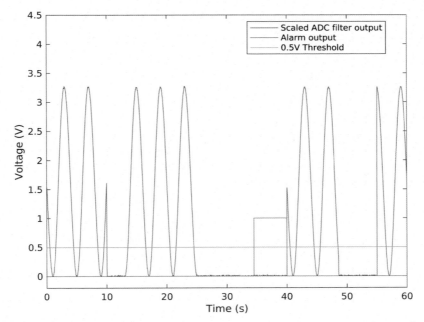

FIGURE 4.19 Timing diagram of a stop-breathing detection circuit shown in Fig. 4.18. The signals include: blue: scaled output of the A/D converter, red: output alarm signal, yellow: the threshold at 0.5 V.

The power in a digital circuit is defined as a sum of dynamic and static power. Static power is the power that a device consumes independent of any activity or task the core is running because even in an inactive state, there is a low "leakage" current path and some static current. Static current is the DC bias current that some circuits need for their correct work. Dynamic power is the power due to switching transistors. It is the power of the active digital components. It is defined as

$$P = A \cdot C \cdot V_{dd}^2 \cdot f_{clk}$$

where A is the switching factor which is the probability of switching from 0 to 1 or from 1 to 0 in a given clock cycle, C is the load, f_{clk} is the clock frequency and V_{dd} is the supply voltage. Therefore, to reduce the dynamic power, one should reduce the supply voltage and the clock frequency. Normally, the clock frequency and the supply voltage can be reduced together with significant power savings. Power savings are normally made by using devices with multiple power modes in which there is a tradeoff between the power consumption and the operating speed or accuracy. The modes range from normal operating modes, where devices operate at full speed, to different power down, standby and sleep modes, depending on the device. The power is computed as the average energy during active and standby modes divided by the total time: $P = \left(P_{active}T_{active} + P_{standby}T_{standby} \right)/T_{total}$. The initial estimate of the power in different modes can be obtained from a datasheet of the component of interest.

4.7.1 An example of estimating the power consumption of an A/D converter

In Chapter 3, we introduced several types of A/D converters. Sigma delta A/D converters perform internally a large number of operations which might be a reason for higher power consumption. Successive approximation A/D converters are typically low power. Low-power A/D converters power down automatically at the end of each conversion phase. In this way, the power consumption is proportional to the sampling rate, as we will show in the following example.

Example 1. Let us consider the power consumption of a 16-bit, 100,000 samples per second (sps) successive approximation A/D converter AD7684 from Analog Devices. The parameters of the converter are given in Table 4.4. Compute power consumption as the sampling frequency increases from 100 sps to 100 ksps.

Solution:. As the A/D converter switches between the operating and standby modes, the energy E_{adc} required to perform one conversion during one sampling period T_s is

$$E_{adc} = V_{dd} \left(I_o \cdot T_c + I_s \cdot (T_s - T_c) \right)$$

The power is then $P_{adc} = E_{adc}/T_s$. The plot of power over the range of sampling frequencies is shown in Fig. 4.20.

4.8 Summary

In this chapter, we introduced several important concepts related to modeling and simulation, including how to:

1. Collect, obtain, or generate data
2. Fit the data to the model
3. Model the noise
4. Estimate the quality of the signal
5. Propagate the uncertainties through a circuit
6. Analyze streaming data by performing block or stream processing
7. Modeling power estimation.

The following applications were simulated in detail while describing the concepts, including:

TABLE 4.4 Operating parameters of A/D converter AD7684 from Analog Devices.

Conversion time, Tc	Supply voltage, Vdd	Operating current at 100 ksps, Io	Standby current, Is
10 µs	2.7 V	0.56 mA	1 nA
	5 V	0.8 mA	1 nA

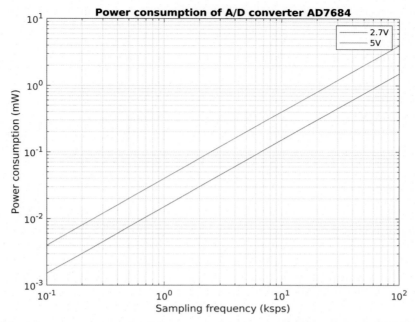

FIGURE 4.20 Power consumption for different sampling frequencies and supply voltages of a successive approximation A/D converter.

1. Windkessel model of the aortic pulse
2. Signal quality analysis of ECG data
3. Uncertainty is propagated through the circuit that converts the signal from the piezo-electric transducer into the voltage to measure acceleration
4. Real-time implementation of the stop-breathing alarm system
5. Power consumption of an A/D converter.

The reader is encouraged to go through the simulations and modify the parameters of the programs to better grasp the concepts. After completing this chapter, the reader should be able to model a biomedical system and its components, including the noise, calculate the signal quality, and perform uncertainty propagation.

4.9 Problems

4.9.1 Simple questions

4.1. *Discuss when white, pink, and brown noises appear in biomedical instrumentation.
4.2. *Give examples of low, middle, and high-frequency noises in biomedical instrumentation.
4.3. Define the differences between noise, interference, and motion artifacts.

4.4. Define concepts related to signal quality analysis, including factors, features, indices, and SQI.

4.5. List advantages of implementation where processing is implemented in software versus implementation in which processing is implemented in analog electronics.

4.9.2 Problems

4.6. Perform uncertainty analysis in Section 4.5.1 using the derivative approach.

4.7. Explain the time and frequency representations of different motion artifacts shown in Fig. 4.7, Fig. 4.8, and Fig. 4.9.

4.9.3 Simulations

4.8. This question is related to the Windkessel model shown in Section 4.3.2.1.

   (a) Perform sensitivity analysis of the circuit parameters of the Windkessel model shown in Fig. 4.3 by modifying the code provided on the book webpage to include a positive correlation between the peripheral and aortic resistance when generating random numbers and a negative correlation between the resistance and compliance. Discuss the difference in the results—if any.

   (b) Modify the cost function in the Windkessel model to be the mean absolute error. Comment on the results.

4.9. This question is related to the quality of the ECG signal described in Section 4.4.3.1.

   (a) Add baseline shift to the ECG signal. Analyze the effect the baseline shift has on the three SQIs. Then filter out the baseline shift and observe the three SQIs again.

   (b) Change SNR for Case 2 of the signal quality example shown in Section 4.4.3.1 and show the effect of different SNRs on all three SQIs.

4.10. Section 4.5.1 shows uncertainty propagation through a charge amplifier. Perform sensitivity analysis and determine what components affect the charge amplifier the most. Select the tolerance of these components to be 0.1% and then perform uncertainty propagation and determine the confidence intervals at the output.

4.11. This question relates to the software-based stop-breathing detection circuit described in Section 4.6.3.3.

   (a) To simulate the effect of aliasing, add 60 Hz interference to the input signal shown in Fig. 4.18. Observe the effects of aliasing at the output. How much is the 60 Hz signal attenuated at the output of the antialiasing filter?

   (b) Run the algorithm that averages the data during the 10 s interval instead of just comparing all data samples against the threshold. Then, compare the generated alarm signal at the output against the one presented in Fig. 4.19. Why is there a delay in processing?

   (c) Add motion artifact of 1 s duration. The artifact should result in a signal amplitude being larger than 1 V. Make sure that the motion artifact occurs during the 15 s stop-breathing intervals. Compare the outputs of the algorithms with and without averaging.

(d) Modify both analog and software-based implementation by adding a push button that will be used to stop the alarm. In this case, the alarm will be stopped only when someone presses the button. This is important to ensure that someone heard and reacted to the alarm.

4.10 Further reading

In this section, we refer to several relevant resources for further reading. However, please note that this list is not comprehensive and that many excellent and relevant resources might have been omitted unintentionally.

In general, modeling physiological control systems and physical systems in Simulink is covered very well in Khoo (2018). Another excellent book that discusses modeling physiological systems using Matlab and Comsol, mainly based on the physical and finite element model, is Dokos (2017).

Modeling of biomedical systems and model-based signal processing is presented in Cerutti and Marchesi (2011).

An excellent description of noise sources in the electric circuits and computation of the sensitivity of the output of the electric circuit to different noise sources is presented in Chapter 8 of Northrop (2012).

Uncertainty propagation for sensors and measurement systems is covered well in Figliola and Beasley (2012). Excellent theoretical discussion about uncertainty quantification is covered in Smith (2014). There are also chapters on sensitivity analysis and uncertainty propagation. One of the main references in the field of global sensitivity analysis is Saltelli et al. (2008).

An excellent reference for real-time implementation of signal processing algorithms is Ackenhusen (1999).

References

12-Bit, 1 MSPS, 2017. Single-Supply, Two-Chip Data Acquisition System for Piezoelectric Sensors, Analog Devices, Circuit Note CN-0350.

Ackenhusen, J.G., 1999. Real-time Signal Processing: Design and Implementation of Signal Processing Systems. Prentice-Hall.

Aew, J., Pollard, T.J., Shen, L., Lehman, L., Feng, M., Ghassemi, M., Moody, B., Szolovits, P., Celi, L.A., Mark, R.G., 2016. MIMIC-III, a freely accessible critical care database. Scientific Data. https://doi.org/10.1038/sdata.2016.35.

Burke, G., Lima, J., Wong, N.D., Narula, J., 2016. The multiethnic study of atherosclerosis. Global Heart 11 (3), 267–268.

Cerutti, S., Marchesi, C., 2011. Advanced Methods of Biomedical Signal Processing. Wiley.

Charlton, P.H., et al., 2019. Modeling Arterial Pulse Waves in Healthy Ageing: A Database for in Silico Evaluation of Haemodynamics and Pulse Wave Indices, fifth ed. 317. American Journal of Physiology-Heart and Circulatory Physiology, pp. H1062–H1085.

Cheng, L., Ivanova, O., Fan, H.H., Khoo, M.C., 2010. An integrative model of respiratory and cardiovascular control in sleep-disordered breathing. Respiratory Physiology & Neurobiology 174 (1–2), 4–28.

Clifford, G.D., Azuaje, F., McSharry, P.E., 2006. Advanced Methods and Tools for ECG Analysis. Artech House Publishing, Boston/London.

Dokos, S., 2017. Modeling Organs, Tissues, Cells and Devices Using MATLAB and COMSOL Multiphysics. Springer.

Elliott, P., 2014. The Airwave health monitoring study of police officers and staff in Great Britain: rationale, design and methods. Environmental Research 134, 280–285.

Ewing, D.J., Martyn, C.N., Young, R.J., Clarke, B.F., 1985. The value of cardiovascular autonomic function tests: 10 years experience in diabetes. Diabetes Care 8 (5), 491–498.

Farago, E., Chan, A.D.C., 2021. Motion artifact synthesis for research in biomedical signal quality analysis. Biomedical Signal Processing and Control 68 (102611).

Figliola, R.S., Beasley, D.E., 2012. Theory and Design for Mechanical Measurements, fifth ed. Wiley.

Khoo, M.C.K., 2018. Physiological Control Systems: Analysis, Simulation, and Estimation, second ed. Wiley.

Kind, T., Faes, T.J.C., Lankhaar, J.W., Vonk-Noordegraaf, A., Verhaegen, M., 2010. Estimation of three and four-element windkessel parameters using subspace model identification. IEEE Transactions on Biomedical Engineering 57, 1531–1538.

K.H. Parker, A practical guide to wave intensity analysis, https://kparker.bg-research.cc.ic.ac.uk/guide_to_wia/00_introduction.html, last updated on August 9, 2013, accessed on June 20, 2021.

Li, Q., Mark, R.G., Clifford, G.D., 2008. Robust heart rate estimation from multiple asynchronous noisy sources using signal quality indices and a Kalman filter. Physiological Measurement 29 (1), 15–32.

Li, Q., Mark, R.G., Clifford, G.D., 2009. Artificial arterial blood pressure artifact models and an evaluation of a robust blood pressure and heart rate estimator. BioMedical Engineering Online 8 (13). https://biomedical-engineering-online.biomedcentral.com/articles/10.1186/1475-925X-8-13. (Accessed 15 November 2020).

Löllgen, D., Müeck-Weymann, M., Beise, R.D., 2009. The deep breathing test: median-based expiration-inspiration difference is the measure of choice. Muscle Nerve 39 (4), 536–544.

Mitra, S.K., 2001. Digital Signal Processing: A Computer-Based Approach, second ed. McGraw-Hill, New York.

Northrop, R.B., 2012. Analysis and Application of Analog Electronic Circuits to Biomedical Instrumentation, second ed. CRC Press.

Orphanidou, C., 2018. Signal Quality Assessment in Physiological Monitoring: State of the Art and Practical Considerations. SpringerBriefs in Bioengineering.

Saltelli, A., et al., 2008. Global Sensitivity Analysis. The Primer. John Wiley and Sons.

Sameni, R., Shamsollahi, M., Jutten, C., 2008. Model-based Bayesian filtering of cardiac contaminants from biomedical recordings. Physiological Measurement 29 (5), 595–613. IOP Publishing.

Sedghamiz, H., Santonocito, D., 2015. Unsupervised detection and classification of motor unit action potentials in intramuscular electromyography signals. In: The 5th IEEE International Conference on E-Health and Bioengineering, EHB 2015, at Iasi-Romania.

Smith, R.C., 2014. Uncertainty Classification: Theory, Implementation and Applications. SIAM.

Spallone, V., Bellavere, F., Scionti, L., et al., 2011. Recommendations for the use of cardiovascular tests in diagnosing diabetic autonomic neuropathy. Nutrition, Metabolism and Cardiovascular Diseases 21, 69–78.

Spencer, M.P., 1987. Normal blood flow in the arteries. In: Spencer, M.P. (Ed.), Ultrasonic Diagnosis of Cerebrovascular Disease. Developments in Cardiovascular Medicine, vol 61. Springer.

Tang, Q., Chen, Z., Ward, R., et al., 2020. Synthetic photoplethysmogram generation using two Gaussian functions. Scientific Reports 10, 13883.

Thakkar, P., 2004. The Removal of Motion Artifacts from Non-invasive Blood Pressure Measurements. Ph.D. thesis, University of Central Florida.

Tri-Council Policy Statement: Ethical Conduct for Research Involving Humans, TCPS2, 2018. Canadian Institutes of Health Research, Natural Sciences and Engineering Research Council of Canada, Social Sciences and Humanities Research Council.

Willemet, M., Chowienczyk, P., Alastruey, J., 2015. A database of virtual healthy subjects to assess the accuracy of foot-to-foot pulse wave velocities for estimation of aortic stiffness. American Journal of Physiology: Heart and Circulatory Physiology 309, H663–H675.

Zhou, S., Lisheng, X., Hao, L., et al., 2019. A review on low-dimensional physics-based models of systemic arteries: application to estimation of central aortic pressure. BioMedical Engineering OnLine 18 (41).

CHAPTER

5

Devices based on oscillometric signal: blood pressure

Pervasive Cardiovascular and Respiratory Monitoring Devices
https://doi.org/10.1016/B978-0-12-820947-9.00009-X

165

Acronyms and explanations

AAMI	The Association for the Advancement of Medical Instrumentation
AM	amplitude modulation
ANSI	The American National Standards Institute
BHS	The British Hypertension Society
CAD	computer-aided design
DBP	diastolic blood pressure
FM	frequency modulation
ISO	The International Organization for Standardization
MAA	maximum amplitude algorithm
MAP	mean arterial blood pressure
ME	mean error
NIBP	noninvasive blood pressure
OMW	oscillometric waveform
OPI	oscillometric pulse indices
PID	proportional-integral-derivative
PP	pulse pressure
PWV	pulse wave velocity
SBP	systolic blood pressure
SDE	standard deviation of error

Variables used in this chapter include:

a, b	model coefficients
C	compliance
C_{cuff}	compliance of the cuff
$env(t)$	envelope of the oscillometric waveform
P_0	maximal inflation pressure of the cuff
$p_a(t)$	arterial pressure
P_{ath}	atmospheric pressure
$p_c(t)$	cuff pressure
$p(t)$	composite or total pressure
$p_t(t)$	transmural pressure
r	rate of deflation
$\hat{t}_{MAP}, \hat{t}_{SBP}, \hat{t}_{DBP}$	estimated time instances of MAP, systolic and diastolic pressures
V_{a0}	volume of the artery for $p_t = 0$
$V_a(t)$	volume of the artery

5.1 Introduction

Blood pressure (BP) is one of five vital signs. It is an important indicator of a patient's cardiovascular health (Taylor et al., 2010). BP is defined as the pressure applied by the blood on the walls of the blood vessels. Blood pressure has different values in different blood vessels in the body. Blood pressure devices normally measure blood pressure in the upper arm (Nichols et al., 2011). The pressure measured invasively in the artery is called the *arterial pressure*. During one cardiac cycle, one can observe one arterial pulse. The minimum of the arterial pulse is called *diastolic pressure* (DBP) and the maximum is called *systolic pressure* (SBP). BP is commonly reported in terms of SBP over DBP and is expressed in millimeters of mercury (mmHg) units. For example, SBP of 120 mmHg and DBP of 80 mmHg are reported as 120/80 mmHg. Please note that mmHg is not an accepted unit for the pressure by the International System of Units (non-SI unit) and that 1 kPa is about 7.5 mmHg.

An *auscultatory method* is an indirect method to measure arterial blood pressure that relies on detecting Korotkoff sounds under an occluding cuff. An *electronic sphygmomanometer* is a device that indirectly measures blood pressure by automatically determining the blood pressure from a signal obtained from the pressure sensors. *Noninvasive blood pressure measurement* (NIBP) is any cuff-based method for measuring BP. *Ambulatory BP monitoring* (ABPM) is a noninvasive, automated, periodic BP measurement over an extended period, such as 24 h. This is different from *home blood pressure monitoring* (HBPM), which typically involves measurements over a longer period, not necessarily at periodic intervals, while the subject is stationary and awake.

Blood pressure measurement at home is commonly performed using an oscillometric method. *Oscillometry* is based on sensing the pressure pulsations within a cuff wrapped around the subject's bicep or wrist. This method was first discovered by Marey (1876). Oscillometry uses a pressure sensor to detect the pressure oscillations within the cuff. Oscillometric methods are normally implemented in electronic sphygmomanometers nowadays.

Other methods for measuring blood pressure, both invasive and noninvasive, require a trained medical practitioner, and they are not described in this book. More information about them can be found in Webster (2010), Baura (2011). They are mentioned in this chapter when we discuss the performance evaluation of the oscillometric devices. Invasive blood pressure involves direct measurement of arterial pressure by inserting an arterial line directly into an artery.

Pulse morphology is the analysis of the shape of the pulse based on its characteristics and parameters. The oscillometric pulse is the oscillometric signal during one cardiac cycle. Oscillometric algorithms are based on extracting features from each oscillometric pulse during cuff deflation and analyzing them to estimate blood pressure. In addition, features extracted from the oscillometric signals can be used to estimate arterial stiffness, central blood pressure, and several other important applications. This chapter will focus on methods and devices for estimating blood pressure. However, we will also show how oscillometric pulse processing can be used for estimating arterial stiffness.

In this chapter, we will try to answer several questions, including:

- Why does an oscillometric waveform have its particular shape?
- How does the oscillometric algorithm work?
- Will the blood pressure estimate depend on the subject's health, including increased arterial stiffness?

5.2 What is measured using a blood pressure device

5.2.1 Definitions related to blood pressure

Fig. 5.1 shows three pulses of an arterial pressure signal obtained invasively. The maximum value of the pulse represents the systolic pressure. SBP is defined as the maximum pressure exerted by the blood against the arterial walls. The minimum value of the arterial pulse represents the diastolic pressure. DBP results from ventricular relaxation during the diastole. The difference between systolic and diastolic pressure is called the *pulse pressure* $PP = SBP - DBP$. Increasing stiffness of arteries very often results in elevated pulse pressure. The *dicrotic notch* represents the interruption of the smooth flow of blood at the moment when the aortic valve closes. The yellow area of the pulses between the diastolic point at the beginning of the pulse and the dicrotic notch is called *systole*, while the green area between the dicrotic notch and the diastolic point of the next cardiac cycle is called *diastole*.

Mean arterial pressure is the mean pressure during the cardiac cycle. It is computed as

$$MAP = \int_0^T p_a(t)dt$$

where $p_a(t)$ is the arterial pressure signal and T is the duration of the cardiac cycle. *MAP* is often approximated as $MAP = (SBP + 2DBP)/3$. This approximation is corrected in several

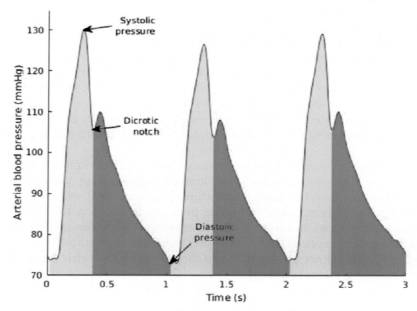

FIGURE 5.1   Arterial pulse waveform $p_a(t)$. The yellow areas correspond to the systole and the green areas to the diastole.

works by including heart rate and other parameters. In general, it should be avoided, and *MAP* should be calculated by the definition shown above. *MAP* of the three pulses shown in Fig. 5.1 is 94 mmHg. *MAP* is important since blood cannot perfuse vital organs when *MAP* is below 60 mmHg (DeMers and Wachs, 2019).

5.2.2 Principles of measuring oscillometric blood pressure

Oscillometric blood pressure measurement requires attaching a cuff to a subject. The measurement starts by initiating the cuff inflation. The cuff is inflated to a point where the blood flow through the artery is ceased. After that, the pressure in the cuff is slowly released. As the blood starts flowing through the artery, the arterial pulses get superimposed on the cuff pressure. The pressure is measured using a pressure sensor, and the oscillometric algorithm is applied to estimate *SBP* and *DBP*.

In oscillometry, intermittent measurements are performed. The oscillometric waveform is obtained through a nonlinear transformation of the arterial pressure signal, where nonlinear transformation includes transform functions of the artery, arm, and cuff. Fig. 5.2A represents the cross-section of the cuff and the upper arm. The pink color represents the skin, and the orange color represents the muscles. The bone is shown as a black circle, and the artery is presented with a red color. Please note that this is a very simplified cross-section of the arm. The outermost gray circle represents the cuff that is wrapped around the arm. Cuff pressure is denoted as $p_c(t)$. The internal pressure in the artery is denoted as $p_a(t)$. The signal at the output of the pressure transducer represents the total pressure obtained during cuff deflation and its waveform is shown on the top of Fig. 5.2B. Small arterial pulses are superimposed on the deflation pressure curve and these pulses are extracted as shown in the middle of Fig. 5.2B. They form a waveform called the *oscillometric waveform* or *oscillometric signal*. The

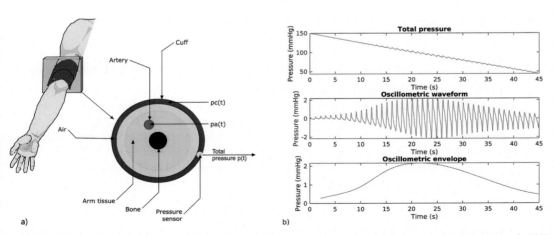

FIGURE 5.2 (A) Simplified cross-section of the cuff and the upper arm. The figure is modified from LifeART Collection Images Copyright © 1989–2001 by Lippincott Williams and Wilkins, Baltimore, MD provided by SmartDraw. (B) The signal called total pressure obtained from the pressure sensor, oscillometric waveform, and oscillometric envelope.

envelope of the oscillometric waveform is called the oscillometric waveform envelope. SBP and DBP are obtained by processing the oscillometric envelope using an algorithm described in Section 5.6.

5.2.3 Issues and controversies in oscillometric blood pressure measurements

Some of the controversial issues regarding oscillometric blood pressure measurements are listed below (Box 5.1).

- The estimation method used: Oscillometric blood pressure is indirectly estimated. A signal from the pressure sensor is obtained during the cuff deflation. Arterial pulses that modulate this cuff pressure signal do not actually carry direct information about the systolic or diastolic blood pressure. The device extracts features from these pressure pulses, analyzes the features, and then maps these features to the cuff pressure signal. Very often, as we will see later, there is no valid model or explanation for why a particular feature or a combination of features should be used to estimate blood pressure. Also, device manufacturers do not report what algorithms they use to estimate blood pressure.
- Frequency of measurements: There is natural variability in arterial blood pressure characterized by continuous fluctuations up to around 20 mmHg over short time intervals, as shown in Fig. 5.3 (Hansen and Staber, 2006). In addition, there are a number of factors that can elevate blood pressure in the short term, including the emotional state, full bladder (e.g., SBP could increase up to 10–15 mmHg when the patient has a full bladder), and outside temperature (Monk, 2010; Adams and Leverland, 1985). Therefore, when the measurement is performed once a day or once a week, the estimated SBP and DBP values might not represent a person's true blood pressure.

BOX 5.1

Controversies regarding blood pressure measurements

Even though blood pressure has been measured for a century, there are still controversies regarding blood pressure measurements. We will add two quotes regarding the importance of measuring blood pressure accurately as well as regarding the fact how poorly the measurements are performed nowadays:

"… 1-million-patient meta-analysis suggests that a 3 to 4 mm Hg systolic increase in blood pressure would translate into a 20% higher stroke mortality and a 12% higher mortality from ischemic heart disease (Lewington et al., 2002)."

"Few measurements in medicine are done as poorly and inconsistently as blood pressure measurement …. Though there is a clear recognition of biological variability, we continue to make decisions largely on measurements taken at random times under poorly controlled conditions (Jones and Hall, 2008)."

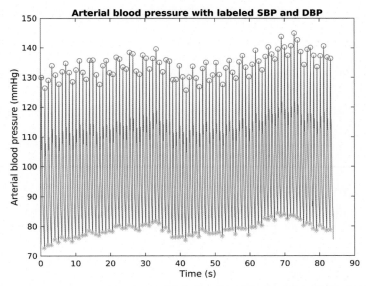

FIGURE 5.3 Arterial waveform over the period of 85 s for a subject under anesthesia diagnosed with cardio-myopathy. Red circles represent systolic pressure, and yellow asterisks represent diastolic pressure. It can be seen that the pressure varies significantly from pulse to pulse.

- Device error: The accepted measurement error based on the ANSI/AAMI/ISO protocol (Non-invasive, 2009) is 5 mmHg or less, with a standard error of 8 mmHg or less. That is a very large error; therefore, making diagnostic decisions based on only one measurement is dangerous. A new statement from the American Heart Association (Muntner et al., 2019) emphasizes the point that a single measurement of BP in the doctor's office is not representative and that ambulatory and home BP monitoring should be done.
- Calibration: Devices are calibrated based on reference measurements from a relatively small sample size (e.g., 85 subjects (Manual, Electronic, or Automated Sphygmomanometers, 2002)). This population includes both healthy people and patients. However, it does not have the number needed to cover "difficult" patients for whom blood pressure measurement using oscillometric methods is unreliable and who actually need to measure blood pressure more often than healthy people. An example includes patients with atrial fibrillation. It was mentioned in O'Brien et al. (2003) that "blood pressure measuring devices vary greatly in their ability to record blood pressure accurately in patients with arrhythmias, indicating that devices should be validated independently in patients with arrhythmias."
- Compliance with the measurement recommendations: Patients and even medical professionals are often unaware or do not adhere to the measurement recommendations. Recommendations include: patients must remain silent during measurements, be seated correctly with back support and legs uncrossed, must have rested at least 5 min before taking the measurement, and should reside in a quiet environment Muntner et al. (2019). Current devices cannot detect incorrect use and nonadherence to measurement procedures. A very important requirement is that the cuff should be at the same vertical

level as the heart to avoid errors due to hydrostatic pressure. Wrist cuff-based devices are particularly vulnerable to this.

- There are no available open hardware and open software devices and no large data-bases of labeled oscillometric signals. Open hardware devices are the devices where one has access to the schematics, logic designs, Computer-Aided Design (CAD) drawings or files, and software of the device and can modify them.

Attempts to address some of these issues are presented later in this chapter.

5.3 Features of the signal and the noise

5.3.1 Why do we analyze each pulse?

Fig. 5.4 shows the total pressure that includes the cuff deflation waveform with superimposed arterial pressure in the graph on the top. The zoomed-in total pressure curve in the regions of the systolic, MAP, and diastolic pressures are shown in the left, middle, and right graphs on the bottom, respectively. It is obvious from these graphs that the pulses have different shapes in these three regions of interest. Therefore, algorithms for estimating blood pressure rely on tracking the specific features extracted from the pulses over time.

When analyzing the features of the pulse, it is common to extract and analyze each pulse separately (Table 5.1). Features related to pulse shape and timing parameters are commonly extracted. Changes in the features can be observed over time. This is important in estimating MAP, SBP, and DBP using oscillometric methods.

In addition to extracting features of each pulse, it is possible to extract features among neighboring oscillometric pulses. This can be used for breathing detection and removal or removing other artifacts that affect multiple pulses, such as motion artifacts.

Features that are commonly extracted in oscillometry include:

- The maximum amplitude of the pulse
- The area under the pulse
- The magnitude: the difference between the minimum and the maximum of each pulse.

5.3.2 Noise and artifacts

5.3.2.1 Motion artifacts

The problem of motion artifacts is common in biomedical engineering and noninvasive blood pressure measurements (Such and Muehlsteff, 2006). Even small, sudden transient motions can result in large signal changes in the oscillometric waveform. What makes estimating blood pressure challenging in the case of motion is that these artifacts might be interpreted as parts of the signal itself. This might degrade the accuracy of the measurement or result in completely erroneous blood pressure estimations (Balestrieri and Rapuano, 2010; Charbonnier et al., 2000). Motion artifacts affect the total pressure curve and might look like short irregular pulses. These artifacts can affect the shape of the oscillometric waveform significantly. Therefore, ways of dealing with motion artifacts need to be developed, and one possible approach is to add an accelerometer to the BP measurement device to detect the motion (Abderahman et al., 2017).

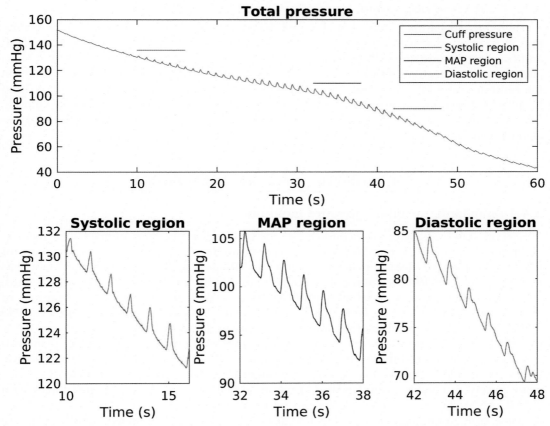

FIGURE 5.4   Total pressure on the top. The top figure also shows the regions of the total pressure where systolic, MAP, and diastolic pressures are estimated. Figures on the bottom show segments of the total pressure in the regions of the systolic, MAP, and diastolic pressures.

TABLE 5.1   Use of the features of pulse to estimate blood pressure or arterial stiffness.

Features of the pulses	Procedure	Estimates
Each oscillometric pulse separately	Track features during deflation	SYS, DIA, MAP
Features among neighboring oscillometric and/or pulses from other transducers.	Track features during deflation	Breathing detection and removal Artifact detection
Each oscillometric pulse separately	Several pulses at a constant cuff pressure	Augmentation index Pulse wave velocity

5.3.2.2 Effects of respiration

There have been many works analyzing the respiration effects on blood pressure estimation. It has been discovered that respiration affects blood pressure through amplitude (AM) and frequency modulation (FM) (Iamratanakul et al., 2003; Bruno and Scalart, 2005). AM accounts for large fluctuations in the amplitude of the signal. FM is evidenced in the changes in the pulse-to-pulse intervals. Deep breathing can result in large fluctuations in SBP of up to 15 mmHg. Breathing at a normal depth can cause SBP fluctuations of up to 3—6 mmHg (Ramsey, 1991). These fluctuations are quite large; hence, several works targeted the suppression of breathing signals (Chen et al., 2010).

5.4 Performance measures and evaluation

5.4.1 Standards for blood pressure monitors

Several standards have been introduced for labeling, safety, and performance requirements of automated BP monitoring devices. A standard published by the Association for the Advancement of Medical Instrumentation (AAMI) and the American National Standards Institute (ANSI) in 1987 is commonly referred to as ANSI/AAMI SP10 (American standard, 1987). This standard was updated in 1993 (American standard, 1993; Manual, Electronic, or Automated Sphygmomanometers, 2002). In 1990, the British Hypertension Society BHS (O'Brien et al., 1990) published a protocol that was updated in 1993 (O'Brien et al., 1993). In 2009, the AAMI and the ANSI, together with the International Organization for Standardization (ISO) introduced their standard for automated BP monitoring devices (Non-invasive, 2009). Another protocol for evaluating BP monitoring devices was developed by the European Society of Hypertension (ESH) (O'Brien et al., 2010). In 2019, ESH joined forces with AAMI/ANSI/ISO to endorse a universal protocol AAMI/ESH/ISO 81060—2; 2019 (Non-invasive sphygmomanometers, 2019). For discussion about the standard AAMI/ESH/ISO 81060—2; 2019, please refer to Stergiou et al. (2019).

Safety requirements are necessary for the devices intended for public, home, or other unsupervised use. They include requirements for maximum cuff pressure (maximum 300 mmHg), release rate, and electrical safety. It should be possible to deflate the cuff by providing rapid exhaust of the pneumatic system from the highest pressure to complete release in less than 10 s. The standard requires that cuff pressure is not maintained above 15 mmHg for longer than 3 min. When used in a normal mode, it shall not be possible to reduce the cuff pressure at a rate lower than 2 mmHg/s for blood pressures between 250 and 50 mmHg (Manual, Electronic, or Automated Sphygmomanometers, 2002).

In Manual, Electronic, or Automated Sphygmomanometers (2002), accuracy and repeatability are defined as follows. The maximum error for measuring the cuff pressure at any point of the scale range shall be ± 3 mmHg (± 0.4 kPa) or 2% of the reading above 200 mmHg. For a given manometer, each repeated reading throughout the graduated range shall agree with one another within 4 mmHg when measured under static conditions at successively lower pressures.

5.4.2 Comparisons of test devices and the reference method

AAMI/ESH/ISO 81060−2; 2019 protocol requires a dataset of BP measurements obtained from at least 85 subjects. Paired measurements are obtained by using the BP device under verification and by two trained observers. Three sets of blood pressure measurements, obtained over a period of 6−30 min, shall be recorded for each subject resulting in a total of 255 measurements. Measurements obtained by a trained observer using the auscultatory technique are regarded as the gold standard. Two trained observers shall make simultaneous, blinded blood pressure measurements on each subject. Any measurements with observer-to-observer differences greater than 4 mmHg should be excluded from the data set.

The general study population for the measurements obtained by the trained observer should include:

- at least 5% of subjects with SBP below 100 mmHg.
- at least 5% of subjects with SBP above 160 mmHg.
- at least 20% of the reference SBP readings above 140 mmHg.
- at least 5% of subjects with DBP below 60 mmHg.
- at least 5% of subjects with DBP above 100 mmHg, and
- at least 20% of the reference DBP readings above 85 mmHg,

The remaining subjects should be distributed between these outer limits.

Another way to evaluate the test device is to compare its accuracy against intra-arterial paired measurements. The clinical patients selected for this study should include only those patients who will undergo surgery that involve the placement of an intra-arterial line. The subclavian artery, axillary artery, and brachial artery are the most desirable sites for intra-arterial measurements. The radial artery is not desirable because of the amount of amplification or attenuation of the blood pressure in this artery. The suggested minimum number of subjects for the study is 15, with a minimum of 5 and a maximum of 10 paired measurements per subject.

Readings by the BP device being verified and the gold standard technique can be taken either sequentially or simultaneously.

The error of BP monitoring devices based on the AAMI/ESH/ISO 81060−2; 2019 is expressed using two measures: the mean error (ME) and the standard deviation of the error (SDE) obtained over the whole 255 measurements. This standard requires an ME to be within plus or minus 5 mmHg and an SDE equal to or less than 8 mmHg. The standard AAMI/ESH/ISO 81060−2; 2019 also introduces an error for individual subjects.

The BHS protocol (O'Brien et al., 1993) classifies the BP monitoring devices into classes A−D, where class A shows the best accuracy, as shown in Table 5.2.

5.4.3 BP simulators

Methods for BP device verification based on data collected from patients require running a pilot study and having trained observers. The studies are normally long and expensive. Objective methods for the calibration of blood pressure devices were proposed (Balestrieri and Rapuano, 2010). They are based on NIBP test simulators that emulate and reproduce the pressure profile of the measurement subject and generate the pressure signal. Two

TABLE 5.2   Classes of BP devices based on british hypertension society protocol.

Class	Percentage of readings with an error less than 5 mmHg	Percentage of readings with an error less than 10 mmHg	Percentage of readings with an error less than 15 mmHg
A	60%	85%	95%
B	50%	75%	90%
C	40%	65%	85%
D	<40%	<65%	<85%

different types of simulators exist: the artificial limb and the waveform generator. Limb simulators consist of an artificial arm incorporating emulated artery with pulsating fluid. These simulators allow for testing the overall BP device, including both mechanical parts and electronics. Waveform generators only generate cuff and oscillometric pressure signals and cannot test the mechanical components of the blood pressure device. Even though they are less comprehensive, waveform generators are more popular due to their simplicity. Ideally, test simulators would have a comprehensive database of physiological waveforms and a protocol that assesses the accuracy of the BP device over a wide range of conditions. Some of the waveform simulators can generate very weak signals and also emulate motion artifacts such as tremors.

In Balestrieri and Rapuano (2010), it was concluded that NIBP simulators could not be used as a substitute for clinical validations to measure the accuracy of NIBP measurement devices for the following reasons:

- The simulators generate waveforms that are significantly different from physiological waveforms.
- There are a number of NIBP simulators, and they differ in the waveforms they provide, the accuracy they report when evaluating the same BP device, and so on. Therefore, there is no standard way of calibrating BP devices based on the NIBP test simulators yet.

However, as the output of the NIBP simulators is very stable, they can be used for evaluating the repeatability of the measurements.

5.5 Oscillometric signal generation model

This section aims to explain the origin of superimposed arterial pulses on the cuff pressure and, therefore, the origin of the oscillometric signal. In addition, we will explore the particular shape of the oscillometric signal and explore changes in the oscillometric signal with increased arterial stiffness and blood pressure. This interpretation would be very difficult without a mathematical model. Therefore, we will model the cuff, arm, and artery to obtain the oscillometric signal. This section is solely based on research presented in Babbs (2012). A detailed review of modeling approaches in oscillometry is presented in Forouzanfar et al. (2015).

5.5.1 Arm-artery system model

In this section, we will discuss arterial volume changes with pressure. These concepts are closely related to compliance and stiffness; therefore, we will introduce them first. The stiffness of an object is a measure of the resistance offered by the object to deformation. The inverse of stiffness is compliance. Compliance and stiffness can be defined for arteries but also other objects such as cuff.

The main factor determining arterial pressure is the stretch of the walls of the arteries by the volume of the blood the arteries contain. This volume increases in systole because of the inflow of blood from the heart. This ability of an artery to expand and increase its volume with increasing pressure is quantified as vessel compliance (C) (or C_a), which is the change in volume (ΔV) divided by the change in pressure measured (Δp) between the inside and the outside of the artery: $C = \Delta V / \Delta p$.

The cross-section of the cuff-arm-artery system is shown in Fig. 5.2. When the cuff is not placed on the arm, the arterial blood pressure is the only pressure in the artery. When the cuff is placed around the upper arm and the pressure is applied, the volume of the artery reduces. The pressure between two sides of the arterial wall is called *transmural pressure*. It is defined as

$$p_t(t) = p_a(t) - p_c(t) \tag{5.1}$$

where $p_c(t)$ is the cuff pressure.

Let P_0 be the maximum pressure when the cuff is inflated at the beginning of the measurement. The maximum inflation pressure is normally set to be about 20–30 mmHg higher than the systolic pressure. The pressure is reduced at the rate r presented in mmHg/s. The common value for the deflation rate is 2–3 mmHg/s. Therefore: $p_c(t) = P_0 - rt$.

To describe the relationship between the volume of the artery and the transmural pressure, we use a model based on two exponential functions: one for negative and one for positive transmural pressure. The volume of the artery is the area of the artery multiplied by the length of the arterial segment under the cuff. We assume the fixed length of 10 cm under the cuff. The area of the artery is called the lumen area. The lumen area gets smaller with increasing external cuff pressure and reduces to zero as the transmural pressure becomes negative. The model is shown below:

$$\frac{dV_a(t)}{dp_t(t)} = \begin{cases} aV_{a0}e^{ap_t(t)}, & \text{for } p_t(t) < 0 \\ aV_{a0}e^{-bp_t(t)}, & \text{for } p_t(t) \geq 0 \end{cases} \tag{5.2}$$

After solving differential equations, we can get

$$V_a(t) = \begin{cases} V_{a0}e^{ap_t(t)}, & \text{for } p_t(t) < 0 \\ V_{a0}\left[1 + \frac{a}{b}\left(1 - e^{-bp_t(t)}\right)\right], & \text{for } p_t(t) \geq 0 \end{cases} \tag{5.3}$$

Here, a and b represent compliance constants in mmHg^{-1}. V_{a0} is the volume of the artery for $p_t = 0$. Constants a and b allow us to simulate different patient conditions. Increasing a and b in proportion results in larger volume changes for a given pressure change, and therefore larger values of a and b can be used to represent a more compliant artery. Decreasing a and b in proportion reduces the volume change for a given pressure change so that smaller values of a and b represent a stiffer artery. Increasing the ratio a/b represents greater maximal distension. Decreasing the ratio a/b represents smaller maximal distension.

Fig. 5.5 shows the relationship between the transmural pressure and the volume of the artery. Simulation is done for the blood pressure of 120/80 mmHg, the artery with a radius of 1.2 mm, and the arterial segment length of 10 cm. Three curves are shown for the normal artery, the stiff artery, and the artery with greater maximal distension. However, we still used the same blood pressure of 120/80 mmHg for both arteries in this example, even though the systolic pressure and pulse pressure would normally increase with increased arterial stiffness. For the normal artery, we selected parameter $a = 0.11 \text{ mmHg}^{-1}$ and $b = 0.03 \text{ mmHg}^{-1}$. We modeled the artery with increased arterial stiffness using $a = 0.076 \text{ mmHg}^{-1}$ and $b = 0.021 \text{ mmHg}^{-1}$. We see that the volume of the artery does not change much between diastolic and systolic pressure in case there is no external pressure. We modeled the artery with greater maximal distension using $a = 0.11 \text{ mmHg}^{-1}$ and $b = 0.0244 \text{ mmHg}^{-1}$.

Aortic compliance gets reduced with age or a disease such as arteriosclerosis (hardening and loss of elasticity of the walls of arteries). When this occurs, the aortic volume–pressure curve is shifted downward compared to the volume-pressure curve for the aorta with normal compliance. This can be observed in Fig. 5.5 if we look only at the pressure regions between 80 and 120 mmHg. It can be observed that the aorta with reduced compliance has a smaller volume (red curve) in the mentioned pressure region than the volume of the normal artery

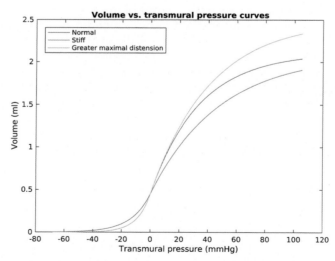

FIGURE 5.5   Volume versus pressure curves for an artery segment of a normal artery with parameters $a = 0.11$ and $b = 0.03$, for the artery with increased arterial stiffness modeled with $a = 0.076$ and $b = 0.021$ and with increased maximal distension with $a = 0.11$ and $b = 0.0244$. All the coefficients are expressed in mmHg^{-1}.

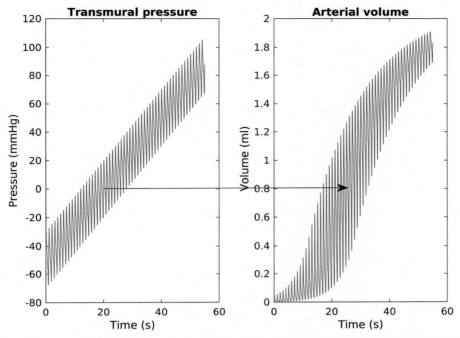

FIGURE 5.6 (A) Transmural pressure over time, (B) arterial volume changes over time. Both figures correspond to the volume–pressure curve shown in Fig. 5.6 for the artery with increased arterial stiffness modeled with parameters $a = 0.076$ mmHg^{-1} and $b = 0.021$ mmHg^{-1}. The arrow shows the position of MAP.

(blue curve). These compliance changes in the aorta are an important cause for the increase in aortic pulse pressure with aging or arterial disease.

The same change in pulse pressure of, for example, 40 mmHg, will result in different changes in the volume of the artery depending on the transmural pressure. It is also interesting to observe that the volume curve is the steepest when the transmural pressure is zero. It was shown experimentally that zero transmural pressure corresponds to the mean arterial pressure. Therefore, the largest amplitude of the oscillometric pulses is observed at the mean arterial pressure. In Fig. 5.6, we presented the transmural pressure and arterial volume changes over time. In the beginning, the transmural pressure is negative, and the volume changes in the artery are very small. When the transmural pressure is around 0 mmHg, the volume changes in the artery become the largest, as shown using an arrow.

5.5.2 Total pressure

We are interested in obtaining the total pressure $p(t)$ that includes the pressure of the cuff itself together with the superimposed pulses. The total pressure at the cuff can be defined as:

$$p(t) = P_0 - rt + P_{ath} + V_a(t)/C_{cuff}(t) \tag{5.4}$$

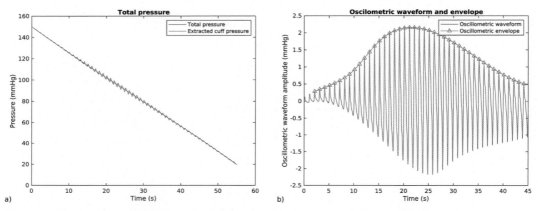

FIGURE 5.7 (A) Simulated pressure over time, (B) Oscillometric waveform representing the extracted and amplified pulses versus cuff pressure from the pressure curve shown in (A).

where P_{ath} is the atmospheric pressure $P_{ath} = 760$ mmHg, and C_{cuff} is the compliance of the cuff. The compliance of the cuff is defined as changes in the cuff's bladder volume versus changes in the cuff pressure. It is obvious that the cuff is stiffer when it is fully inflated and less stiff as the pressure in the cuff decreases; therefore, the compliance of the cuff is time-dependent. The compliance of the cuff can be modeled as $C_{cuff}(t) = V_0 / (P_0 - rt + P_{ath})$ (Babbs, 2012), where V_0 is the volume of the cuff after inflation in ml. The simulated total pressure over is shown in Fig. 5.7A. The extracted oscillometric waveform is shown in Fig. 5.7B.

The assumption in the model (5.4) is that the arm tissue is mostly incompressible; therefore, the arterial volume changes are superimposed on the volume of the cuff bladder. Therefore, the changes in the artery volume are proportional to the oscillations observed on the total curve. This is, in most cases, consistent with what is observed with the recorded oscillometric pulses (Forouzanfar, 2014).

Several works addressed the modeling assumption of incompressible arm tissue. In Lan et al. (2011), an upper arm model based on finite elements was developed to study the effect of mechanical properties of arm soft tissue and muscles on oscillometric measurements. It was shown that the arterial lumen area increases from the center to the edge of the cuff. Therefore, when the artery is closed at the center of the cuff, it can remain open underneath the rest of the cuff. This effect is not considered in the model presented in this section.

5.6 Oscillometric algorithm

The traditional algorithm for estimating blood pressure based on the oscillometric method includes several steps shown in Fig. 5.8. They are:

1. Extracting the oscillometric waveform (OMW),
2. Extracting features from the pulses as described in Sections 5.3.1 and 5.3.2 these features over time are called *oscillometric pulse indices* (OPI),

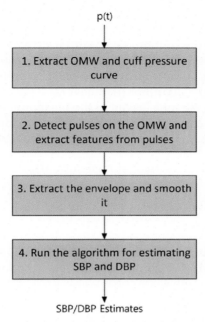

p(t)

1. Extract OMW and cuff pressure curve

2. Detect pulses on the OMW and extract features from pulses

3. Extract the envelope and smooth it

4. Run the algorithm for estimating SBP and DBP

SBP/DBP Estimates

FIGURE 5.8 Block diagram of the oscillometric blood pressure estimation algorithm.

3. Interpolating OPIs to obtain the envelope and smoothing the envelope. The *envelope* is defined as a smooth curve outlining the extremes of the signal.
4. Estimating MAP, SBP, and DBP using the oscillometric algorithm.

Existing methods for steps 1−4 are quite diverse and varying. The input to the algorithm is the total pressure shown in Fig. 5.7A. The figure shows the decreasing cuff pressure waveform starting from about 150 mmHg and going down to about 25 mmHg. This pressure curve is used in step 1 to extract the OMW and in step 4 for estimating SBP and DBP.

The first step is extracting the arterial pressure signal from the cuff deflation signal. This extracted signal represents the OMW. The downward trend relating to the cuff deflation should be removed to obtain the OMW. This can be performed by digital filtering (band pass (Geddes et al., 1982; Sorvoja et al., 2005) or high pass (Moraes et al., 1999)), analog filtering (Ramsey, 1991), or detrending (Jazbinsek et al., 2005). The extracted OMW is shown in Fig. 5.7B.

From the OMW, OPIs are then required as part of step 2. The OPIs require that each pulse in the OMW is separately processed to extract the features. The features include the height from baseline-to-peak, peak-to-peak, or area of each pulse (Ng and Small, 1994; Jazbinsek et al., 2005; Ball-llovera, 2003). The features are then recorded and used in step 3.

The OPIs are then interpolated to form an envelope in step 3. This step is also aimed at cleaning the envelope from any possible disturbances caused by noise or interference, which is necessary for the algorithms used in step 4 that estimate SBP and DBP values. Methods for cleaning the envelope include frequency domain filtering (Geddes et al., 1982), moving average filtering (Baker et al., 1997), and curve fitting (Lin et al., 2003).

In step 4, points that correspond to the SBP and DBP are found on the cleaned envelope. We will show only the most commonly used algorithms for estimating MAP, SBP, and DBP, which are called the maximum amplitude algorithm (MAA) (Ramsey, 1991) and the slope change algorithm (Jilek,). Other traditional algorithms for estimating blood pressure based on the envelope include the linear approximation algorithm (Medero, 1996) and the points of maximum/minimum slope algorithm (Ball-llovera, 2003).

MAA is based on finding the time instant of the maximum of the envelope and then mapping the time instant to the cuff deflation waveform and reading the MAP. If the cuff deflation waveform is denoted as $p_c(t)$ and the envelope is denoted as $env(t)$, then

$$\widehat{t}_{MAP} = \underset{t}{argmax}\ env(t)$$

$$\widehat{MAP} = p_c\left(\widehat{t}_{MAP}\right)$$

where $\widehat{t}_{MAP}$ represents the time instant where the envelope is maximum and $\widehat{MAP}$ is the estimate of MAP using the MAA algorithm. It has been explained in Section 5.5.1 that the maximum of the oscillometric envelope corresponds to MAP. SBP and DBP are estimated using empirical coefficients as follows:

$$\widehat{t}_{SBP} = env^{-1}(\max(env(t)) * c_{SBP}),\ \ \widehat{t}_{DBP} = env^{-1}(\max(env(t)) * c_{DBP})$$

$$\widehat{SBP} = p_c\left(\widehat{t}_{SBP}\right),\ \ \widehat{DBP} = p_c\left(\widehat{t}_{DBP}\right)$$

where $\widehat{t}_{SBP}$ represents the time instant where the amplitude of the envelope is equal to $max(env(t))c_{SBP}$ and the estimated systolic pressure $\widehat{SBP}$ is the pressure at the cuff deflation waveform at the same time instant. $\widehat{t}_{DBP}$ represents the time instant where the amplitude of the envelope is equal to $max(env(t)) * c_{DBP}$ and the estimated diastolic pressure $\widehat{DBP}$ is the pressure at the cuff deflation waveform at the same time instant. Coefficients c_{SBP} and c_{DBP} are determined during the calibration phase, and they are commonly fixed. They are empirical coefficients obtained based on data collected from pilot studies against the reference measurements. Fig. 5.9 shows mapping from the OMW to MAP, SBP, and DBP.

Even though the maximum amplitude algorithm is commonly used in practice, there is significant evidence that the coefficients should not be fixed. In Chandrasekhar et al. (2019), it has been shown that the true ratios vary with the pulse pressure and the widths of the arterial compliance curve (coefficients a and b). A very complex relationship between the systolic and diastolic coefficients and the parameters (PP, a and b) was given in Chandrasekhar et al. (2019). This is shown in Fig. 5.9A which shows the cuff pressure and the envelope for a person with normal arterial stiffness. Fig. 5.9B shows the envelope for a person with increased arterial stiffness but with the same blood pressure. Simulation is run so that the systolic and diastolic blood pressures are the same for both cases: 120/80 mmHg. It is obvious that the fixed coefficients cannot be used in both cases to estimate blood pressure. As we can see in this case, there is a significant error in estimating the diastolic blood pressure for the situation

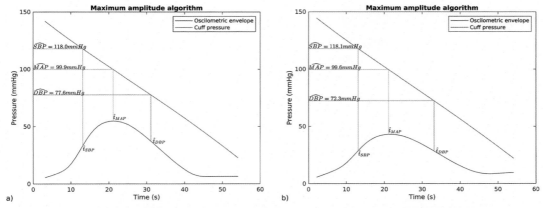

FIGURE 5.9 The results of the oscillometric algorithm for the simulation shown in Fig. 5.5 for SBP = 120 mmHg and DBP = 80 mmHg. The coefficients are fixed to $c_{SBP} = 0.6$ and $c_{DBP} = 0.65$. (A) Estimates of the blood pressure for the normal artery with parameter $a = 0.11$ mmHg^{-1} and $b = 0.03$ mmHg^{-1}. Estimated pressures are: MAP = 100 mmHg, SBP = 118 mmHg and DBP = 78 mmHg. (B) A estimates of the blood pressure for the artery with increased arterial stiffness modeled with $a = 0.076$ mmHg^{-1} and $b = 0.021$ mmHg^{-1}. Estimated pressures are $MAP = 100$ mmHg, $SBP = 118$ mmHg, and $DBP = 72$ mmHg.

when the subject has stiff arteries. The errors do not occur for DBP only; SBP and MAP are also affected. The situation also gets much worse when we introduce noise and motion artifacts.

Unlike the maximum amplitude algorithm, the maximum slope algorithm considers the slope of the envelope rather than the heights. Here, the systolic point is found as the point of the maximum slope of the envelope, and the diastolic point is the point of the minimum slope of the envelope. These two points can be found by taking the derivative of the envelope. The point where the slope is maximum corresponds to the maximum of the derivative signal, and the point where the slope is minimum also corresponds to the minimum of the derivative signal. When the maximum slope algorithm is run on the same data shown in Fig. 5.9, it correctly estimated systolic blood pressure for both normal and stiff arteries, but it also underestimated the diastolic pressure.

5.7 Sensors and circuits

5.7.1 Pressure sensors

Pressure is defined as force per unit area. The SI unit of pressure is the Pascal (Pa), where $1\ Pa = 1\ \mathrm{Nm}^2$. Pascal is a relatively small unit; therefore, for most biomedical applications, the pressure is expressed in kPa. The unit still in use in blood pressure measurements for historical reasons is mmHg.

Two types of transducers have been introduced in Chapter 2: strain gauges and piezoelectric sensors. Both of these transducers are used in designing pressure sensors. However, piezoelectric sensors can be used only for dynamic and not static pressure measurements. *Pressure sensors* operate in one of three modes of measurement: absolute, gauge, and

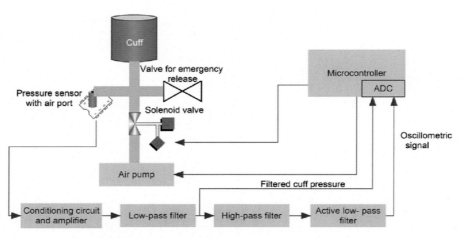

FIGURE 5.10   Hardware block diagram of the BP device.

differential. The *absolute* pressure sensor is designed with one port that allows air or fluid to enter and exert pressure on the transducer (see the pressure sensor in Fig. 5.10). The output of the sensor is the signal whose magnitude is proportional to the pressure applied. *Differential* and *gauge* sensors measure pressure at the input port relative to the pressure on another input port or to the atmospheric pressure, respectively. In comparison with the absolute pressure sensors, differential sensors have two air or fluid ports.

Sensing is normally performed by converting the force applied by the air or fluid into displacement. This is done by placing the transducer on a material that can be bent, compressed, or extended by force. Pressure sensing diaphragm, capsule, or expanding bellows are commonly used for this purpose. The *pressure-sensing diaphragm* is a circular plate fixed around the edges and placed below the air or fluid pressure port so that that is exposed to the air or fluid. The picture of a pressure sensor with a single air pressure port is shown in Fig. 5.10. In the case of the absolute pressure sensor, the other side of the diaphragm is a sealed chamber with a fixed pressure. In the case of the differential pressure sensor, the other side of the diaphragm might be exposed to the air or fluid on the other input port. When pressure is applied, the diaphragm gets deflected to the extent that is proportional to the magnitude of the applied pressure. In the case of resistive strain gauges, up to four foil strain gauges can be bonded to the surface of the diaphragm. These strain gauges are then connected to the Wheatstone bridge and then to the amplifiers. Diaphragms can be made from metal or ceramic materials; in this case, the pressure sensor can withstand considerable pressure. For MEMS sensors, the diaphragm is normally made of silicone.

Absolute pressure sensors are normally used for invasive blood pressure measurements. Gauge pressure sensors are used for oscillometric blood pressure, which means the blood pressure is measured relative to the atmospheric pressure. A tube from the cuff is connected to the air pressure port on the sensor's package. The sensors normally contain additional circuitry. In the case of strain gauges, the electronics that are a part of today's pressure sensors include the Wheatstone bridge only or the whole analog processing chain, including amplifiers and filters or the analog processing chain together with the A/D converter producing the digital output.

An example of a sensor commonly used for research and education is the TruStability board mount pressure series of transducers by Honeywell (Honeywell Datasheet, 2014) that have both digital and analog outputs. The sensors are temperature compensated and can measure both absolute and differential pressure. Parameters of that transducer include a maximum pressure range of up to 1550 mmHg, an accuracy that is normally better than ± 1 mmHg, combined linearity and hysteresis error: ± 0.25%, response time: 1 ms, and the sampling rate of the digital output up to 2 ksamples/s. Another example includes a resistive pressure sensor 2SMPP-02 from Omron with sensors already connected to the Wheatstone bridge. The pressure medium for this sensor is air. The bridge should be driven with a constant current of 100 µA DC, and the sensing range is between 0 and 37 kPa (277 mmHg).

5.7.2 Operation of the device

The following are the mechanisms that a blood pressure device needs to implement:

- Inflation mechanism: During inflation, the pressure on the artery is increased by inserting air into the bladder. During that time, the systolic blood pressure is roughly estimated, and the cuff is inflated to a pressure of about 20—30 mmHg higher than the estimated systolic pressure.
- Deflation mechanism: deflation rate, which is defined as $dp_c(t)/dt$ is kept constant in the majority of devices. This can be done by controlling the air release in the bladder and tubing by opening and closing the valve. Opening and closing the solenoid valve can be controlled using pulse width modulation. The pressure is monitored during deflation, and the pulse width can be controlled using a proportional-integral-derivative (PID) controller to maintain a constant deflation rate.
- Sampling and filtering are performed in hardware. Analog electronics normally includes the Wheatstone bridge, an amplifier, an antialiasing filter, and an A/D converter.
- Processing: It is required to perform the operations of inflation, deflation, and data collection in order, and this is done using a state machine that is implemented in software on a microcontroller. In addition, the microcontroller runs the oscillometric algorithm to estimate blood pressure.
- Safety mechanisms: Requirements for accuracy and safety are extracted from the standards (Manual, Electronic, or Automated Sphygmomanometers, 2002). Regarding safety, the standard requires that cuff pressure is not maintained above 15 mmHg for longer than 3 min. The maximum cuff pressure is 300 mmHg and, if required, the complete release of pressure must happen in less than 10 s. Watchdog timers are used to check if the time intervals for the safety mechanisms have expired.

5.7.3 Block diagram

The following components are commonly needed:

- Cuff of the appropriate size.

- The air pump that moves the air from the inlet and delivers it to the outlet. It is commonly a motor-driven diaphragm pump. When the diaphragm is compressed, the air within the diaphragm gets removed. When the diaphragm is decompressed, the chamber refills with air.
- The solenoid valve that connects the air circuit between the pump and the cuff allowing the air to flow toward the cuff during inflation and to the atmosphere during deflation. The rate of inflation and deflation can be changed by controlling the opening and closing of the valve.
- Valve for emergency release that can be implemented as a push-button valve or controlled by a safety circuit.
- Tubes that are used to connect the cuff, air pump, and valve. They need to be connected tightly to avoid any air leaks. A latex T-tube with a diameter of 3 mm is commonly used. T tubes are used so that it is possible to connect a cuff on one side with the pressure transducer and with a third port to either connect to the air pump or release air.
- Pressure sensors.
- Voltage regulators: very often, there are two voltage regulators: one for providing DC voltage for the electronics and one for providing higher voltages for the air pump and the motor.
- A microcontroller that is used to perform processing and implement safety mechanisms as described in Section 5.7.2.
- A communication unit that is used to connect the BP device to a smartphone or a computer using a wired or wireless connection.

The block diagram of a potential design is shown in Fig. 5.10. The design comprises components explained previously, including the motor, valve, cuff, microcontroller, pressure transducer, and communication unit.

5.7.4 Simulation of the electronics

The circuit is shown in Fig. 5.11. The sensor is modeled in a way that change in the resistance of $\Delta R = 1\,\Omega$ corresponds to 300 mmHg change in blood pressure. The bridge resistance is 120 Ω. Here, $V_+ = 5$ V and all operational amplifiers and the A/D converters operate in the range of 0–5 V. Maximum strain that can be then applied to this strain gauge

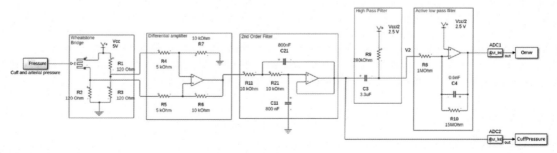

FIGURE 5.11 A circuit model of the oscillometric system. This model does not include mechanical components, such as the air pump, cuff, and valves.

is $\varepsilon = \Delta R / RG = 0.0041$ where G is the gauge factor which value is set to 2. Therefore, we converted the pressure at the input into the strain to model the sensor using a linear transformation. The output of the Wheatstone bridge is connected to the differential amplifier that amplifies the signal twice. This amplifier is connected to a second-order low-pass filter with a cutoff frequency of 19.89 Hz. At this stage, the cuff pressure is acquired and digitized using an A/D converter ADC2. A/D converters are set to 12 bits and the sampling rate of 200 samples/s.

The oscillometric signal is further extracted by using a high-pass filter that removes frequencies below 0.17 Hz. Therefore, this filter removed the cuff pressure, and the remaining signal is the oscillometric waveform. The filter is implemented in a way so that the oscillometric waveform is centered around $V_+/2$. It needs to be amplified and filtered further. For that, an active low-pass filter is used. Components $C_4 = 0.6$ nF and $R_{10} = 15$ MΩ are selected to achieve $f_c = 17.6$ Hz. The gain of the filter at the DC level is 15. The oscillometric signal at the output is sampled at the same sampling rate of 200 samples/s using the A/D converter ADC1.

After the cuff pressure and oscillometric signals are obtained, they are in the range between 0 and $2^{12} - 1 = 4095$. Since we need only relative values of the oscillometric pulses, we do not need to rescale the signal in general. However, the cuff pressure needs to be calibrated, which is done in software by the manufacturer. After the signals are obtained, they are processed using an oscillometric algorithm.

5.8 Simulation of the overall system

5.8.1 Why model an end-to-end system?

In this section, we will model the complete system that includes the generation of cuff pressure and oscillometric waveform, the circuit for processing blood pressure, and the algorithm for estimating blood pressure.

Since we know the systolic and diastolic blood pressures at the input of our model, we can compute the mean absolute error for the estimated systolic and diastolic pressure as well as for the mean arterial pressure. Therefore, the system's performance is evaluated in terms of the mean absolute error for SBP and DBP. The block diagram of the overall simulation system is shown in Fig. 5.12. Fig. 5.12 shows the cuff/artery/arm model described in Section 5.2.2, the oscillometric circuit described in Section 5.7, as well as the oscillometric algorithm

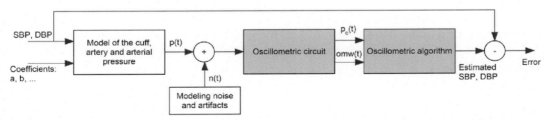

FIGURE 5.12 A model of the complete blood pressure system. The mechanical components, including the air pump, valve and cuff, are not modeled.

presented in Section 5.6. The output of the cuff/artery/arm model is the composite pressure $p(t)$ that includes the cuff pressure with superimposed arterial pulses. We can add Gaussian noise, 60 Hz interference signal and/or motion artifacts to it. This signal is connected to the pressure sensor, where it is filtered and amplified to extract the cuff pressure $p_c(t)$ and the oscillometric signal $omw(t)$. These two signals are used as an input to the oscillometric algorithm that produces estimated systolic and diastolic blood pressure. The mean square error is calculated in the end.

The model can be used to:

- Fit the parameters of the algorithm, such as c_{SBP} and c_{DBP}.
- Determine the error in ideal conditions without noise.
- Determine the performance of the system under different noise sources, including Gaussian noise as well as impulse noise that we will use to model motion artifacts.
- Determine the performance of the system under deep respiration.

In addition, the model can be used to understand better the effect of the model parameters on the output. This is done through sensitivity analysis, where the parameters of the cuff/arm/artery model, as well as the parameters of the circuit and the algorithm, are modified to determine the most influential parameters regarding accuracy.

Please note that we will do only partial analysis here and the rest will be assigned as problems to the reader. Also, for readers interested mainly in algorithm development, the simulation can be run without the oscillometric circuit by connecting the model of the cuff/artery directly to the oscillometric algorithm.

We will follow the framework introduced in Fig 1.8. The analysis is divided into multiple steps. In step B, the parameters of the models are fitted based on the data. In step C, the important parameters that affect the performance are selected through performance analysis and uncertainty propagation is done. Finally, the error analysis is done. Please note that the sensitivity analysis of the circuit will not be performed here. The MAA and other algorithms for blood pressure estimation introduce large errors by themselves; therefore, the tolerance of the parameters of the circuit does not affect the result much.

5.8.2 Fitting the coefficients

When fitting the coefficients of the algorithms, it is important to have a representative data set. As explained in Section 5.4.2, the data set should include subjects with very low and very high systolic and diastolic blood pressure values.

In our case, data is simulated. However, it should be observed that systolic and diastolic pressures are correlated with coefficients a and b. Therefore, data is generated from the multivariate Gaussian distribution in which the correlation between the variates is set. The mean of the Gaussian distribution for systolic and diastolic blood pressure and coefficients a and b are $\mu = \left[130 \text{ mmHg}, 80 \text{ mmHg}, 0.09 \text{ mmHg}^{-1}, 0.027 \text{ mmHg}^{-1}\right]^T$ and standard deviation $\sigma = \left[10 \text{ mmHg}, 5 \text{ mmHg}, 0.02 \text{ mmHg}^{-1}, 0.004 \text{ mmHg}^{-1}\right]^T$. We also assumed that a and b are correlated with a correlation coefficient of 0.75 and that systolic pressure is also negatively correlated with the coefficients a and b with a correlation coefficient of -0.6. The system is simulated without the oscillometric circuit. The coefficients of the MAA algorithm

TABLE 5.3 Accuracy of the algorithms for blood pressure estimation without including the oscillometric circuit.

Algorithm	Mean absolute error and standard error of algorithms without any noise or motion artifacts	Mean absolute error and standard error of algorithms with simulated motion artifacts and Gaussian noise
MAA	SBP error $= 4.1 \pm 1.4$ mmHg, DBP error $= 6.4 \pm 4.6$ mmHg	SBP error $= 3.8 \pm 1.9$ mmHg, DBP error $= 5.8 \pm 5.5$ mmHg
Maximum slope	SBP error $= 2.2 \pm 0.6$ mmHg, DBP error $= 5.5 \pm 5.2$ mmHg	SBP error $= 3.4 \pm 4.4$ mmHg, DBP error $= 11 \pm 13.5$ mmHg

are selected as the mean value of the obtained coefficients in case the results are known. They are, in our case, for the set that is generated $c_{SBP} = 0.65$ and $c_{DBP} = 0.61$.

5.8.3 Error analysis without artifacts

First, the error analysis is performed without any artifacts. Data for the test set is generated from a distribution in which some correlation coefficients have different values than the coefficients in the training data set used for fitting. In addition, the mean value for systolic blood pressure is increased to 140 mmHg. The errors are shown in the second column of Table 5.3. It can be seen that these algorithms are not optimal even for an ideal case without the oscillometric circuit, without noise and motion artifacts. The reason is that the algorithms do not take into account the knowledge about the model and data generation process. The maximum slope algorithm has a very small error in estimating systolic blood pressure. However, this algorithm is very sensitive to the noise in the envelope. When the circuit is included in the simulation, the accuracy is similar. The results and the details of the simulation are shown in Section 5.12.

5.8.4 Error analysis with motion artifacts

Impulse artifact can be simulated by the *sinc* function (Li et al., 2009) in the discrete domain as $a_{imp}(n) = m \cdot \sin(\pi n T_s / \eta) / (\pi n T_s)$, as described in Chapter 4. The central lobe of the *sinc* function was added to the cuff pressure signal. The duration of the artifact is determined by η. We selected $\eta = 1.5$, resulting in about 5 s long movement artifacts. The amplitude of the movement artifact is m. This type of artifact could be due to the motion of an elbow, for example, during the oscillometric measurement.

The signal and the envelope with noise and motion artifacts are shown in Fig. 5.13. Movement artifact is visible between the interval of 18–22 s. The noise is white Gaussian noise with a standard deviation of 0.4 mmHg. Extracted and filtered envelopes are shown in Fig. 5.13B. Smoothing and median filtering are performed to remove the motion artifacts from the envelope and obtain a smoother envelope. It is obvious that even if we work with the smoother envelope, the maximum slope algorithm will fail because the slope of the envelope is

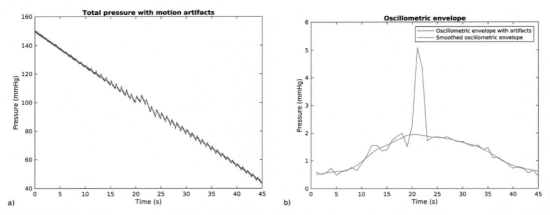

FIGURE 5.13   (A) Total pressure with simulated noise and the motion artifact. (B) Oscillometric envelope (blue line) that shows the effects of noise as well as motion artifacts around the second 20. The red line is the envelope after applying median and low pass filters.

modified. MAA algorithm fails when MAP is incorrectly detected because of the movement artifact or noise in the signal.

The results of the blood pressure estimates are shown in column 3 of Table 5.3. It is obvious that the diastolic pressure estimation accuracy is significantly affected. Many solutions have been published regarding more efficient noise and movement artifact removals. We will not discuss them here. The reader is referred to Forouzanfar et al. (2015) for the short review on motion artifact removal solutions.

5.9 Improving oscillometric methods and devices

5.9.1 Technology

There have been several innovations regarding the electronic and mechanical components of oscillometric devices. The advances are related to reducing the size of the device, innovations in the cuff design, reducing the level of noise in the system, and so on. Several of these advantages have recently been achieved with the air pump called piezoelectric micropump. TTP Ventus has recently designed the micropump for ambulatory blood pressure monitoring (TTP Ventus n.d). The micropump is small (coin size), silent, and produces no vibrations since it works on a different principle (piezoelectric) than traditional motor-driven diaphragm pumps. The micropump changes the way blood pressure is measured in several ways as it is integrated into the cuff and does not require tubing. It is silent and, therefore, can be used for ambulatory measurements during the night, and it measures blood pressure during inflations since it does not create any vibrations that can affect the oscillometric signal during the inflation.

A novel cuff with ECG dry electrodes attached to the inner side of the cuff was developed (Chen et al., 2010). It allows for the joint acquisition of BP and ECG and the use of pulse transit time to assist the oscillometric blood pressure measurements. A new cuff in the form of a wristband was developed by Omron (OMRON Healthcare, 2023). It is an inflatable cuff within the wristband, so the blood pressure monitor looks like a smartwatch. We are witnessing a number of devices with a smaller form factor, including smartphone-based devices as well as miniature wrist-based oscillometric watches.

5.9.2 Methods and processing

Potential innovations based on when during the measurement cycle the improvements can be performed are shown in Fig. 5.14. Improvements in the blood pressure monitoring methods can be made before, during, and after blood pressure measurement is performed. Before the measurement is performed, we can monitor the patient's status, including the heart rate, breathing rate, and motion. This can lead to initiating the start of the measurements not only when the device is connected and the start button is pressed, but when the device determines that the measurement can be performed based on the patient's conditions.

Additional transducers can be added to monitor adherence to the measurement recommendations. For example, they can monitor motion, heart rate, whether the patient speaks before and during the measurement, and so on. In Wagner et al. (2012), a system was developed to monitor activity and posture, and classify audio data as speech or silence during blood pressure measurements. In addition, it is possible to detect if the cuff is properly applied by developing the mathematical model for generating the signal when the cuff is tight and comparing the model parameters obtained after fitting the model by using the acquired data with the stored model parameters (Mersich and Jobbágy, 2009).

It is possible to detect motion artifacts and remove them during blood pressure measurement in real-time. This is very important because it would allow measurements while people

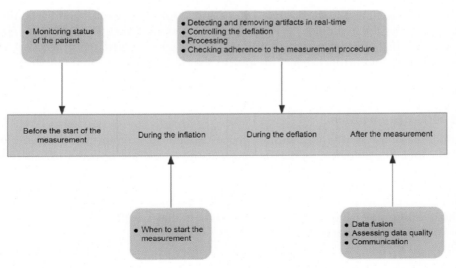

FIGURE 5.14 Potential ways to improve blood pressure measurements by analyzing the blood pressure measurement cycle.

are active and allow for performing measurements without adherence to the measurement recommendations as it is commonly done at home.

Automation of the measurements can be done by doing one of the following:

- Identifying the subject either explicitly using, for example, Near Field Communication technology on the subject's phone or algorithmically by recognizing the subject based on collected pulses and adjusting the algorithms based on the subject.
- Detecting a patient medical condition (e.g., atrial fibrillation) and adjusting the algorithm based on the conditions.
- Determining when to start the measurement.
- Checking adherence to measurement recommendations.
- Automated real-time control of the cuff.

5.9.2.1 Algorithmic innovations

The major innovation on the algorithmic side is the development of model-based algorithms for blood pressure estimation. These algorithms rely on modeling arterial pulse, tissue, and cuff and estimating parameters of the model based on acquired data (Forouzanfar et al., 2015). The oscillometric waveform does not have a uniform shape or pattern across all population groups and is affected by age and arterial stiffness. Therefore, algorithms can be developed for specific patient groups, and the parameters of the models can be learned separately for each group (Forouzanfar, 2014). An approach was proposed to adapt the algorithm to respond to the effects of increased arterial stiffness at higher pressures (Drzewiecki et al., 1994).

Further improvements on the modeling side are based on including personal information about the subject, such as age, sex, body mass index, medical history, and previous measurements.

Other sensors could potentially be used to detect patient conditions. A potential approach is shown in Fig. 5.15. In that approach, the arrhythmia is detected and potentially classified using an ECG signal, and then different processing algorithms are applied based on the type of arrhythmia. A similar problem is recognized in Amoore (2012). The following approaches can be applied if the artifacts or arrhythmias are detected: adjust the processing algorithm based on the detected artifact, reject the pulses that fall outside of predefined limits, correct the pulses corrupted by arrhythmia, or perform real-time detection of the arrhythmia and adjustment of the deflation curve.

5.9.2.2 Other innovations

An open database that would include the oscillometric waveforms, invasive blood pressure, and other physiological signals is necessary to accelerate research in this field. Many research groups collected data, but data is not shared.

Better ways to calibrate the devices need to be developed. Methods based on NIBP test simulators are not widely used yet, because the simulators are not standardized and do not contain comprehensive databases of the signals that would cover different population groups and different artifacts and arrhythmias.

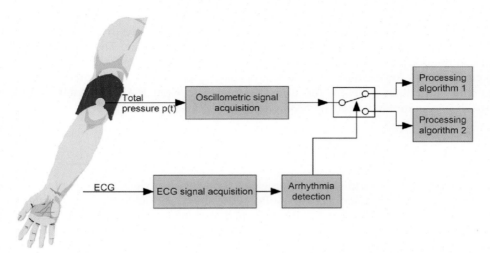

FIGURE 5.15   Potential ways to improve blood pressure measurements by selecting algorithms based on patient's conditions.

5.10 Oscillometry and arterial stiffness

This section briefly discusses arterial stiffness estimation based on oscillometric pulses. Other techniques based on tonometry, ultrasound, and magnetic resonance imaging will not be considered here. Measurements of arterial stiffness based on oscillometric methods are still not done at home and require trained technicians or specialized equipment to do the measurements. Therefore, we will introduce the method for arterial stiffness estimation but will not elaborate on it. Even though the measurement still requires expertise, we expect that this will change in the near future because it has been recognized that pharmacological interventions and lifestyle modifications reduce blood pressure and affect arterial function; therefore, it is important to monitor both blood pressure and arterial stiffness (Chen et al., 2017).

Arterial stiffness is related to the extent of hardening of the artery's inner wall. The stiffness of arteries directly affects the effort the heart makes to pump blood through the body. Large arteries in our body are the most elastic; they retain the blood during systole and then release it within diastole. In that way, they provide steady blood flow toward smaller arteries. Therefore, measuring aortic stiffness is more important than measuring stiffness in other arteries. Arterial stiffness is influenced by factors like age, high blood pressure, and atherosclerosis and is quantified through various indices as a function of pressure, diameter, volume, and pulse wave velocity (PWV).

Pulse wave velocity is defined as a velocity of a pressure wave propagating between two points in the arterial system. It is usually measured over several beats and then averaged. Higher PWV normally means higher arterial stiffness. The general consensus is that *Carotid-Femoral Pulse Wave Velocity* (CFPWV) is accepted as the most established technique for the assessment of arterial stiffness (Townsend et al., 2015; Van Bortel, 2012). CFPWV is the propagation speed at which the pressure wave propagates between the carotid and femoral arteries. CFPWV greater than 12 m/s for middle-aged hypertensive patients is regarded as an indicator of high cardiovascular risk and damage. The pulses from the two points on the body can be obtained in different ways, and one way is to use the blood pressure cuff around, for example, the neck and the leg, collect the oscillometric signals at the same time, and then measure the time difference between the oscillometric pulses. This approach was commercialized by SMT Medical (Vicorder, n.d).

5.11 Summary

In this chapter, we introduced oscillometric blood pressure devices. To explain algorithms and hardware, we first described the oscillometric signal and its features and different types of noise and artifacts. Also, we briefly described standards for oscillometric blood pressure measurements and identified requirements related to safety and accuracy. Then we introduced the basic algorithm for oscillometric blood pressure estimation and the hardware and software of a typical blood pressure monitoring device.

Modeling a blood pressure device encompasses modeling the overall system, including the oscillometric signal generation, blood pressure hardware, and algorithms. Cuff and mechanical components have not been modeled at this point. The model is implemented in Simulink and can be used for sensitivity analysis and analysis of accuracy. It can also be used for research on improving algorithms or reducing the effects of different types of artifacts on the accuracy of BP estimation.

Research trends were presented as well. They included research on hardware such as transducers, pumps and cuffs, methods when to do measurements and how, and algorithms for oscillometric blood pressure.

We also noticed that features extracted from the oscillometric pulse could be used for other applications. Therefore, at the end of the chapter, we discussed using oscillometric measurements to estimate pulse wave velocity and augmentation index that are used to infer arterial stiffness.

5.12 Appendix A

Simulation parameters and some other results of simulations for Sections 5.8.1–5.8.3 are shown here. The parameters included in the model are shown in Table 5.4.

The values of pressure and coefficients a and b are generated from a multivariate distribution of dimension 4. Here, $x = [SBP, DBP, a, b]^T$ and $\mu_x = [140 \text{ mmHg}, 80 \text{ mmHg}, 0.09 \text{ mmHg}^{-1}, 0.027 \text{ mmHg}^{-1}]^T$ and standard deviation $\sigma_x = [10 \text{ mmHg}, 5 \text{ mmHg}, 0.02 \text{ mmHg}^{-1}, 0.004 \text{ mmHg}^{-1}]^T$. Correlation coefficients are defined as follows $\rho_{12} = 0.2$, $\rho_{13} = -0.6$, $\rho_{14} = -0.6$, $\rho_{23} = -0.3$, $\rho_{24} = -0.3$, $\rho_{34} = 0.75$. These values are arbitrary and research needs to be done to select the appropriate values.

TABLE 5.4 Parameters used in simulations of blood pressure signals and other parameters.

Parameter	Definition	Value	Units
P0	Cuff pressure at the onset of deflation	SBP+30 mmHg	mmHg
r	Cuff pressure decay rate	2.5	mmHg/s
Va0	Artery segment volume at zero pressure	0.45	ml
V0	The volume of the cuff at the beginning in ml	200	ml
σ	The standard deviation of the white Gaussian noise added to the pressure signal	0.4	mmHg
$a_{imp}(n)$	Motion signal parameters	m = 5	mmHg

TABLE 5.5 Error of the algorithms when the oscillometric circuit is also included in the simulation.

Algorithm	Mean absolute error and standard error of algorithms without any noise or motion artifacts with the simulated circuit	Mean absolute error and standard error of algorithms with noise and motion artifacts with the simulated circuit
MAA	SBP error = 6.4 ± 2.0 mmHg, DBP error = 4.6 ± 3.3 mmHg	SBP error = 6.3 ± 2.2 mmHg, DBP error = 4.3 ± 3.8 mmHg
Maximum slope	SBP error = 3.2 ± 1.6 mmHg, DBP error = 3.4 ± 1.3 mmHg	SBP error = 4.2 ± 3.3 mmHg, DBP error = 10.6 ± 14 mmHg

The simulation results, including the Simscape simulation of the oscillometric circuit, are shown in Table 5.5.

5.13 Problems

5.13.1 Simple questions

5.1. What are vital signs? What is their normal range?
5.2. *Why is increasing blood pressure dangerous?
5.3. *What is blood pressure? What are systolic and diastolic BPs? Compare blood pressure in different vessels.
5.4. What is the cardiac cycle? What does the pulse look like (pulse morphology), and what features can be extracted from it?
5.5. What is a sphygmomanometer?
5.6. What is Auscultation?
5.7. Describe the process of measuring blood pressure using the cuff.

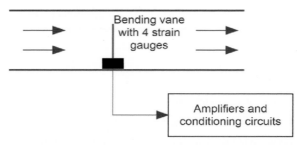

FIGURE 5.16   An example of an implementation of transducers for measuring flow that can be used to measure blood pressure by placing it in the air tube.

5.8. How does the oscillometric algorithm work?

5.9. What are the common sources of errors in oscillometric devices?

5.10. What is the acceptable error for BP devices?

5.13.2 Problems

5.11. Four strain gauges are placed on the bending vane shown in Fig. 5.16.

    (a) Please answer the following:

        i. Draw how one would place these four strain gauges on the vane.

        ii. Label these four strain gauges using labels R_1, R_2, R_3, and R_4. Then, draw the Wheatstone bridge and properly place the strain gauges R_1, R_2, R_3, and R_4 in the bridge.

        iii. Derive the sensitivity of the bridge.

    (b) Now, assume that this tube represents a tube for air release from the blood pressure cuff. Instead of the bending vane, a flexible diaphragm is used. Strain gauges are connected in the same way as in part (a). The maximum resistance change is 2 Ω. The nominal resistance is 120 Ω. The range that needs to be covered is 300 mmHg. The signal needs to be amplified after the Wheatstone bridge and connected to the A/D converter, whose input range is between 0 and 5 V. Compute the gain of the amplifier.

    (c) Draw the circuit of the oscillometric BP device that includes the components from part (b).

    (d) *Now, let us take a look at one oscillometric pulse. Draw a single pulse and explain its morphology. Below that pulse, draw an ECG pulse. Explain how these two pulses are related.

5.14. Assume that we would like to measure the blood pressure of a person whose blood pressure is commonly 140/90. Inflating the cuff takes 3 s, and deflation takes 17 s. His/her heart rate is 60 beats per minute. The signal $p(t)$ is processed using the oscillometric blood pressure device so that the oscillometric waveform is extracted.

    (a) Draw the waveforms for $p_a(t)$, $P(t)$ and extracted oscillometric waveform using Fig. 5.17. For $p_a(t)$, $p(t)$ identify minimums and maximums and explain the shape

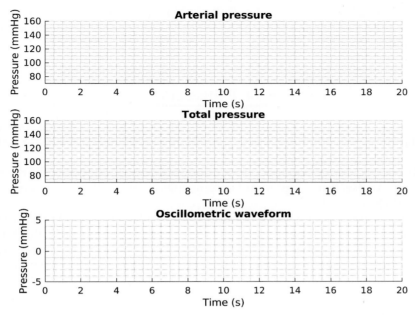

FIGURE 5.17 Fill the oscillometric, cuff pressure, and arterial pressure waveforms.

of the pulses. For the oscillometric waveform, show approximately where the maximum will occur in time, assuming that the measured blood pressure is again 140/90 mmHg.

(b) Draw a block diagram of an oscillometric blood pressure monitoring device and explain the function of each block.

5.13.3 Simulations

5.15. Start with the blood pressure system that includes the Simscape circuit on the book webpage. Assume that we are interested in simulating the blood pressure of 120/80 mmHg.

(a) Find coefficients that need to be used to generate the arterial signal with that blood pressure.

(b) Perform sensitivity analysis and discuss what parameters of the circuit affect the accuracy of the overall system the most.

(c) Simulate the effects of motion artifacts on the blood pressure estimation accuracy. You can model the motion as a 1–2 s pulse superimposed on the cuff pressure. This pulse will, in most cases, cause an error in estimating blood pressure.

5.16. Determine the sensitivity of the blood pressure estimation algorithm to

(a) Respiration, where the respiration signal modulates the pulse arterial signal. For example, it can be simulated as a sine wave with a frequency of 0.3 Hz.

(b) Change in the precision of the pressure transducer.

5.17. Change coefficients a and b in Eq. (5.3) while keeping the blood pressure constant. Explore how the changes in compliance affect the accuracy of the blood pressure algorithm.

5.14 Further reading

In this section, we refer to several relevant resources for further reading. However, please note that this list is not comprehensive and that many excellent and relevant resources might have been omitted unintentionally.

A detailed description of different types of pressure sensors is given in the report (Pressure sensors, 2020). An excellent chapter on instrumentation for oscillometric blood pressure can be found in Webster (2010).

A high-level description of models developed for oscillometric blood pressure is described in Forouzanfar et al. (2015). We recommend (Balestrieri and Rapuano, 2010) for analysis of the calibration of biomedical devices. Excellent work on understanding blood pressure algorithms and adjusting them based on physical models is presented in Chandrasekhar et al. (2019).

References

References related to blood pressure

Abderahman, H.N., Dajani, H.R., Bolic, M., Groza, V.Z., 2017. An integrated blood pressure measurement system for suppression of motion artifacts. Computer Methods and Programs in Biomedicine 145, 1—10. Elsevier.

Adams, C.E., Leverland, M.B., 1985. Environmental and behavioral factors that can affect blood pressure. Nurse Practical 10, 39—40.

Amoore, J.N., 2012. Oscillometric sphygmomanometers: a critical appraisal of current technology Blood pressure monitoring. Blood Pressure Monitoring 17 (2), 80—88.

American Standard for Electronic and Automated Sphygmomanometers, 1987. ANSI/AAMI SP10 Standard, Washington, DC.

American Standard for Electronic and Automated Sphygmomanometers, 1993. ANSI/AAMI SP10 Standard, Arlington, TX.

Pressure Sensors: The Design Engineer's Guide, 2020. Avnet-Abacus, Report.

Babbs, C.F., 2012. Oscillometric measurement of systolic and diastolic blood pressures validated in a physiologic mathematical model. BioMedical Engineering Online 11 (56).

Baker, P.D., Westenskow, D.R., Kuck, K., 1997. Theoretical analysis of non-invasive oscillometric maximum amplitude algorithm for estimating mean blood pressure. Medical and Biological Engineering 35, 271—278.

Ball-llovera, A., 2003. An experience in implementing the oscillometric algorithm for the non-invasive determination of human blood pressure. In: Proceeding of the 25th Annual International Conference of the IEEE Engineering in Medicine and Biology Society, Cancun, Mexico.

Balestrieri, E., Rapuano, S., 2010. Instruments and methods for calibration of oscillometric blood pressure measurement devices. IEEE Transactions on Instrumentation and Measurement 59 (9), 2391—2404.

Baura, G., 2011. Medical Device Technologies: A Systems Based Overview Using Engineering Standards. Academic Press.

Bruno, S., Scalart, P., 2005. Estimation of cardiac and respiratory rhythms based on an AMFM demodulation and an adaptive eigenvector decomposition. In: 13th European Signal Processing Conference, Antalya, Turkey.

Chandrasekhar, A., Yavarimanesh, M., Hahn, J.O., Sung, S.H., Chen, C.H., Cheng, H.M., Mukkamala, R., 2019. Formulas to explain popular oscillometric blood pressure estimation algorithms. Frontiers in Physiology 10. https://www.frontiersin.org/articles/10.3389/fphys.2019.01415/full. (Accessed 12 November 2020).

Charbonnier, S., Siche, J.P., Vancura, R., 2000. On-line Detection of Motion Artifacts to Improve Ambulatory Blood Pressure Monitoring". IEEE Instrumentation and Measurement Technology Conf, pp. 693—700.

Chen, S., Bolic, M., Groza, V.Z., Dajani, H.R., Batkin, I., Rajan, S., 2010. Improvement of Oscillometric Blood Pressure Estimates through Suppression of Breathing Effects. I2MTC.

DeMers, D., Wachs, D., 2019. Physiology, mean arterial pressure [Updated 2019 Feb 24]. In: StatPearls [Internet]. StatPearls Publishing, Treasure Island (FL). Available from. https://www.ncbi.nlm.nih.gov/books/NBK538226/.

Drzewiecki, G.M., Hood, R., Apple, H., 1994. Theory of the oscillometric maximum and the systolic and diastolic detection ratios. Annals of Biomedical Engineering 22, 88—96.

Forouzanfar, M., 2014. A Modeling Approach for Coefficient-free Oscillometric Blood Pressure Estimation. Ph.D. thesis, University of Ottawa, Canada.

Forouzanfar, M., Dajani, H.R., Groza, V.Z., Bolic, M., Rajan, S., Batkin, I., 2015. Oscillometric blood pressure estimation: past, present, and future. IEEE Reviews of Biomedical Engineering 8, 44—63.

Geddes, L.A., Voelz, M., Combs, C., Reiner, D., Babbs, C.F., 1982. Characterization of the oscillometric method for measuring indirect blood pressure. Annals of Biomedical Engineering 10, 271—280.

Hansen, S., Staber, M., 2006. Oscillometric blood pressure measurement used for calibration of the arterial tonometry method contributes significantly to error. European Journal of Anaesthesiology 23, 781787.

HeartGuide™. Blood pressure anytime, anywhere. https://omronhealthcare.com/products/heartguide-wearable-blood-pressure-monitor-bp8000m/. (Accessed 02 April 2023).

Honeywell Datasheet, 2014. TruStability® board mount pressure sensors: SSC series—standard accuracy. Compensated/Amplified. https://prod-edam.honeywell.com/content/dam/honeywell-edam/sps/siot/en-ca/products/sensors/pressure-sensors/board-mount-pressure-sensors/trustability-ssc-series/documents/sps-siot-trustability-ssc-series-standard-accuracy-board-mount-pressure-sensors-50099533-a-en-ciid-151134.pdf. (Accessed 02 April 2023).

Iamratanakul, S., McNames, J., Goldstein, B., 2003. Estimation of respiration from physiologic pressure signals. In: Proceedings of the 25th Annual International Conference of the IEEE Engineering in Medicine and Biology Society, Cancun, Mexico, pp. 2734—2737.

Jazbinsek, V., Luznik, J., Trontelj, Z., 2005. Non-Invasive Blood Pressure Measurements: Separation of the Arterial Pressure Oscillometric Waveform from the Deflation Using Digital Filtering. IFBME Proceedings of EMBEC'05, Prague, Czech Republic.

Jilek, J., "Physiology of Oscillometric Blood Pressure Measurement," (unpublished).

Jones, D.W., Hall, J.E., 2008. Hypertension: pathways to success," Hypertension. Editorial 51 (5), 1249—1251.

Lan, H., Al-Jumaily, A.M., Lowe, A., Hing, W., 2011. Effect of tissue mechanical properties on cuff-based blood pressure measurements. Medical Engineering & Physics. 33, 1287—1292.

Lewington, S., Clarke R, R., Qizilbash, N., Peto, R., Collins, R., 2002. Prospective Studies Collaboration. Age-specific relevance of usual blood pressure to vascular mortality: a meta-analysis of individual data for one million adults in 61 prospective studies. Lancet 360, 1903—1913.

Li, Q., Mark, R.G., Clifford, G.D., 2009. Artificial arterial blood pressure artifact models and an evaluation of a robust blood pressure and heart rate estimator. BioMedical Engineering Online 8 (13). https://biomedical-engineering-online.biomedcentral.com/articles/10.1186/1475-925X-8-13. (Accessed 15 November 2020).

Lin, C.T., Liu, S.H., Wang, J.J., Wen, Z.C., 2003. Reduction of interference in oscillometric arterial blood pressure measurement using fuzzy logic. IEEE Transactions on Biomedical Engineering 50, 432—441.

Manual, Electronic, or Automated Sphygmomanometers, 2002. ANSI/AAMI SP10 Standard, Arlington, TX.

Marey, E.J., 1876. Pression et vitesse du sang. Physiologie Experimentale 2, 307—343.

Medero, R., 1996. Determination of oscillometric blood pressure by linear approximation. US Patent 5 577 508.

Mersich, A., Jobbágy, Á., 2009. Identification of the cuff transfer function increases indirect blood pressure measurement accuracy. Journal of Physiological Measurement 30 (3), 323—333.

Monk, S., 2010. 10 factors that can affect blood pressure readings. Blog at SunTech Medical. https://www.suntechmed.com/blog/entry/bp-measurement/10-factors-that-can-affect-blood-pressure-readings (Accessed 02 April 2023).

Moraes, J.C.T.B., Cerulli, M., Ng, P.S., 1999. Development of a New Oscillometric Blood Pressure Measurement System. IEEE Computers and Cardiology, Hannover, Germany, pp. 467—470.

Muntner, P., et al., 2019. Measurement of blood pressure in humans: a scientific statement from the American heart association. Hypertension 73 (No. 5), e35–e66.

Ng, K.G., Small, C.F., 1994. Survey of automated noninvasive blood pressure monitors. Journal of Clinical Engineering 33, 452–475.

Nichols, W.W., O'Rourke, M.F., Vlachopoulos, C., 2011. McDonald's Blood Flow in Arteries: Theoretical, Experimental and Clinical Principles, sixth ed. Hodder Arnold Publishers, London, UK.

Non-invasive Sphygmomanometers — Part 2: Clinical Validation of Automated Measurement Type, 2009. ANSI/AAMI/ISO 81060-2 Standard, Arlington, TX.

Non-Invasive Sphygmomanometers — Part 2: Clinical Investigation of Automated Measurement Type. American National Standards Institute. ANSI/AAMI/ISO 81060–2:2019, 2019. International Organization for Standardization, Geneva.

O'Brien, E., Petrie, J., Littler, W., de Swiet, M., Padfield, P.L., O'Malley, K., Jamieson, M., Altman, D., Bland, M., Atkins, N., 1990. The British Hypertension Society protocol for the evaluation of automated and semi-automated blood pressure measuring devices with special reference to ambulatory systems. Journal of Hypertension 8, 607–619.

O'Brien, E., Petrie, J., Littler, W., de Swiet, M., Padfield, P.L., Altman, D., Bland, M., coats, A., Atkins, N., 1993. The British Hypertension Society protocol for the evaluation of blood pressure measuring devices. Journal of Hypertension 11, S43–S62.

O'Brien, E., et al., 2003. European Society of Hypertension recommendations for conventional, ambulatory and home blood pressure measurement. Journal of Hypertension (issue 2), 821–848.

O'Brien, E., Atkins, N., Stergiou, G., Karpettas, N., Parati, G., Asmar, R., Imai, Y., Wang, J., Mengden, T., Shennan, A., 2010. European society of hypertension international protocol revision 2010 for the validation of blood pressure measuring devices in adults. Blood Pressure Monitor vol 15, 23–38.

Ramsey III, M., 1991. Blood pressure monitoring: automated oscillometric devices. Journal of Clinical Monitoring and Computing 7, 55–67.

Sorvoja, H., Myllyla, R., Karja-Koskenkari, P., Koskenkari, J., Lilja, M., Kesaniemi, A., 2005. Accuracy comparison of oscillometric and electronic palpation blood pressure measuring methods using intra-arterial method as a reference. Molecular and Quantum Acoustics 26, 235–260.

Stergiou, G.S., Palatini, P., Asmar, R., et al., 2019. Recommendations and practical guidance for performing and reporting validation studies according to the universal standard for the validation of blood pressure measuring devices by the association for the advancement of medical instrumentation/European society of hypertension/international organization for standardization (AAMI/ESH/ISO). Journal of Hypertension 37 (3), 459–466.

Such, O., Muehlsteff, J., 2006. The challenge of motion artifact suppression in wearable monitoring solutions. In: IEEE Engineering in Medicine and Biology Society, International Summer School on Medical Devices and Biosensors, pp. 49–52.

Taylor, C.R., Lillis, C., LeMone, P., Lynn, P., 2010. Fundamentals of Nursing: The Art and Science of Nursing Care, seventh ed. Lippincott Williams and Wilkins, Baltimore, MD.

TTP Ventus, "Disc Pump: a better solution for Ambulatory Blood Pressure Monitoring," https://www.ttpventus.com/applications/ambulatory-blood-pressure-monitoring, accessed on October 31, 2019.

Wagner, S., et al., 2012. Context Classification during Blood Pressure Self-Measurement Using the Sensor Seat and the Audio Classification Device. Pervasive Health Workshop.

Webster, J.G. (Ed.), 2010. Medical Instrumentation: Application and Design, fourth ed. John Wiley and Sons.

References related to arterial stiffness

Chen, Y., Shen1, F., Liu1, J., Yang, G.Y., 2017. Arterial stiffness and stroke: de-stiffening strategy, a therapeutic target for stroke. Stroke and Vascular Neurology 2 (Issue 2), 62–72.

Townsend, R.R., et al., 2015. Recommendations for improving and standardizing vascular research on arterial stiffness: a scientific statement from the American Heart Association. Hypertension 66, 698–722.

Van Bortel, L.M., 2012. Expert consensus document on the measurement of aortic stiffness in daily practice using carotid-femoral pulse wave velocity. Journal of Hypertension 30, 445–448.

Vicorder, Cardio-vascular and Peripheral-vascular Testing, https://www.smt-medical.com/, last accessed on November 12, 2019.

Devices based on photoplethysmogram and pulse oximetry

Acronyms and explanations

APG	acceleration plethysmogram
bpm	beats per minute
LED	light-emitting diode
PPG	photoplethysmogram
pSQI	spectral power ratio signal quality index (SQI)
PWM	pulse width modulation
SaO$_2$	arterial blood oxygen saturation
SDDVP	second derivative of digital volume pulse
SDR	spectral distribution ratio
SpO$_2$	pulse oximeter-based oxygen saturation

Variables used in this chapter include:

Ab	absorbance
B_0	total blood fraction
B_i	fraction of ith substance in the tissue
c	concentration
D	pulse width offset
d_i	the thickness of the ith layer of the tissue
I	the intensity of the detected light
$I_{dia}^{(\lambda)}, I_{sys}^{(\lambda)}$	the intensity of light for a particular wavelength detected during systole or diastole
l	optical path
P	power
$P(f)$	power spectral density
R	ratio of AC and DC components at the two wavelengths
$s_{PPG}(t)$	PPG signal
T	PWM period
W	content of water in the tissue

λ wavelength
μ_a absorption coefficient
μ_s scattering coefficient
$\varepsilon(\lambda)$ molar extinction coefficient

6.1 Introduction

In this chapter, we analyze systems based on a photoplethysmography signal. These systems require photons of light to penetrate tissue and interact with the tissue through absorption and scattering. Then, the results of the interaction are observed as the photons propagate out of the tissue. Normally, light emitting diodes (LEDs) are used to generate light of certain intensity and wavelength, while photodetectors are used to detect the light's intensity at the skin's surface. As the light propagates through the tissue, it is differently absorbed depending on if the blood is oxygenated or not during the cardiac cycle. The time-varying component of the signal at the output of the photodetector is proportional to the blood volume changes in the skin under the probe containing the light source and the detector. This signal (that is further amplified and filtered) is called a *photoplethysmogram* (PPG) *signal*. A PPG device called a *photoplethysmograph* measures the PPG signal. PPG systems are commonly used nowadays in smartwatches to obtain the heart rate.

PPG signals obtained using two or more LEDs of different wavelengths can be used to infer an oxygen saturation level in blood. A *pulse oximeter* is a device that noninvasively measures and estimates the level of oxygen-saturated hemoglobin in a subject's arterial blood. It is, in general, based on Beer—Lambert law, which states that the intensity of light transmitted through a solution depends on the concentration of the elements in the solution.

Acquired PPG signal can be used for a number of other applications besides measuring heart rate and oxygen saturation. PPG systems were integrated into a finger cuff and used for continuous measurement of blood pressure in the finger. There has also been a surge of research works on using only a PPG sensor for cuff-less estimation of blood pressure. In addition, pulse wave velocity is then used to estimate arterial stiffness, which is an important indicator of cardiovascular health. Pulse wave velocity can be measured by observing the time difference between signals obtained from two distant sites on the body. Multisite PPG measurements on the body have been proposed for peripheral vascular disease detection. PPG probe can also be placed directly on the top of a particular blood vessel, such as the jugular vein, to obtain the signal from it. Important observations can be made by observing the morphology of the PPG signal.

In this chapter, we first introduce PPG devices and pulse oximeters. Then, we present our models and show simulation results for (1) light propagation through the tissue and (2) an electronic circuit for PPG devices. It is very important to understand light propagation through the tissue when designing a new PPG device to know what tissue layers are illuminated by the LED, how to position the emitter and the detector depending on the measurement site, and what kind of signal is expected at the photodetector. This knowledge could also contribute to developing robust algorithms that do not depend on a subject's skin color and health conditions—however, this is not covered in this chapter.

The major difficulty when using PPG devices is motion artifacts. Besides motion artifacts, pulse oximeters are inaccurate in poorly oxygenated patients. This is because there is a lack of data for fitting the parameters of the models used in pulse oximeters for patients whose oxygen saturation levels are below 70%.

In this chapter, we will try to answer several questions, including.

- How does the light propagate through the tissue?
- How much light is picked up by the photodetector?
- What does a pulse oximeter measure?
- What does a photoplethysmograph measure?

The material the author covers in a basic bioinstrumentation course includes Sections 6.2.1–6.2.3, 6.3, 6.6, 6.7.2, and 6.8. Therefore, the reader can skip some parts of the chapter depending on her interest. For example, the simulation of light propagation through the tissue in Section 6.5 can be skipped without significantly impacting the understanding of the rest of the chapter. The same is with performance measures in Section 6.4. The other material should be useful for readers interested in understanding how the overall system works and/ or in simulating and designing the overall system.

6.2 What is measured using pulse oximeters and PPG devices?

In this section, we start with the definitions of SpO_2, followed by a high-level explanation of how PPG and pulse oximeter devices function. Beer–Lambert law is introduced, showing how one can compute SaO_2. The fact that multiple layers of human tissue are not considered and that the scattering of light is not included in Beer–Lambert law makes it not directly applicable in reflectance pulse oximetry. We introduce terminology needed to discuss light propagation through the tissue. We also present the layers of human tissue and introduce the optical properties of tissue.

6.2.1 PPG and oxygen saturation

6.2.1.1 Oxygen saturation

Hemoglobin is the main light absorber in human blood at the wavelengths used in pulse oximetry. Hemoglobin binds oxygen in the pulmonary capillaries and releases it in the systemic capillaries. When hemoglobin is fully saturated with oxygen (carrying four oxygen molecules), it is called *oxyhemoglobin (HbO_2)*. If it is not fully saturated with oxygen, it is called *reduced* or *deoxygenated hemoglobin (Hb)*. An oxygenated hemoglobin molecule is bright red, while a deoxygenated hemoglobin molecule is dark red. This color change is an important feature used in pulse oximetry to measure hemoglobin oxygen saturation. Each red blood cell contains approximately 265 million molecules of hemoglobin (Webster, 1997).

Arterial blood's oxygen saturation, SaO_2, is the percentage of hemoglobin in arterial blood bound with oxygen. It is normally measured in vitro on blood samples using a device called a co-oximeter.

SpO$_2$ represents the percentage of hemoglobin in arterial blood that is bound with oxygen, and it is measured using a pulse oximeter. A *pulse oximeter* is a noninvasive device that indirectly measures SpO$_2$ by generating light and then observing the light on the surface of the skin after interacting with the tissue. Oxygen saturation is normally larger than 95% (Box 6.1).

6.2.1.2 PPG

PPG measures relative changes in blood volume in blood vessels under the probe. This definition assumes that the light will propagate deep enough to get absorbed by the pulsating arterioles or arteries in our body. Our simulation in Section 6.5 assumes that PPG devices measure the blood volume changes. However, it has been pointed out recently that displacement of capillaries could be the origin of the PPG signal—please see Box 6.2.

The basic components of a PPG device are a light source or emitter and a photodetector. The light source is commonly an LED. The light that the LED produces should have narrow bandwidth and high-enough power so that the PPG device can acquire a good quality signal at a photodetector. The photodetector should have a fast response time and be sensitive to the wavelength of the LED.

PPG devices operate in transmittance or reflectance modes. In the *transmittance mode*, the photodetector and the LED are on the opposite sides of the tissue that is illuminated, and the light is transmitted through the tissue. In *reflectance mode*, the photodetector and the LED are on the same side of the tissue and the origin of the detected signal is from the scattered and reflected light. There are only several measurement sites on the body, such as the finger and the ear, where the transmittance mode can be used because these sites need to be thin enough so that the light is not completely absorbed in the tissue. In this chapter, we will focus on the reflectance mode PPG. Although transmittance mode PPG devices and pulse oximeters are mostly used in clinical applications, reflectance mode-based devices have an advantage because they can be attached almost anywhere on the body. Measurement sites can be external and internal. External sites include the forehead, wrist, arm, chest, and others, while

---

BOX 6.1

Oxygen saturation and COVID-19

Normal range for oxygen saturation is more than 95%. Low oxygen levels in the body are known as hypoxia. Hypoxia can lead to shortness of breath, and it is one of the known symptoms of COVID-19. In a recent study (O'Carroll et al., 2020), 18 COVID-19 patients, who were discharged from a hospital, measured their oxygen saturation levels at home for about 12 days. The system was set to alarm when the oxygen saturation is below 94% which happened 5.5% of the time for this group of patients. The National Institutes of Health recommended maintaining oxygen saturation levels at 92%—96% for COVID-19 patients (COVID-19 Treatment Guidelines Panel).

BOX 6.2

Controversies regarding the PPG origin

Recent works question the origin of the PPG signal, especially the signal measured in the reflectance mode. It was pointed out in Bashkatov et al. (2005) that red light can penetrate up to 1.8 mm. Up to that distance, most of the light is either absorbed or scattered by the capillaries. On the other hand, it was shown that capillaries do not change their volume much during the cardiac cycle (Fung et al., 1966). Therefore, it is claimed in Volkov et al. (2017) that volume change cannot be the reason for generating the PPG signal in many sites on the body. It was

proposed in Volkov et al. (2017) that displacement of capillaries during the cardiac cycle causes the AC component of the signal.

It was also discovered that the absorbance change between systole and diastole is due to a change in the axis of red blood cells. They are aligned parallel to the direction of blood flow during diastole and perpendicular to the direction of flow during systole, which causes variation in the absorbance and reflectance of blood in motion within the cardiac cycle (Webster, 1997).

---

internal sites include the mouth mucosa and gums. Smartwatches only use reflectance mode for acquiring the PPG signal (Box 6.3).

Fig. 6.1 shows the results of our simulation of the light propagating through the tissue. Details of the simulation are explained, and simulation results are shown in Section 6.5. Fig. 6.1A presents the fluence rate (defined in Section 2.5.1) in the vertical z dimension divided by the

---

BOX 6.3

Clinical applications of PPG

PPG has been applied in many different clinical settings, including physiological monitoring, vascular assessment, and evaluation of autonomic functions (Allen, 2007). In this chapter, we are focusing on physiological monitoring where PPG has been used for monitoring heart rate, blood pressure, and respiration. Blood pressure can be measured by using a finger-based device that uses an inflatable finger cuff with an

integrated PPG sensor (Peñáz, 1973). We will mention a new application of PPG for continuous cuff-less monitoring of blood pressure in Section 6.9.5.1. Vascular assessment includes the assessment of arterial disease, arterial compliance and aging, endothelial function, and venous assessment. Autonomic function includes the estimation of blood pressure and heart rate variability.

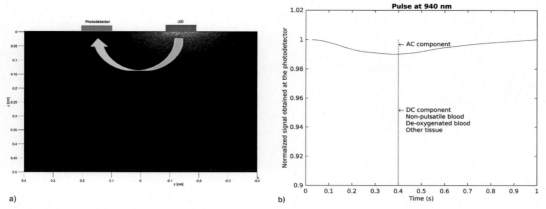

FIGURE 6.1 (A) Simulated normalized fluence rate $[(W/cm^2)/W]$ through the tissue (the dark area). LED and photodetector are placed on the skin surface. Their positions and dimensions correspond to the dimensions used in the simulation. Z-axis shows the depth of light illumination starting from zero at the top and y-axis shows the relative position of the components in cm. (B) Normalized and smoothed pulse obtained at the photodetector from our simulations. The signal is normalized between 0 and 1; however, we present only the range between 0.9 and 1 to make AC component of the signal visible. The figure is obtained by using the software package MCMatlab (Marti et al., 2018) and running a modified example provided by MCMatlab.

incident light power. This is why it is referred to as the normalized fluence rate. A yellow arrow points in the direction of scattered light. A very small percentage of light is actually picked up by the photodetector in the reflectance mode. Out of the light collected by the photodetector, more than 98% is the DC component that does not change during the cardiac cycle. The *DC component* of the light is caused by venous pulsations that are weak, as well as scattering and reflections from all other tissues except for the arterial blood. DC component is relatively constant for a particular wavelength. The *AC component* represents the pulsating change mainly caused by blood volume changes during the systole. The amplitude of the AC component is generally up to 1%–2% of the DC component, and the AC component is superimposed on the DC component, as shown in Fig. 6.1B. Sometimes, the DC component is referred to as $I_{dia}^{(\lambda)}$ where λ represents the wavelength of interest, *dia* refers to the diastole part of the cardiac cycle, and I is the intensity of the detected light. AC components is then computed as $I_{sys}^{(\lambda)} - I_{dia}^{(\lambda)}$ where *sys* represent systole part of the cardiac cycle.

The block diagram of a PPG device is shown in Fig. 6.2. Besides the LED and the photodetector, the system includes amplifiers and filters to process the signal from the photodetector. In addition, it includes driving and timing control circuits for the LED. These components will be explained in detail in later sections.

Monitoring sites for reflectance mode PPG have been analyzed by Nilsson et al. (2007) based on the ability to obtain heart and breathing rates. It was shown that an accurate heart rate could be obtained by placing the probe on the shoulder; a moderate accuracy heart rate can be obtained from the wrist and forehead, while low accuracy heart rate can be obtained from the forearm.

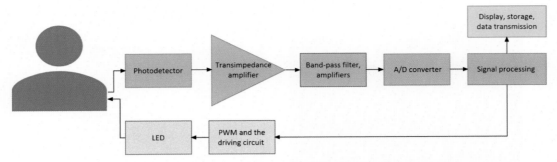

FIGURE 6.2   Block diagram of a PPG device. PWM refers to pulse width modulation circuit.

6.2.1.3 *Pulse oximetry*

Pulse oximetry indirectly measures oxygen saturation SpO_2 (Box 6.4). It is based on using at least two emitters (LEDs). Commonly used emitters in practice illuminate light in the red (R) and near-infrared (IR) spectrums. One photodetector is normally used, requiring that LEDs are alternately turned on and off in a time-multiplexed fashion. Red and infrared lights are absorbed differently by HbO$_2$ and Hb; therefore, the AC signals obtained after illuminating the tissue with these two emitters will differ. In addition, at these wavelengths, the DC components of light are different too. Therefore, SpO_2 is determined by computing first the ratio R of AC and DC components at the two wavelengths:

$$R = \frac{1 - \dfrac{I_{sys}^{(\lambda_R)}}{I_{dia}^{(\lambda_R)}}}{1 - \dfrac{I_{sys}^{(\lambda_{IR})}}{I_{dia}^{(\lambda_{IR})}}} = \frac{\dfrac{AC_R}{DC_R}}{\dfrac{AC_{IR}}{DC_{IR}}} \tag{6.1}$$

Very often, R is called the *ratio of ratios*, where the ratios are $(AC_R)/(DC_R)$ and $(AC_{IR})/(DC_{IR})$. After R is computed, SpO_2 is estimated mainly using regression. An example of linear regression is shown below:

$$SpO_2 = C_A + C_B \cdot R \tag{6.2}$$

BOX 6.4

C l i n i c a l   a p p l i c a t i o n s   o f   p u l s e   o x i m e t e r s

Oxygen saturation is considered to be one of five vital signs. It has been used nowadays in many clinical applications, including evaluating the depth of anesthesia, neonatal care, and sleep research.

---

where C_A and C_B are empirically determined constants. The reasoning behind Expressions in Eqs. (6.1) and (6.2) will be explained next.

6.2.2 Oxygen saturation computed based on the Beer–Lambert law

The *Beer–Lambert law* has been a foundation for developing pulse oximeters. The Beer–Lambert law is detailed in biomedical instrumentation textbooks, including Webster (2010). The goal of this section is to show how one can compute the ratio of ratios R presented in Eq. (6.1) and what the physical relationship between the ratio of ratios and oxygen saturation is.

Beer–Lambert law describes how the light is attenuated by a nonscattering absorbing medium. According to Beer–Lambert law, after the monochromatic light with the light intensity I_0 enters the medium, one part is absorbed by the medium while the other part is transmitted through the medium. The transmitted light intensity I decreases with the distance l as

$$I = I_0 e^{-\varepsilon(\lambda)cl} \tag{6.3}$$

where $\varepsilon(\lambda)$ is called the molar extinction coefficient (also called molar absorptivity). The unit of the molar extinction coefficient is $l \cdot \text{mol}^{-1} \cdot \text{cm}^{-1}$. The larger values indicate a stronger absorptive capacity of the material at the specific wavelength of light. Variable c is the molar concentration of the medium, and it is expressed in mol/l; l is the optical path (distance) that the light traveled in the medium.

The light intensity decreases with the distance because of the absorbance. Absorbance (Ab) in this process is defined as

$$Ab(\lambda) = \ln\left(\frac{I_0}{I}\right) = \varepsilon(\lambda)cl \tag{6.4}$$

The absorbance is unitless. The absorbance of 0 means that all the light passes through the medium (no absorption). The absorbance of 1 means that about 10% of the light was transmitted through the medium.

In a complex medium, each of n components of the substance will affect the absorbance. In this case, the absorbance is computed as

$$Ab(\lambda) = \sum_{i=1}^{n} \varepsilon_i(\lambda)c_i l \tag{6.5}$$

where $\varepsilon_i(\lambda)$ and c_i are extinction coefficient and the concentration of the ith component of the medium material, respectively.

Let us consider blood as a complex medium that contains only hemoglobin and reduced hemoglobin, and let us illuminate the blood with red (R) and infrared (IR) light. The absorbance is then:

$$Ab(\lambda_R) = \ln\left(\frac{I_0^{(\lambda_R)}}{I^{(\lambda_R)}}\right) = \varepsilon_{Hb}(\lambda_R)c_{Hb}l + \varepsilon_{HbO_2}(\lambda_R)c_{HbO_2}l \tag{6.6}$$

and

$$Ab(\lambda_{IR}) = \ln\left(\frac{I_0^{(\lambda_{IR})}}{I^{(\lambda_{IR})}}\right) = \varepsilon_{Hb}(\lambda_{IR})c_{Hb}l + \varepsilon_{HbO_2}(\lambda_{IR})c_{HbO_2}l \tag{6.7}$$

where c_{Hb} and c_{HbO_2} are the concentration of reduced hemoglobin and oxyhemoglobin in blood, respectively. Here we ignored all other components in the blood and considered only hemoglobin and reduced hemoglobin. This is one of the approximations that will make this technique inapplicable in pulse oximeters.

We also know, based on the definition of oxygen saturation in Section 6.2.1.1 that

$$SaO_2 = \frac{c_{HbO_2}}{c_{Hb} + c_{HbO_2}} \tag{6.8}$$

From there, we can find that

$$c_{Hb}/c_{HbO_2} = -1 + 1/SaO_2 \tag{6.9}$$

Let us now take a look at the change in the transmitted light intensity over time. We will start by taking the first derivative of Expression (6.3).

$$\frac{dI}{dt} = I_0 e^{-\varepsilon(\lambda)cl} \cdot (-\varepsilon(\lambda))c\frac{dl}{dt} = -I\varepsilon(\lambda)c\frac{dl}{dt} \tag{6.10}$$

From Eq. (6.10), we can see that the ratio of change of intensity over time versus the intensity of light at the output can be computed as $(dI/dt)/I = -\varepsilon(\lambda)c\, dl/dt$.

Let us also define the ratio of the ratios as

$$R = \frac{\dfrac{dI^{(\lambda_R)}}{dt}/I^{(\lambda_R)}}{\dfrac{dI^{(\lambda_{IR})}}{dt}/I^{(\lambda_{IR})}} \tag{6.11}$$

In systole, the arteries contain more blood than the amount of blood during the diastole. The vessel diameter of arteries also increases because of the rise in blood pressure. The increase in skin tissue absorbance is primarily due to increased red blood cells in the arteries. We can further approximate the derivatives in some time instances T_1 and T_2 as $dI^{(\lambda_R)}/dt = \left(I_{T_2}^{(\lambda_R)} - I_{T_1}^{(\lambda_R)}\right)/(T_2 - T_1)$ and $dI^{(\lambda_{IR})}/dt = \left(I_{T_2}^{(\lambda_{IR})} - I_{T_1}^{(\lambda_{IR})}\right)/(T_2 - T_1)$. Now, let us assume that T_2 is the instant of the minimum of the light intensity during the systole and T_1 is the maximum of the light intensity during the diastole. In this case, $I_{T_2}^{(\lambda_R)} = I_{sys}^{(\lambda_R)}$ and $I_{T_1}^{(\lambda_R)} = I_{dia}^{(\lambda_R)}$. Therefore the difference between systolic and diastolic components represents the AC components of the pulse. In this case, we obtain that the ratio presented in Eq. (6.11) is the same as the ratio we defined in Eq. (6.1):

$$R = \frac{\dfrac{I_{dia}^{(\lambda_R)} - I_{sys}^{(\lambda_R)}}{I_{dia}^{(\lambda_R)}}}{\dfrac{I_{dia}^{(\lambda_{IR})} - I_{sys}^{(\lambda_{IR})}}{I_{dia}^{(\lambda_{IR})}}} = \frac{\dfrac{AC_R}{DC_R}}{\dfrac{AC_{IR}}{DC_{IR}}} \tag{6.12}$$

Here, $(T_2 - T_1)$ in the nominator and the denominator canceled out. Several assumptions are made during this derivation. First of all, the approximation of the first derivative with the difference in intensities divided by $(T_2 - T_1)$ is valid only is $(T_2 - T_1)$ is small, which is not the case if we select these points to be the minimum and the maximum of the pulse. In addition, we assumed that dI/dt is the same for both wavelengths, which might not be the case in practice. $I^{(\lambda_R)}$ and $I^{(\lambda_{IR})}$ are replaced with the light intensities at the diastolic instances.

By computing intensities of light from Eqs. (6.6) and (6.7), taking the first derivative like in Eq. (6.10) and plugging it in Eq. (6.12) and assuming that dI/dt is the same for both wavelengths, we get

$$R = \frac{\varepsilon_{Hb}(\lambda_R)c_{Hb} + \varepsilon_{HbO_2}(\lambda_R)c_{HbO_2}}{\varepsilon_{Hb}(\lambda_{IR})c_{Hb} + \varepsilon_{HbO_2}(\lambda_{IR})c_{HbO_2}} \tag{6.13}$$

After dividing both the nominator and denominator with c_{HbO_2} and using Expression (6.9) to replace the ratio of concentrations with SaO$_2$, we obtain the relationship between the ratio R and the saturation SaO$_2$ as

$$SaO_2 = \frac{\varepsilon_{Hb}(\lambda_R) - R\varepsilon_{Hb}(\lambda_{IR})}{R(\varepsilon_{HbO_2}(\lambda_{IR}) - \varepsilon_{Hb}(\lambda_{IR})) - (\varepsilon_{HbO_2}(\lambda_R) - \varepsilon_{Hb}(\lambda_R))} \tag{6.14}$$

6.2.2.1 Challenges in directly applying the Beer–Lambert law in reflectance-mode pulse oximetry

The Beer–Lambert law considers only absorption by the medium and, therefore, cannot be directly applied to the reflectance-mode pulse oximetry. In addition, the absorbance of light is not simply proportional to the concentration of hemoglobin or the length of the optical path. Multiple scattering causes the light to take paths of different lengths. This is shown in Fig. 6.3, where the skin is simulated using six layers. The simulation setup is shown in Fig. 6.1 and further explained in Section 6.5. Photon paths are shown using black lines. It is obvious that multiple scattering events occur and that different photons take paths of different path lengths. Several works expanded the Beer–Lambert law to include multiple scattering

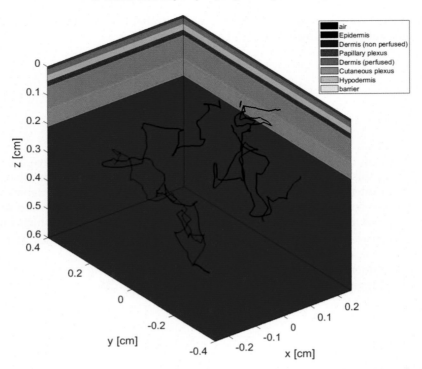

FIGURE 6.3 Simulated skin layers and photon paths of five photons for the simulation setup shown in Fig. 6.1 and further explained in Section 6.5. The figure is obtained by using the software package MCMatlab (Marti et al., 2018) and running a modified example provided by MCMatlab.

effects (Davies et al., 1989). In addition, the Beer–Lambert law assumes that the tissue is homogeneous—however, the tissue has different layers with distinct optical properties.

Therefore, in pulse oximeters, SpO$_2$ is not computed using Expression (6.14) but using regression formulas such as Expression (6.2).

Development of computational resources and simulations of light through nonhomogeneous tissues based on Monte-Carlo methods allowed for computational approaches in determining more truthful propagation of light through different tissue. One of the first works analyzing light propagation through homogeneous and nonhomogeneous tissue includes Tuchin (1997). More recent works include Reuss and Siker (2004), Reuss (2005), and Chatterjee and Phillips (2017). These works are discussed in Section 6.5.

6.2.3 Optical properties of tissues

The material for this section is mainly extracted from Jacques (2020). The tissue will absorb a fraction of incident light traveling within the medium. Absorption is described using an absorption coefficient that depends on tissue type. The *absorption coefficient, μ_a* $\left[\mathrm{cm}^{-1}\right]$, is the product of the number density of molecules in a medium and the absorption cross-section, and it describes the ability of a molecule to absorb light. $1/\mu_a$ determines the mean free path until absorption. The absorption spectrum shows the absorption coefficient over the wavelengths, and it is shown in Fig. 6.4 for oxygenated $\mu_{a,oxy}$ and deoxygenated blood $\mu_{a,deoxy}$. The isosbestic point is the wavelength at which the absorption of HbO_2 and Hb is the same—one isosbestic point is around 800 nm.

Please note that molar absorptivity is different from the absorption coefficient and that the absorption coefficient is proportional to molar absorptivity and concentration. The absorption coefficient can be described using the molar absorptivity and concentration, or it can be presented as the volume fraction of each absorbing component of the medium and the absorption coefficient of that component. The total absorption coefficient for a particular tissue is based on the content of several components, including water, blood, bilirubin, fat, and melanin (Jacques et al., 2017). The simplified formula that includes only contributions of the absorption coefficients of oxygenated and deoxygenated blood and water in a specific tissue is shown below:

$$\mu_a = BS\mu_{a,oxy} + B(1-S)\mu_{a,deoxy} + W\mu_{a,water} \tag{6.15}$$

where B represents the average blood volume fraction, W is the fraction of water volume in the tissue, and S is tissue oxygen saturation. Here, $BS\mu_{a,oxy}$ is the absorption due to oxyhemoglobin, $B(1-S)\mu_{a,deoxy}$ is the absorption due to deoxygenated hemoglobin, and $W\mu_{a,water}$ is the absorption due to water content in the tissue. The quality of PPG signals relies on the fact that blood is a high absorber compared to other tissues.

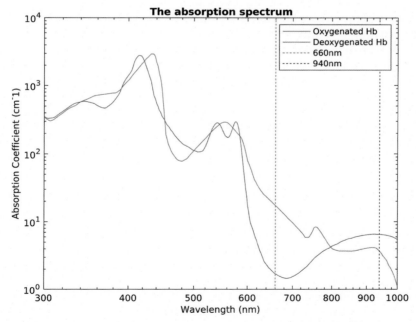

FIGURE 6.4 Absorption coefficient versus wavelength in the visible and near-infrared spectrum for oxygenated and deoxygenated blood. The figure is obtained by using the software package MCMatlab (Marti et al., 2018).

Scattering causes photons to change the directions of their trajectories. Photons are strongly scattered by those structures whose size matches the photon wavelength. Scattering of light by structures on the same size scale as the photon wavelength is described by Mie theory and mainly causes scattering in the forward direction. Scattering of light by structures much smaller than the photon wavelength is called Rayleigh scattering. The *scattering coefficient*, μ_s $[\text{cm}^{-1}]$, depends on the scattering cross-section and the number density of the scatterers, and it varies for different tissues and with the light wavelength. The scattering coefficient determines the frequency of scattering, and therefore $1/\mu_s$ determines the mean free path between scattering events. The *anisotropy*, g [dimensionless] looks at the projection $\cos(\theta)$ of the scattering beam that scattered at the angle θ from the original trajectory. g is defined as the average value of the projection. It is a measure of the amount of forward direction retained after a single scattering event. It is normally in the range of 0.7—0.9, indicating mainly forward scattering. Scattering is of significant interest in reflectance-mode pulse oximetry as we are interested in observing the fraction of the light that has been scattered and ends up at the surface of the detector.

Refraction occurs when the propagating light encounters a change in the refractive index. Boundaries between skin layers can cause the refraction of light.

6.2.4 Human tissue

Layers of human skin are shown in Fig. 6.5. The skin is composed of the epidermis, dermis and hypodermis. Simulated layers are shown in Fig. 6.3. The epidermis is an outer layer of skin up to 1.5 mm thick. It protects the body from the environment. The stratum corneum is the outermost layer of the epidermis and comprises 10—30 thin layers of dead cells. Besides the stratum corneum, there are several layers of live cells in the epidermis. The basal layer is the innermost layer that contains cells that produce melanin, which gives the skin its color and helps protect the skin's deeper layers from the harmful effects of the sun. The dermis is the layer below the epidermis, and it is 1.5 mm up to 4 mm thick. The network of capillaries is visible in the dermis. The dermis also contains lymph vessels, hair follicles, and sweat glands. Hypodermis serves as an isolator conserving body temperature and as an additional protector of internal organs. The cutaneous plexus is a network of arteries and arterioles present along the border of the hypodermis. These arteries supply blood to the fatty tissue of the hypodermis and deep parts of the dermal layer.

6.2.5 Light penetration through the tissue and the choice of wavelength for oxygen saturation estimation

Light penetration depth in biological tissue is dependent on the wavelength. Generally, the longer wavelength, the deeper the penetration depth. For example, the penetration depth of red light is about 0.5—1.8 mm. On the other hand, blue light penetrates only a quarter of the depth of red light.

In Fig. 6.4, vertical dashed lines represent red (660 nm) and infrared (940 nm) light. This region between these two wavelengths is called the diagnostic and therapeutic window. The absorption of HbO_2 is higher than that of Hb in the infrared wavelengths, while the

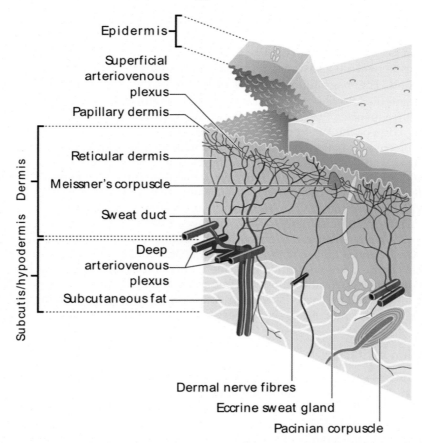

Epidermis

Superficial
arteriovenous
plexus

Papillary dermis

Reticular dermis

Meissner's corpuscle

Sweat duct

Deep
arteriovenous
plexus

Subcutaneous fat

Dermis

Subcutis/hypodermis

Dermal nerve fibres

Eccrine sweat gland

Pacinian corpuscle

FIGURE 6.5   Layers of the human skin. This figure is from Wikimedia Commons, the free media repository, last modified on October 9, 2012. *Source: https://en.wikipedia.org/wiki/File:Skin_layers.png. The file is licensed under is licensed under the Creative Commons Attribution-Share Alike 3.0 Unported license.*

situation is reversed in the range of red light. Therefore, a good choice for a wavelength smaller than the wavelength of the isosbestic point (at about 800 nm) is 660 nm because of a large difference in the absorption coefficients between Hb and HbO_2. Another issue for the wavelength choice is the relative flatness of the absorption spectra shown in Fig. 6.4 around the chosen wavelengths in the red (660 nm) and infrared (940 nm) range. If the absorption spectra are not flat, shifts in the peak wavelength of the LEDs will result in a larger error in estimating oxygen saturation level (Webster, 1997).

Pulse oximeters in transmittance mode might require LEDs with smaller beam angles than those operating in reflectance mode. The reason is that pulse oximeters operating in the reflectance mode rely on the scattering of light occurring in the tissue, and therefore there is more chance for the light to arrive at the detector if the beam angle is wider.

6.3 Signal properties and noise characterization

PPG signal $s_{PPG}(t)$ during two cardiac cycles is shown in Fig. 6.6A. The pulse is normalized to amplitudes between 0 and 1 and time between 0 and 1 s. Therefore, the heart rate is one beat per second. The magnitude spectrum of the signal $|S_{PPG}(f)|$ is shown in Fig. 6.6B. It is computed using the Discrete Fourier Transform (DFT) with 4092 points. The spectrum shows harmonics at frequencies that are multiples of the heart rate.

The rising edge of the PPG pulse corresponds to the systolic phase, and the PPG peak is sometimes referred to as the systolic peak. The falling edge of the pulse corresponds to the diastolic phase, and the minimum is called onset. The time difference between peaks of the consecutive pulses is called a *peak interval*, while the time difference between onsets is called an *onset interval*. When we talk about *pulse intervals*, we refer to either peak or onset intervals.

The morphology of the PPG pulse depends on several factors, including age and health conditions. Changes in pulse morphology with age or health conditions are of clinical importance. It was shown in Allen and Murray (2003) that the duration of the systolic rising edge of the PPG pulse becomes longer with age as well as the notch becomes less prominent.

The signal can be corrupted differently due to ambient light, 50 Hz or 60 Hz interference, and motion artifacts. In addition, both the frequency and the amplitude of the PPG signal are modulated by breathing. These issues can cause difficulties in recognizing the pulses and determining the heart rate. Next, we will explain the problem with motion artifacts.

6.3.1 Motion artifacts

Motion artifacts can be caused even by small movements. Examples of small movements are typing on the keyboard if the probe is placed on the wrist or breathing if the probe is placed on the shoulder. Motion artifacts can corrupt the signal in different ways. They can cause baseline drift and affect the morphology of the signal. Motion artifacts result mainly from the changes in the lengths of the air gap between the probe and the tissue with motion.

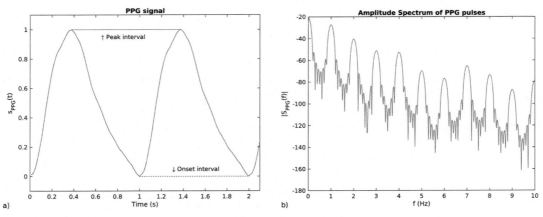

FIGURE 6.6   (A) Two pulses of the PPG signal. (B) Amplitude spectrum of the PPG signal.

This causes changes in the amount of light absorbed in the tissue as well as the amount of light scattered and reflected from the tissue.

Motion artifacts are mainly low-frequency interference. They can be modeled in different ways. Changes in the air gap can be included in the model presented in Section 6.5. The work that included air gap changes in the Monte Carlo model assumed three types of motion: fast arm swing, slow arm swing, and typing (Zhou et al., 2020). The sensor is used in reflectance mode and placed on the wrist. Analytical approaches for including motion artifacts were also developed. They assume that the signal scattered from nonpulsatile and pulsatile tissue is amplitude modulated by the motion artifact signal (Hayes and Smith, 1998). Recently, several works have been done on characterizing the type of noise or motion artifacts of PPG signals obtained when people perform different activities such as sitting, walking, and running (Cajas et al., 2020). That research direction has been further intensified because there are publicly available databases of these signals (Zhang et al., 2015). A significant amount of work has been done on algorithms for motion artifact removal, which is summarized in Pollreisz and TaheriNejad (2019).

The effect of motion artifacts on the illumination of the photodetector is simulated in Section 6.5.3—the results are shown in Fig. 6.7. Simulation is performed for three different values of the air gap: 0.001 mm, 0.01 mm, and 0.2 mm. The first two data points correspond to the situation without an air gap, and the air gap of 0.2 mm corresponds to a small air gap of a smartwatch. The barrier is simulated between the emitter and the photodetector, so there is no direct optical path between them. It is clear that the ratio of the AC component to the DC component of the signal reduces with increasing the air gap depth—see Fig. 6.7A. This significantly affects the ratio of ratios and, therefore, the SpO_2 estimation, as shown in Fig. 6.7B. We should obtain the same ratio of ratios for the same oxygen saturation level (we simulated 95%), which is not the case here. The devices are normally calibrated for a fixed air gap depth. However, when the air gap depth changes due to motion artifacts, it causes significant changes in the ratio of ratios and errors in estimating SpO_2.

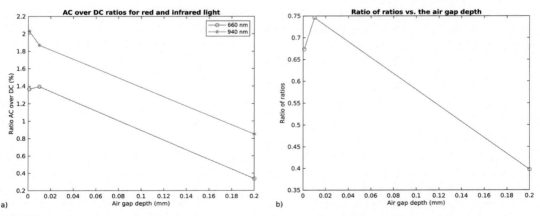

FIGURE 6.7 (A) Simulated AC over DC ratios at red and IR wavelengths for the air gap of 0.001 mm, 0.01, and 0.2 mm with simulated barrier placed in between the emitter and the detector. (B) The ratio of ratios for the same air gap values. Oxygen saturation was kept at 95% during the simulations.

6.4 Performance measures

6.4.1 Performance of PPG devices for heart rate measurements

Using PPG to measure heart rate has become a standard for medical and nonmedical applications. The widespread use of smartwatches with embedded PPG sensors allows for heart rate measurement at home. Wristband-based reflectance mode PPG is implemented in commercial wearables (such as Apple watch (Apple, 2020)) as well as in research-grade wearables (such as the Empatica watch https://www.empatica.com/). Almost 500 trials have been reported using the Fitbit device on PubMed (Bent et al., 2020).

Sources of errors include motion artifacts due to the motion of the sensor or skin deformations, and they are manifested as missing heartbeats that affect the heart rate. It was also reported that the devices are less accurate when attached to people with darker skin, even though the work of Bent et al. (2020) showed that this is not the case. The mean absolute error of the heart rate relative to the heart rate obtained from a reference device (such as ECG) is normally reported. In addition, manufacturers normally report the range for heart rate measurements, for example, 30−235 beats per minute (bpm), as well as the resolution of measurements (e.g., 1 bpm). Other parameters shown are the number of levels of the PPG signal amplitude (for example, 0 to 100 with a resolution of 1%).

6.4.2 Signal quality metrics of PPG signals

Signal quality assessment is detailed in Chapter 4. The signal quality is usually described by the signal-to-noise ratio (SNR). However, calculating SNR is difficult when it comes to physiological signals because it is impossible to distinguish the signal from noise. Signal quality assessment is commonly done by analyzing morphological features of the signal and comparing them with the known reference features. Morphological features can include peak intervals, the ability to detect cardiac pulses, the spectral distribution of the signal, and so on. Signal quality is summarized by using signal quality indices (Li et al., 2008).

The method based on spectral density estimation was introduced in Section 4.4.3 for the example of the ECG signal. Let us assume that the main frequency components of the PPG signal are in the range between 0.8 and 5 Hz, and the total bandwidth is from 0.5 to 30 Hz. The spectral distribution ratio (SDR) can be represented as

$$\text{SDR} = \int_{f=0.8}^{f=5} P(f)df \Big/ \int_{f=0.5}^{f=30} P(f)df \tag{6.16}$$

where $P(f)$ is the power spectrum of the PPG signal. SDR has values between 0 and 1. It is normally compared against a fixed threshold to come up with the spectral power ratio SQI (pSQI) that can be either 0 or 1.

Another way to compute the quality of each pulse is by detecting similarities between the pulse and a reference pulse. This was introduced in Section 4.4.3 under the title "Template matching using average correlation coefficient" and is summarized here for the PPG signal.

As the reference pulse is commonly unavailable, it is computed as an ensemble average of the detected pulses in the time interval T. Ensemble averaging is performed by aligning the pulses so that their peaks occur at the same time and then averaging sample by sample of all the pulses (see Fig. 6.21 in Section 6.11.2). After the ensemble-averaged template pulse is computed, we can proceed in several ways to determine what pulses are of satisfactory quality. One way is to compute the correlation coefficient between each pulse and the template pulse and then consider pulses with a correlation coefficient above a predefined threshold as pulses of satisfactory quality. This is used in the PulseAnalyse toolbox described in Section 6.7.3. Another way is to compute the Euclidean distance between each pulse and the reference pulse and use an empirical threshold to select pulses of satisfactory quality (Sukor et al., 2011). It is also possible to combine the signal quality indexes obtained per pulse in a composite index that can determine the overall quality of pulses in the interval of interest T. For example, we can average the Euclidean distance measure obtained for each pulse in the interval T.

6.4.3 Standards related to pulse oximeters

Pulse oximeters normally need to meet requirements set by the following standards: IEC 60601-1, IEC 60601-1-2, and ISO 80601-2-61. ISO 80601-2-61 is the standard that defines requirements for the safety and performance of pulse oximeters (ISO 80601-2-61, 2011). The ISO pulse oximeter standard requires that the root-mean-square error is smaller than 4% compared with the co-oximeter blood gas readings for the SpO_2 range between 70% and 100%. The standard permits validation against a single approved pulse oximeter as the secondary standard.

Pulse oximetry devices must adhere to the safety standard (Medical Electronic Equipment, 2012) and the electromagnetic compatibility standard (IEC 60601-1-2, 2020). The device maker must show evidence that the device has passed the safety and functional tests defined in this standard. Pulse oximetry devices are classified as class II.

FDA guidance document for pulse oximeters (Pulse Oximeters, 2013) suggests that the study that is used for device configuration should include 10 or more healthy subjects that vary in age and gender, out of which at least 15% (or two subjects, whatever is larger) need to be darkly pigmented subjects. The data should include at least 200 paired observations from the pulse oximeter and co-oximeter. A typical value for root mean square error is up to 3% for the transmittance probe and up to 3.5% for the reflectance probe for SpO_2 in the range between 70% and 100%.

6.5 Simulating the propagation of light through the tissue in reflectance-mode pulse oximetry

Simulating light propagation through the tissue is important since it allows us to understand issues in designing pulse oximeters operating at different monitoring sites. The simulation also allows us to predict the shape of the signal expected from an arbitrary measuring site, the effects of motion artifacts on the signal, and the response to different degrees of perfusion and oxygenation. It is very important to simulate heterogeneous skin tissue because only then we can obtain a realistic picture of the propagation of light through the tissue.

6.5.1 Simulating the propagation through the tissue

The propagation of light through the tissue is simulated using the Monte Carlo simulation-based library called MCMatlab (Marti et al., 2018). This library allows for modeling the light interaction with the tissue. Similar software packages exist for addressing the same problem, such as mcxyx (Jacques et al., 2017). However, we decided to use MCMatlab because the complete simulation can be performed within a Matlab environment.

The first step in simulating light propagation through the tissue is to describe the experiment, which includes the modeled tissue and its properties, the light source, and the detector. In the next step, the light is simulated as photons that are absorbed or randomly scattered as they travel through the medium. The rate of absorption and scattering depends on the absorption and scattering coefficients of each layer of the skin. This light distribution is described by the solution to the radiative transfer equation, which is solved using Monte Carlo (MC) methods (Wilson and Adam, 1983). Using tools such as MCmatlab, the implementation of the solver is provided so that the user only needs to describe the experiment and interpret the simulation results.

The simulation is mainly based on the work described in Reuss and Siker (2004) and Reuss (2005). The simplified model of a reflectance PPG probe consists of an emitter and a detector. The wavelengths of 660 and 940 nm were used in the simulation. Emitter and detector separation was set to 3 mm. The emitter is centered at the coordinates (-0.15 cm, 0 cm, 0 cm) while the detector is placed at (0.15 cm, 0 cm, 0 cm) and normal to the tissue surface, as can be observed in Fig. 6.1A. At the emitter side, a Gaussian beam of a radial width of 4 mm was simulated. The Gaussian beam was selected because it approximates noncollimated LED (the light does not have parallel rays). The detector was simulated with a diameter of 2 mm. Several parameters are set differently than those in Reuss and Siker (2004) and Reuss (2005). They include the level of melanin set to the value of 0.3% as in Chatterjee and Phillips (2017), which is 10-fold below what is commonly found in lightly-pigmented human skin. The reason was to enable the simulation to run in a reasonable time. The existence of melanin in the skin significantly increases the absorption of light and reduces the fraction of illuminated light received at the detector. The absorbance of 660 nm light by melanin is hundreds of times greater than the absorbance of the same wavelength by blood- and melanin-free tissues. A simulation of the effect of melanin on the ratio of ratios is given as an exercise in the end of this chapter.

Tissue is modeled using several layers and presented as a cuboid in a 3D Cartesian coordinate system. The thicknesses and blood volume distribution in the tissue layers are adapted from Reuss (2005). Human skin is essentially divided into epidermis, dermis, and hypodermis, as described in Section 6.2.4. A multilayer skin tissue model can be seen in Fig. 6.3. As each layer is uniformly thick, their relative volumes are expressed by their thicknesses alone. Modeled thickness d_i of skin layer i and the relative volume of blood B_i in each layer are presented in Table 6.1.

TABLE 6.1 Sublayers of the skin together with their thickness and diastolic blood volume. The meaning of * is that the blood volume fractions remain the same during the systolic cycle except for the blood volume fraction in the cutaneous plexus, which changes from 0.2037% to 0.2454%.

Layer number (i)	Layer names	Sublayers thickness (mm)	Diastolic and systolic* blood volume (%)
1	Epidermis	0.2	0
2	Dermis (nonperfused)	0.2	0
3	Papillary plexus	0.2	0.0556
4	Dermis (perfused)	0.8	0.0417
5	Cutaneous plexus	0.6	0.2037 in diastole and 0.2454 in systole
6	Hypodermis	8	0.0417

The total blood fraction B_0 is defined as the mean concentration of blood in the total tissue volume during the diastolic stage, and it is assumed to be 5%. Therefore, the values of the diastolic blood volume and the depth of the tissue are selected in a way that the total blood fraction ends up being 5% by computing the total blood fraction as

$$B_0 = \left(\sum_{i=1}^{6} d_i B_i \right) / \left(\sum_{i=1}^{6} d_i \right)$$

This example is simulated for 660 nm wavelength, with a 5 min simulation duration, which allows for simulating about 10^7 photons. Fig. 6.8A shows normalized absorbed power in different layers. The figure shows that most light is absorbed in the dermis layers. $X-Y$ plane is placed at about 0.1 cm depth below the skin's surface at the boundaries of the hypodermis. $X-Z$ plane shows that light gets absorbed by the barrier on the top and the dermis layer. Fig. 6.8B shows the normalized fluence rate of the collected light at the photodetector. The $X-Z$ plane is placed to cut the middle of the photodetector. We can see a nonuniform distribution of the light where more light is collected on the side of the photodetector that is closer to the LED.

6.5.2 Modeling changes in blood volume

Software packages for simulating light propagation through the tissue do not automatically include information about the parameters that change over time. Therefore, modifications must be done to address the variables that change over time, such as arterial or venous blood volume. In this model, venous blood is completely ignored. The systole was simulated by increasing blood volume only in the skin layer that contains larger blood vessels. The increase in blood volume per different sublayers was based on the work of Reuss (2005). The arterial pulsation was simulated by blood volume increase in the cutaneous plexus layer only by adding the arterial blood. The pulse fraction B_p is defined in Reuss (2005) as the fraction of the total

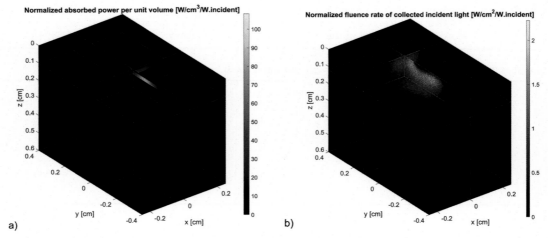

FIGURE 6.8 Simulation setup in case of 95% saturation and 0.01 mm air gap. (A) Normalized absorbed power per unit volume, (B) normalized fluence rate of collected light. The figure is obtained by using the software package MCMatlab (Marti et al., 2018) and running a modified example provided by MCMatlab.

tissue volume displaced by the arterial pulse, which is assumed to be 0.25%. The relative volume in systole in the cutaneous plexus layer is then computed as $B_5' = B_5 + \frac{B_p d_0}{d_5} = 0.2037 + 0.0025 \cdot 10 \text{ mm}/0.6 \text{ mm} = 0.2454\%$. In this way, the relative volume of blood in the cutaneous plexus changed from $B_5 = 0.2037\%$ to $B_5' = 0.2454\%$. In all other layers, the relative volume of blood is the same during systole and diastole. However, arterial pulsation can occur in other layers as well. If we simulate the arterial pulsation in the layers closer to the skin surface, we can say that the depth of pulsation is lower.

In this simulation, about 84% of incident light hits the cuboid boundaries, and about 16% of incident light is absorbed within the cuboid. Out of the total incident light hitting the boundaries of the cuboid, about 2.5% is detected by the detector placed on the skin's surface. Selected simulation parameters for dermis layers include the water content $W = 0.65$, Mie scattering coefficient 1.0, and anisotropy $g = 0.9$.

6.5.3 Simulation results

Fig. 6.9 was obtained by simulating the interaction of light with the tissue for the experimental setup described previously when oxygen saturation was changed from 50% up to 100% in steps of 5%. Fig. 6.9A shows the illumination fraction at the detector side for red and near-infrared light during systole and diastole. It shows the relative values of $I_{dia}^{(660nm)}$, $I_{dia}^{(940nm)}$, $I_{sys}^{(660nm)}$, and $I_{sys}^{(940nm)}$. Estimated SpO$_2$ versus the ratio of ratios R is shown in Fig. 6.9B for given simulation parameters.

The curve SpO$_2$ versus the ratio of ratios shown in Fig. 6.9B moves horizontally with (A) changes in the blood volume fraction during the systolic period, (B) the depth of the

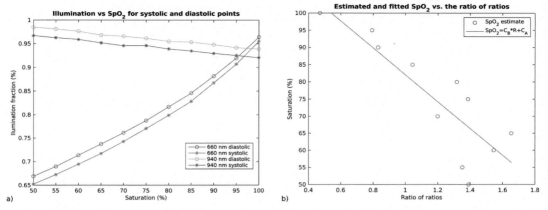

FIGURE 6.9 (A) Illumination fraction of light vs. oxygen saturation level obtained at the detector for 660 and 940 nm wavelengths. The blood volume was changed in the cutaneous plexus layer to simulate the increase of blood volume during the systole. Therefore, there are two different curves for each wavelength, one obtained during systole and another obtained during the diastole period of the cardiac cycle. (B) The ratio of ratios computed using Eq. (6.1) and SpO$_2$ fitted using Eq. (6.2). Coefficients C_A and C_B are estimated to be $C_A = 121.2$, $C_B = -39.2$.

pulsation, and (C) change in distance between the emitter and the detector. It was shown in (Reuss and Siker, 2004) that for small distances between the emitter and the detector, the curve for a smaller depth of pulsation during the systolic period is on the right relative to the curve obtained with the nominal volume fraction. The situation gets reversed when the distance between the emitter and detector increases. Therefore, the pulse oximeter design needs to be adjusted to the site where the probe is attached since the SpO$_2$ value depends on the depth of the pulsation.

Fig. 6.10 shows the simulated illumination fraction at 38 discrete points obtained by running simulation $2.62 \cdot 10^7$ photons at a rate of $2.62 \cdot 10^6$ photons per minute. As can be seen from the simulations, the ratio of AC versus DC component is about 1.3% when the air gap is very small (0.01 mm). This is realistic and has been confirmed in several experimental studies. Curve fitting is done using a fourth-degree polynomial.

When the air gap is increased to 0.2 mm, some photons are reflected directly from the epidermis without entering the tissue. The barrier is simulated in the midrange between the emitter and the photodetector to prevent direct coupling. This increases the DC level but decreases the magnitude of the AC component. For the same setup used to obtain Fig. 6.10, except for the depth of air gap of 0.2 mm, the ratio of AC versus DC component decreases to 0.54%.

This section showed a powerful way of simulating light propagation through the tissue. The major problem with Monte Carlo simulations is that it is slow. However, it allows us to evaluate different scenarios that would otherwise be very difficult. They include the following:

- Analyzing the effects of wavelength and melanin on the pulse shape and estimated SpO$_2$.
- Analyzing the effect of blood volume changes during the systole at different depths.

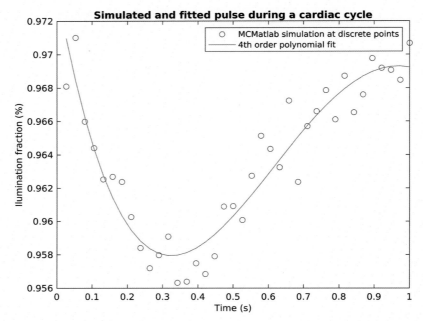

FIGURE 6.10   Simulated illumination fraction versus time of one cardiac pulse at SpO$_2$ level of 95% with air gap of 0.01 mm.

- Analyzing the effect of placement of the probe by modeling the skin properties at the probe position.
- Analyzing the effect of motion artifacts and changing the air gap size on the shape of the pulse.

Some of these scenarios are given as practice problems at the end of the chapter.

6.6 Sensors and circuits

This section will introduce circuits for driving LEDs and for converting photodiode current into voltage. Driving LEDs is switch-based, and switching is commonly done using the pulse width modulation (PWM) technique introduced in Section 3.6.2. In addition, the transimpedance amplifier is used for current-to-voltage conversion. It was previously introduced in Section 3.3.6.4.

6.6.1 LED driver circuits

LEDs can be switched on and off at a high rate and are not commonly turned on all the time. If the LED is switched on for only a small percentage of the time, this will reduce the power consumption of the LED significantly. Switching LEDs on and off is done by using PWM. PWM systems normally have several parameters to tune, including:

- PWM frequency is the frequency of switching LEDs on and off, and it is in the order of 1 kHz. PWM period T is equal to the reciprocal of PWM frequency.
- Pulse width offset D can be given in seconds or in the percentage of the PWM period. It can also be defined in some implementations as
  $D = 100\% \cdot (V_{REF+} - V_{REF-})/(V_{max} - V_{min})$ where V_{REF+} is the positive input voltage in the range between the minimum and maximum voltages (V_{min}, V_{max}), and V_{REF-} is the negative input voltage in the range between the minimum and maximum voltages. In Fig. 6.11, the PWM block is shown with the following parameters: $V_{min} = 0$ V, $V_{max} = 5$ V, $V_{REF-} = 0$ V and V_{REF+} is equal to voltage set in the PWM control voltage source. Therefore, the pulse width offset is obtained as $D = (V_{REF+}/5 \text{ V}) \cdot 100\%$.
- Pulse delay time represents the shift in time compared to the pulse starting at 0 s. The values can be given in the percentage of the PWM period or in seconds.
- Output voltage represents the amplitude of the voltage signal at the output.

The simulated circuit shown in Fig. 6.11 contains an LED, a current source for driving the LED, a PWM block for controlling the timing of the LED and the constant voltage sources (PWM Control) for setting up V_{REF+} of the PWM circuit. Blocks i and w represent the Ammeter and Wattmeter that measures the radiated power. We used the PWM module already provided by Simscape. The voltage–current characteristics of an LED shown in Fig. 2.5A implies that small variations in the forward voltage cause large fluctuations in the forward current and, therefore, in the radiant power. Therefore, LEDs are commonly driven using current sources (Dong and Xiong, 2017). There have been a number of current source designs for driving LEDs efficiently (Li et al., 2016). In this book, we will not consider the details of these designs. We will assume that the LED is driven using a voltage-controlled current source, as shown in Fig. 6.11. The current source is set to produce current $I = (V^+ - V^-)/R$ where R is selected to be 200 Ω.

The circuit presented in Fig. 6.11 was used to obtain the timing diagrams shown in Fig. 6.12. Fig. 6.12A represents forward current and radiated power over time. PWM frequency is set to be 1 kHz, and the duty cycle is 30%. Fig. 6.12B represents a case where V_{REF+} at the input of the PWM block is changed from 0 to 5 V and therefore, the duty cycle of the PWM is changed from 0% to 100%. Averaged radiant power of the LED is computed

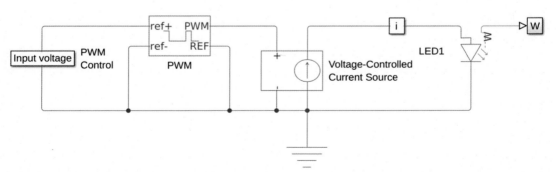

FIGURE 6.11 Voltage-controlled current source for driving an LED that relies on switching based on the PWM module.

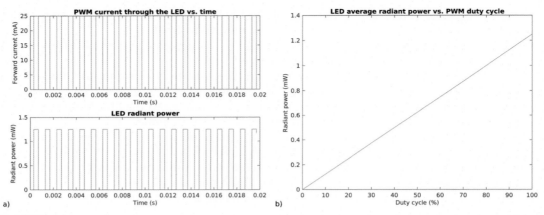

FIGURE 6.12   (A) Figure on the top shows current pulses at a PWM frequency of 1 kHz and a duty cycle of 30%. The figure at the bottom shows radiated power versus time. Of course, the LED will emit only when the forward current is above a certain threshold; therefore, it emits when the forward current is high. (B) Radiant average power output versus the PWM duty cycle.

and presented versus the PWM duty cycle. This figure shows that it is possible to control the luminous intensity of the LED by controlling the PWM duty cycle.

6.6.2 Connecting a photodiode to an amplifier

As pointed out in Chapter 3, a transimpedance amplifier is used to amplify the reverse-biased current of the photodiode and generate voltage. The transimpedance amplifier is shown in Fig. 6.13. The operational amplifier is a single-rail amplifier with $V_{cc} = 3.3$ V. The voltage at the output of the amplifier is proportional to the product of the resistor R_1 and the photodiode current i_p. The capacitor C_1 is used to provide low-pass filter characteristics. The divider circuit formed by resistors R_2 and R_3 ensures that the voltage at the output of the operational amplifier is higher than the low output voltage specification (in our case, 0 V) when there is only a small dark current flowing through the photodiode.

Input to the photodiode is irradiance measured in mW/cm^2. The LED produces this input signal with a PWM frequency of 1 kHz with a duty cycle of 30%. Simulated irradiance is computed in the following way. The radiant power of LED was considered to be 1 mW, and the illuminance fraction is considered to be 2% based on our simulations in Section 6.5. The area of the photodiode is considered to be 4 mm^2. Therefore, the irradiance is computed as $0.02 \cdot 1$ mW/4 mm$^2 = 5$ W/m^2. Based on the photodiode current versus the reverse voltage curve from Fig. 2.7, the current through the photodiode is 5 µA. Fig. 6.14A shows the current through the photodiode during the first 20 ms.

The transimpedance amplifier converts the current generated by the photodiode into voltage to be filtered and processed. Voltage V_0 is calculated as $V_0 = i_p \cdot \frac{R_1}{1+j\omega C_1 R_1} + Vcc \cdot \frac{R_3}{R_2+R_3}$

For this particular example, voltage V_0 can take values between 0.35 V (3.3 V$\cdot$1.2 kΩ/11.2 kΩ) and 3.3 V. In Fig. 6.14B, it is shown that the output takes values around 0.557 V for the given current input and that the high PWM switching frequency is

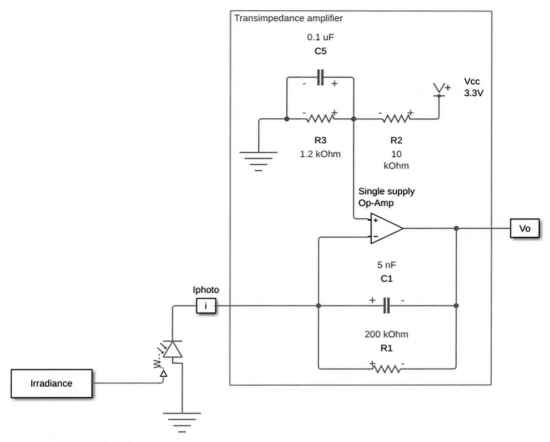

FIGURE 6.13 Transimpedance amplifier for converting photodiode current into voltage.

filtered out. The resistance and capacitance are selected to set the filter cutoff frequency to $f_c = 1/(2\pi R_1 C_1) = 160$ Hz. This is important since the PWM frequency is 1 kHz, and therefore this amplifier and filter will smooth the signal initially. The signal needs further processing and amplification to sample it using an A/D converter. These further processing stages will be shown in Section 6.7.

For information about stability issues in transimpedance amplifiers, we are referring the reader to Baker (2017).

6.7 PPG system

Next, we will describe a PPG system that includes an LED and its driver circuit, a model of the tissue and motion artifacts, and a photodiode and conditioning circuit. The system is shown in Fig. 6.15. The LED driver circuit has already been presented in Section 6.6.1. Feature extraction and pulse processing are not shown in Fig. 6.16, but we will discuss them in Section 6.7.3.

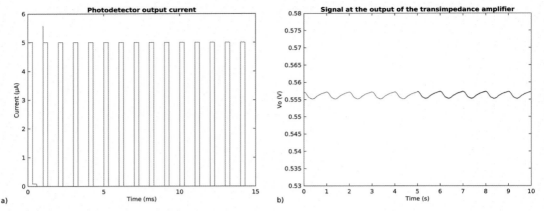

FIGURE 6.14 Timing diagrams for transimpedance amplifier in Fig. 6.13. (A) Current at the photodetector output, and (B) Voltage at the output of the transimpedance amplifier.

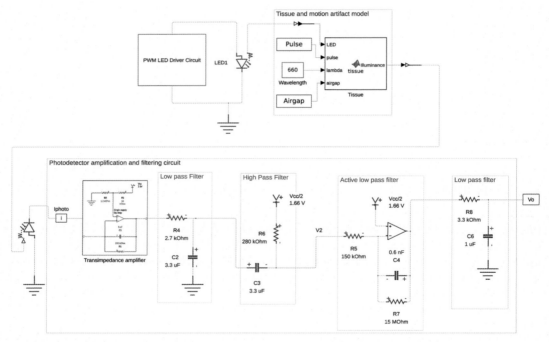

FIGURE 6.15 Simulated PPG system that includes an LED and its driver circuit, a simplified model of the light propagation through the tissue and motion artifacts, and a photodiode and conditioning circuit.

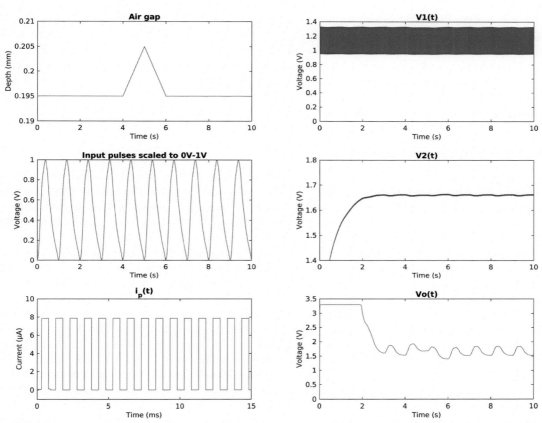

FIGURE 6.16 Timing diagrams for the PPG system simulating the movement. Top left: airgap between the photodiode and the tissue, left in the middle: pulses at the input scaled to the range of 0–1V, bottom left: photodiode current, top right: signal at the output of the transimpedance amplifier, right in the middle: signal at the output of the highpass filter at the point V_2, bottom right: output signal V_o.

6.7.1 Modeling tissue

Modeling the light propagation through the tissue is performed based on simulations described in Section 6.5. The tissues with air gaps that simulate motion artifacts are modeled based on the model used in Fig. 6.7. We used that model to obtain the fraction of illumination (%I) detected by the photodiode versus the air gap between the photodiode and the tissue. The curves illumination fraction versus air gap depth for 660 and 940 nm were fitted using second-order polynomial regression. The air gap is simulated to be 0.195 mm except during the time interval between 4 and 6 s, where we simulated movement that corresponded to extending the air gap and then reducing it to its original depth, shown in Fig. 6.16 on the top left. Fig. 6.16 in the middle left shows the input signal *pulse(t)*. This input signal is added to the signal generated by the LED. The output power of the signal from LED is simulated to be $P = 1$ mW. Photodiode irradiation was modeled as $P \cdot (\%I(\lambda, airgap) + pulse(t) \cdot k)/A$. We assume that photodiode had the area of $A = 4$ mm^2. The parameter k is a constant used to scale down input signal $s_{PPG}(t)$ to be in the range of 2% of the signal from the LED.

It would be possible to run the Monte Carlo light propagation model explained in Section 6.5. However, it takes a very long time to obtain each point; therefore, it will take too long to simulate for 10 s. This is the reason for modeling the light propagation through the tissue and motion artifacts in the way it was described earlier.

6.7.2 Photodetector circuit

The complete circuit that amplifies and filters the signal from the photodiode in order to extract the AC component of the PPG signal is shown in Fig. 6.15. Please note that multiple filtering stages are needed here, especially since the filters are simple and designed as first-order filters. The transimpedance amplifier amplifies the PWM-modulated signal from the photodiode and acts as an initial lowpass filtering stage. The first lowpass filter extracts the noisy AC component from the signal. Next, the highpass filter removes the DC component. As we know, the AC component is small (about 1%) relative to the DC component and therefore needs to be amplified and further filtered, which is done in the next two processing stages.

The components of the system include the transimpedance amplifier, two passive lowpass filters, and a high pass filter, as well as an active lowpass filter. The design is based on Lopez (2012). The transimpedance amplifier and its output have already been shown in Fig. 6.13 and Fig. 6.14. The current through the photodetector, as well as the signal at the output of the transimpedance amplifier $V_1(t)$ are shown in Fig. 6.16 on the bottom left and on the top right, respectively. The first lowpass filter after the transimpedance amplifier is set to the cutoff frequency of 17.9 Hz. In this way, it will further smooth the signal. The high pass filter has a cutoff frequency of 0.17 Hz. This filter removes the DC component of the signal. Also, as the power supply voltage is nonsymmetric (0 V to Vcc = 3.3 V), this highpass filter produces the output centered at Vcc/2. The signal V_2 (t) after the highpass filter is shown in Fig. 6.16 in the middle on the right side. Please note that the AC signal level at this stage is low (in the order of mV) and the AC signal needs to be amplified. For that purpose, an active low pass filter is used, which amplifies the signal 100 times. The cutoff frequency of the filter is 16.6 Hz. The last filter has a cutoff frequency of 48 Hz.

The output signal $V_o(t)$ is shown in Fig. 6.16 on the bottom right. As can be seen, it takes about 3 s for the signal to stabilize. In the case of motion artifacts during seconds 4 and 5, the output signal is corrupted. In addition, due to filtering, the shape of the signal is modified and not all the features in the input signal are preserved.

This system should be followed by an A/D converter and processing circuit. An alternative solution would be to place an A/D convertor immediately after the first lowpass filter. In this case, both DC and AC components of the signal will be captured, and filtering will be done in the digital domain algorithmically. However, this will also require a high-resolution A/D converter since the AC component is only up to 2% compared to the DC component.

This section showed that it is possible to simulate a complete PPG system. In addition, it is possible to analyze the effects of motion artifacts on the quality of the signal. This model can be extended in multiple ways, including the following:

1. Replacing the tissue model with a more feasible model that will take into account the light propagation through the tissue.
2. Removing all analog stages after the first lowpass filter and adding the high-resolution A/D converter to acquire the signal from the first lowpass filter.
3. Including different noise and motion artifacts presented in Chapter 4 that simulate various activities.

6.7.3 Pulse analysis

Processing a PPG signal normally includes heart rate estimation and morphology analysis. Analysis of the pulse morphology or *pulse analysis* is composed of two main steps: detecting important points on the pulse (*fiducial points*) and computing *indices* based on the fiducial points. This section is based on the PulseAnalyse toolbox for Matlab (Charlton et al., 2019). The algorithm for morphology analysis includes the following steps.

1. Detecting each pulse and recording the time and amplitudes of the beginning as well as the minimum and maximum of the pulse.
2. Determining what pulses are of satisfactory quality for further processing.
3. Filtering the pulses.
4. Computing the derivatives of the pulses.
5. Identifying fiducial points and calculating fiducial point times and amplitudes.
6. Calculating indices.

The easiest way to detect pulses is to use ECG R peaks as the reference points for pulses. This will be discussed in Chapter 8. If an ECG signal is unavailable, simple detection of maximums and minimums could be used to determine the beginning, systolic peak, and end of the pulse. A more advanced way to detect pulses is presented in Karlen et al. (2012). The algorithm is based on the following steps: preprocessing, line segmentation, classification of the segments, and artifact removal. The signal is first segmented so that it is approximated using lines of predetermined length. Then, the segments are classified as up-slope, down-slope, or horizontal lines. Next, pulse peaks are identified as endpoints of the previously detected up-slopes. The obtained features are pulse magnitude, maximum and minimum of each pulse, and pulse period. After this, artifact detection is performed based on physiological parameters such as the duration of the pulse.

Pulse quality can be performed by computing signal quality indicators introduced in Section 6.4.2 (Charlton et al., 2019).

PPG pulse and its derivatives are shown in Fig. 6.17. The following fiducial points are detected in the figure on the top: time instant and the amplitudes of the first peak (identified as *s* for systolic), the second peak (identified as *dia* for diastolic in Fig. 6.17, and the notch (identified as *dic* for dicrotic notch). Maximum slope (ms) is obtained by finding the maximum of the first derivative of the pulse. Several points are extracted from the second derivative of the pulse, sometimes called acceleration plethysmogram (APG), or the second

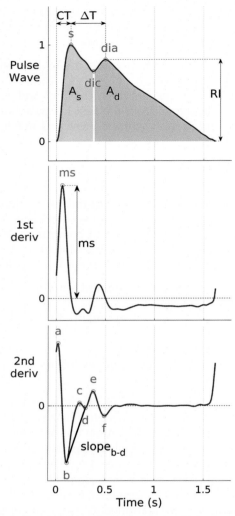

FIGURE 6.17 Fiducial points (*red*) and indices (*black*) of a PPG pulse and its first and second derivatives. The graph is obtained by running the program PulseAnalyze.m described in Charlton et al. (2019) and available at https://peterhcharlton.github.io.

derivative of the digital volume pulse (SDDVP). The second derivative allows us to determine where the pulse is convex up or down. The ratios of the values of parameters a, b, c, d, e (e.g., b/a or c/a) extracted from APG are sometimes used for the analysis of arterial stiffness (Elgendi, 2012). The reflection index (RI) is defined as the ratio of the magnitudes of the second peak and the first peak. Two timing parameters are shown: crest time (CT) which is the time between the beginning of the pulse and the systolic peak and the time delay between the systolic peak and the second peak (ΔT). ΔT decreases with age. Also, the areas under systolic and diastolic peaks are labeled as A_s and A_d. Some parameter names are derived from blood pressure terminology that deals with the reflected waves.

6.8 Pulse oximeters

The Block diagram of the pulse oximeter circuit is shown in Fig. 6.18. It contains only one photodetector and two LEDs. Therefore, the signals from LEDs need to be time multiplexed. That means that when the red LED is on, the infrared LED is off and vice versa. Actually, there is an off period between the periods when the two LEDs are turned on. After the photodetector, the current is converted to voltage using the transimpedance amplifier. Then, the signal is fed into two processing subsystems. The signal from the red LED is passed through the sample and hold block, and the signal value is kept constant until the new sample from the red LED is obtained. That signal is filtered and amplified. The same steps are performed in the second subsystem that processes the IR signal. Two input A/D converter is used to collect digital data from the two subsystems. This design requires obtaining both DC and AC components of red and infrared signals to compute the ratio shown in Expression (6.1). If the system is implemented in a way to keep DC components constant for both LEDs using a circuit called automatic gain control, then only AC components from red and infrared signals will be required. Automatic gain control is not shown in Fig. 6.18.

Pulse oximeters can be implemented in many different ways. There are implementations with two photodiodes, so multiplexing is not required. Also, there are solutions with multiple LEDs and photodiodes.

6.9 New developments

In Fig. 6.19, we identified several drivers and challenges for improvements in SpO_2 technology. They include long-term monitoring of heart rate and SpO_2, improving the accuracy of

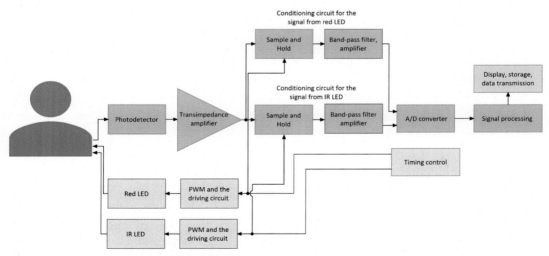

FIGURE 6.18   Block diagram of a pulse oximeter.

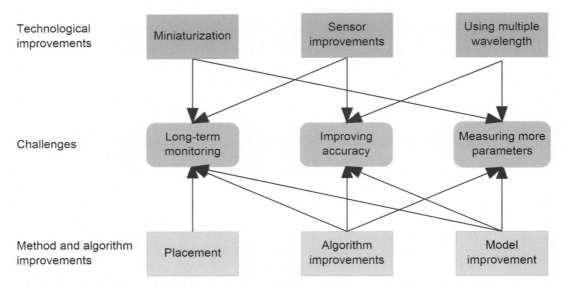

FIGURE 6.19 Research directions related to PPG and SpO$_2$ devices and technologies. *Green* blocks represent challenges and drivers for improving the technology; *blue* blocks are related to improving the technology, and *cyan* blocks to improving methods, measurements, or algorithms.

the devices for patients with different conditions (patients with very low SpO$_2$), as well as measuring more parameters than what standard SpO$_2$ devices measure.

To address these challenges, several research and development directions are identified, including:

- Miniaturizing the system
- Improving sensors
- Reducing the effects of tissue and other parameters on the performance by using more than two wavelengths
- Algorithm improvements through addressing motion and other artifacts
- Determining the proper placement of the device and then designing the device fitted for that location
- Improving mathematical models.

Fig. 6.19 shows these research directions in blue and cyan blocks and challenges in green blocks. Arrows indicate what research directions address what challenges.

6.9.1 Probe placement

Long-term monitoring of oxygen saturation has become very important during the COVID-19 pandemic—please check Box 6.1. Long-term monitoring of SpO$_2$ requires that the pulse oximeter is small and nonobtrusive. Therefore, long-term monitoring is impossible if the transmittance-mode PPG probe is placed on the finger because it presses the finger and prevents people from performing daily activities, including washing hands and typing on the keyboard. Several recent research projects explored the feasibility of placing the device on the wrist as a smartwatch, in the ear canal, and on the neck or forehead. The issue with many of these placements is that there is a smaller capillary density in the tissue under the probe and possibly poorer sensor contact with the skin. Therefore, obtained AC signal has a lower amplitude and is noisier. An example of that kind of work is Davies et al. (2020), where the researchers developed a miniature SpO_2 device placed in the ear canal. The device shows very promising results in determining SpO_2. The location of reflective PPG also affects the morphology of the pulse. It was shown in Hartmann et al. (2019) that the systolic peak of the pulse shifts forward in time for pulses obtained from the forehead and the ear compared to the systolic peak of the pulses measured at the finger.

There are also industrial solutions for long-term monitoring. For example, Masimo (www.masimo.com) announced in 2019 their wearable solution for pulse oximetry monitoring that is FDA-approved and can be used for continuous long-term monitoring of SpO_2.

A number of research works discussed the pressure that needs to be applied to the reflectance-mode PPG probe to obtain the PPG signal of maximum amplitude. The AC pulse amplitude and the AC to DC amplitude ratio were maximum when the probe was slightly pressed to the skin with the contact force of 0.2–0.8 N (Teng and Zhang, 2004).

6.9.2 Developments in miniaturization and sensor technologies

Long-term monitoring is also directly related to the *miniaturization* of the system by integrating multiple components on a chip. A large number of integrated single-chip solutions for photoplethysmography and pulse oximetry are available from major manufacturers, including Texas Instruments, Maxim Integrated, AMS, and Osram Opto Semiconductors. We can classify these solutions into several categories, including:

- Integrated LEDs and photodetectors on a chip,
- Integrated electronics solution that requires only LEDs and photodetectors to be added,
- Completely integrated pulse oximeters,
- Integrated PPG/SpO_2 solution with other sensors such as ECG.

Integrated LEDs and photodetectors became popular with the increased use of PPG technology in smartwatches for measuring heart rate. Solutions include several LEDs in a single chip together with one or more photodiodes. These packages include a light barrier to reduce optical cross-talk between the LEDs and the photodiode. Smartwatches normally contain green LEDs and one photodetector. An example of the industrial solution is SFH7060 by Osram Opto Semiconductors, which includes three green, one red, one infrared emitter, and one photodetector. Therefore, it can be used for heart rate and SpO_2 estimation in smartwatches (https://www.osram.com/os/).

There have been a number of analog front-end integrated chips for optical biosensing applications. These chips require attaching additional LEDs and photodiodes and a microcontroller or a processor for signal processing. An example is Texas Instruments AFE4410 which allows for attaching up to four LEDs and up to three photodiodes (https://www.ti.com/). The chip contains a transimpedance amplifier for converting current into voltage and for amplification, followed by a noise reduction filter, antialiasing filter, and A/D converter.

So, AC and DC components of the signal are not extracted in hardware but in software. Therefore, the A/D converter used has a high resolution of 24 bits. There are also integrated solutions designed specifically for pulse oximeters (for example, TI AFE4403). They commonly include a configurable timing controller for generating the PWM signal and LED drivers.

Fully integrated pulse oximeters contain LEDs, photodiodes, drivers, conditioning electronics and filters, and timing control in a single chip. As such, they represent an integration of two approaches presented earlier ((1) LEDs and photodetectors in a chip, and (2) analog front-end in a chip). They are used for smartwatches, smartphones, and tablets. An example chip is Maxim's MAX30102 PPG sensor that requires only a processor or microcontroller to be attached for processing (https://www.maximintegrated.com/). The chip contains red and infrared LEDs and one photodiode. Regarding electronics, it contains ambient light cancellation, an A/D converter, and a discrete time filter, as well as an LED time control circuit and drivers. A/D converter has a resolution of 18 bits and a programmable sampling rate in the range between 50 samples/s and 3.2k samples/s. In the majority of the integrated solution, the frequency and pulse width can be programmed to achieve a trade-off between accuracy and power consumption.

In addition, integrated pulse oximeter solutions are further integrated with other sensors, such as temperature sensors and ECG. An example of that type of integration is Maxim's MAX86150 which integrates ECG and pulse oximeter. Red and infrared LEDs are integrated into the chip together with the photodiode. To measure ECG, two electrodes need to be attached.

We presented mainly industrial solutions up to this point. Academic researchers have also been very active in this field; however, we will limit our description to only one work here for the interest of space. New research on PPG sensors that are very thin, soft, skin-like electronic patches was described in Chung et al. (2019). The main application is for monitoring the vital signs of babies. Therefore, the design's focus was to have skin-like electronics that is easily attachable to the baby's body without wires and batteries. The power was supplied using near-field radio-frequency technology in a contactless fashion.

6.9.3 Multiwavelength systems

Up to this point, we discussed systems that contain a single LED for PPG measurements or two LEDs of different wavelengths for pulse oximetry. Model-based analysis of multi-wavelength pulse oximetry was performed in Aoyagi et al. (2007). Several sources of errors were recognized in two-wavelength systems, including the effect of venous blood, which volume changes during the cardiac cycle, the effect of different tissue layers and different depths of tissue, as well as the effect of motion artifacts. It was shown that the five-wavelength system could eliminate the effects of tissue and venous blood as well as improve motion artifact detection and removal.

A prototype device was developed with multiple finger probes used simultaneously (Urpalainen, 2011). The project aimed to develop an eight-wavelength pulse oximeter that measures fractional hemoglobins and total hemoglobin. The work showed promising results for measuring methemoglobin, while the measurement of carboxyhemoglobin was more challenging. Methemoglobin is important because if the body produces too much methemo-globin, it can lead to insufficient oxygen getting to your cells. Carboxyhemoglobin (HbCO) forms in red blood cells after contact with carbon monoxide (CO). Therefore, there is significant interest in measuring it to detect signs of poisoning with CO. Moreover, 5—12 wavelengths of light were used in the Masimo Rainbow Pulse CO-Oximeter, which allows for measuring HbCO and methemoglobin.

An increase in the number of wavelengths led to the development of *pulse spectrometers*. An example of the device is V-Spec Sensors by Senspec, Germany (https://www.sentec.com/). These devices have broadband LEDs. The spectroscopy sensor collects light at 350 nm wavelength. Estimation of fractional hemoglobins is performed. Studies show high accuracy in measuring HbCO and methemoglobin.

6.9.4 Processing and modeling

We have already mentioned in Section 6.3.1 that significant work has been done on modeling motion artifacts and removing the effects of motion artifacts. That work was summarized in Pollreisz and TaheriNejad (2019) and in Tamura (2019). Besides reducing motion artifacts algorithmically, artifacts can also be reduced by adding LEDs that operate at multiple wavelengths (Alharbi et al., 2018) or by adding the accelerometer next to the pulse oximeter to detect the motion.

There has been a surge of research papers on extracting features from the PPG signal and using the features for continuous blood pressure estimation. This research is also important for the PPG field because it requires a detailed analysis of the morphology of the PPG signal and morphological changes with blood pressure. For example, almost a 100 features were extracted from the PPG signal in Chowdhury et al. (2020).

Modeling can be divided into models that deal with the interaction of light with the tissue and models used to simulate or generate the PPG signal itself. In this chapter, we focused only on the first type of models. Regarding PPG signal generation modeling, we can broadly divide models into those that take into account the hemodynamics and the model that are just focused on the shape of the PPG pulse. These models normally include blood flow velocity, luminal area, volume flow rate, and arterial pressure waves. Models are built for a particular segment of the body or the whole body. In Charlton et al. (2019), whole body simulation is performed and PPG is also simulated by assuming that the PPG depends on the volume of arterial blood in a tissue.

A number of parametric models have been developed where the PPG signal is fitted to a particular model that does not take into account hemodynamics. An example is a model composed of the superposition of two waves presented as two Gaussian curves. Different models are proposed to handle different signal lengths, morphologies, and beat-to-beat variability. This approach is shown in Tang et al. (2020), where the PPG signal is modeled for regular and irregular heartbeats. Noise and motion artifacts are also modeled. The pulse is represented as a sum of two Gaussian functions. The simulation is performed in Matlab, and the code was made publicly available.

Models allow for a better understanding of PPG, detecting and removing motion artifacts because the artifacts "look" different than the signal, reconstructing corrupt or missing epochs of the signal, evaluating the effect of physiological variables on the accuracy of the device, and so on.

6.9.5 Addressing health problems in a new way

6.9.5.1 PPG signal analysis for continuous blood pressure monitoring

PPG is used for cuff-less blood pressure monitoring by extracting indices (Section 6.7.3) from the PPG pulses and using these indices as features in a regression-type machine learning model. Models are commonly trained with data from publicly available databases. Currently, two databases have been commonly used. The Medical Information Mart for Intensive Care (MIMIC) database includes PPG and ECG data for ICU patients (Saeed et al., 2011). The issue with the database is reported in the literature—for some subjects, PPG and ECG are not properly synchronized. In addition, patients are from the intensive care unit, so medications and comorbidities are most likely present, but information about them is not included in the database. Another available database is called PPG-BP Database, and it contains PPG signals collected along with BP readings from 219 patients admitted to the Guilin People's Hospital in Guilin, China (Liang et al., 2018).

Recent works based on machine learning show very high accuracy in blood pressure measurements only based on PPG signals. The mean absolute error for systolic blood pressure is 3 mmHg and for diastolic is 1.7 mmHg (Chowdhury et al., 2020). More details about estimating blood pressure noninvasively are given in Chapter 8.

6.10 Summary

In this chapter, we discussed the design of PPG devices and pulse oximeters. We simulated the propagation of light through the tissue for different wavelengths. We modified the simulation to show the effect of the blood flow and of the distance between the sensor and the skin on the illumination fraction and the ratio of ratios. We showed the circuits for driving the LED and conditioning the photodiode's signal. Finally, we showed the design of the overall PPG circuit. We also introduced a conceptual diagram of a pulse oximeter. Trends and research directions are pointed out as well.

What was not covered in this chapter is the detailed block diagram of the pulse oximeter device. Instead, this is given as an exercise. In addition, we do not describe improvements in materials and transducer/sensor technologies.

6.11 Problems

6.11.1 Questions

6.1. *What is the average density of capillaries in human tissue?
6.2. What is the penetration depth of green, blue, and red light through the tissue if the LEDs are placed directly on the skin?
6.3. Explain why different wavelengths of LEDs are used in biomedical applications and what these wavelengths are.
6.4. *What is the photoelectric effect?
6.5. What does a pulse oximeter measure?

6.6. *Why is it important to monitor an anesthetized patient with a pulse oximeter?

6.7. What is the difference between transmittance and reflectance modes of operation?

6.8. Explain, in general terms, the operating principles behind a pulse oximeter. What is the Beer–Lambert absorption law, and how is it used for pulse oximetry?

6.9. *Why is it that only the arterial blood gases get measured? If a pulse oximeter measures just two light wavelengths, why will it get confused by the presence of another gas in the blood (such as CO)?

6.10. What does photoplethysmography measure?

6.11. Describe clinical applications of PPG.

6.11.2 Problems

6.12. We designed a device and defined the SpO_2 ratio of ratios R as

$$R = \frac{I_{AC}^{(\lambda_R)}}{I_{AC}^{(\lambda_{IR})}}$$

Assume the AC components at both red and infrared lights are measured with 2% uncertainty at a 95% level. What is the uncertainty of R?

6.13. In photo-plethysmography in a smartwatch, the pulsatile signal at the output of the photodetector and conditioning circuit (transimpedance amplifier and the filter) has a maximum amplitude of 1 mV while the DC term is 1 V. This conditioning circuit is connected to a filter, amplifier, antialiasing filter and 8-bit A/D converter that works in the range of 0–5 V.

    (a) Smartwatches normally use green LED. What is the propagation depth of the green LED compared to the propagation depth of the red and infrared LEDs?

    (b) Is this device operating in transmittance or reflectance modes? Why? What are the advantages and disadvantages of these two modes?

    (c) Why do we need to filter out the DC term?

    (d) After removing the DC component, the AC signal is amplified. What should be the gain of the amplifier? Why?

    (e) You need to design an antialiasing filter for the sampling rate of $fs = 50$ Hz. Chose components of the Sallen-Key second order lowpass active filter so that all the components above $f_s/2$ are attenuated at least five times.

6.14. Pulse oximeters work by passing red light (660 nm) and infrared light (940 nm) through a body part (such as a finger).

    (a) Why do pulse oximeters use two different wavelengths to measure SpO_2?

    (b) What is the origin of the DC component of the detected signal? What is the origin of the AC component?

(c) A sample of the raw pulse oximeter data is shown in Fig. 6.20. The SpO_2 can be estimated by Expressions (6.1) and (6.2). Compute the ratio of ratios as well as estimate the coefficient C_B if the real SpO_2 value is 96% and $C_A = 121$.

(d) What is the patient's heart rate, according to the raw pulse oximeter data in Fig. 6.20?

6.15. The pulse oximeter device is shown in Fig. 6.18.

(a) Why is there only one transimpedance amplifier and two sample and hold circuits, bandpass filters, and amplifiers?

(b) Signal for both red and IR LEDs is amplified, processed and then sampled using an A/D converter and collected by the processor. What kind of processing needs to be done to estimate SpO_2? How many pulses need to be collected to estimate SpO_2? How would the oximeter know how many samples to collect to have enough to estimate SpO_2?

(c) Redesign the circuit so that the amount of analog processing is minimal. What kind of A/D converter would you use in this case? Hint: A/D conversion needs to be done as soon as possible in the circuit pipeline. Please make sure that the antialiasing filter is used.

6.16. Ensemble averaging is widely used to extract repetitive pulses corrupted by noise. These pulses are detected as they come and added to the previous pulses in the

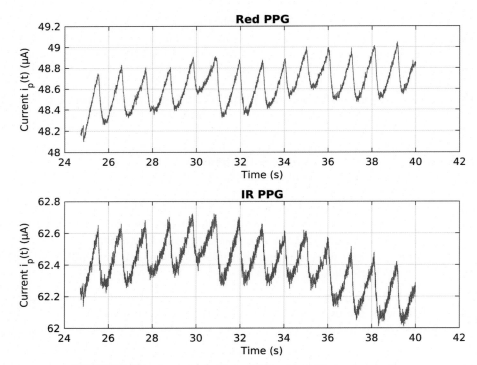

FIGURE 6.20   An example of raw PPG red and IR signals over time.

following way—the first sample of the first pulse is added to the first sample of the second pulse, the second sample of the first pulse is added to the second sample of the second pulse and so on for M samples in the pulse and N pulses. In the end, we get only one pulse with reduced SNR. Your goal is to do ensemble averaging of PPG pulses in real time as they are measured. Your design should provide one PPG pulse at the output for every $N = 64$ pulses.

(a) Explain the sources of noise of the PPG signal and how you would address/reduce the noise for each of these sources. How would you reduce the noise if you observe a noisy PPG signal, but you are not able to identify the source of the noise?

(b) Design the circuit (on paper) that perform ensemble averaging of PPG signals and explain your design. Please do NOT store all your pulses and then add them. You would need to perform addition as the pulses arrive in real-time. In one possible implementation, you would need a comparator to detect the beginning of the pulse and a memory element to store the pulse (and the sum) (Fig. 6.21).

6.11.3 Simulations

6.17. Using provided MCmatlab simulation, change the level of melanin in the epidermis to (a) 3% and (b) 12% and observe changes in the slope of the ratio of ratios. How would you design a robust pulse oximeter to handle different skin colors?

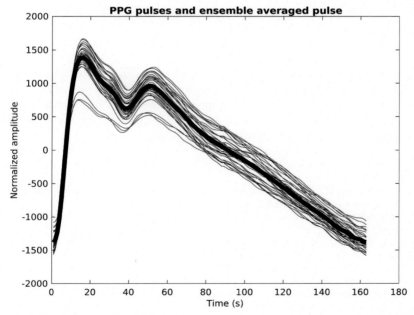

FIGURE 6.21 Several PPG pulses presented in a way so that the foot of each pulse occur at the same time instance. The black line shows the averaged pulse.

6.18. Using provided MCmatlab simulation, add the contribution of venous blood by following Reuss and Siker (2004) or Chatterjee and Phillips (2017).

6.19. In the provided simulation, it was assumed that the blood volume increases only in the cutaneous plexus layer. Let us assume that the blood volume changes in the papillary plexus layer instead. How would that change affect the ratio of ratios R?

6.20. Perform optimization of the PPG system shown in Fig. 6.15. Modify the circuit so that the distortion to the PPG signal is the smallest. This is very important if we need to extract features from the PPG signal and not only to estimate the heart rate or measure the signal's amplitude. The optimization of the components of the circuit can be done in the optimization application in Simulink.

6.21. Modify the system in Fig. 6.15 to model a pulse oximeter. Assume that red and IR LEDs are used and that only one photodetector is used.

6.12 Further reading

In this section, we refer to several relevant resources for further reading. However, please note that this list is not comprehensive and that many excellent and relevant resources might have been omitted unintentionally.

Excellent material on biomedical optics can be found at https://omlc.org/classroom/index.html. Similar material was very well presented in the slides by Jacques (2020). We relied mainly on work by J.L. Reuss in modeling the propagation of light through the tissue (Reuss and Siker, 2004; Reuss, 2005). For the simulation, we used an excellent simulator based on Matlab, which details are explained in Marti et al. (2018). Detailed covering of pulse oximeter technology is given in Webster (1997). That book (Webster, 1997) is strongly recommended to someone who wants to design a new pulse oximeter. Excellent coverage of PPG applications is given in Allen (2007).

To perform pulse processing, excellent resources regarding simulating pulse morphology at different sites and performing pulse processing can be found at https://peterhcharlton.github.io. For pulse generation that does not include physiology but focuses only on the pulse morphology and modeling of noise and arrhythmias, please see Tang et al. (2020) and https://github.com/Elgendi/PPG-Synthesis/tree/master/code.

References

Alharbi, S., Hu, S., Mulvaney, D., et al., 2018. Oxygen saturation measurements from green and orange illuminations of multi-wavelength optoelectronic patch sensors. MDPI Sensors 19 (1).

Allen, J., 2007. Photoplethysmography and its application in clinical physiological measurement. Physiological Measurements 28 (3), 1−39.

Allen, J., Murray, A., 2003. Age-related changes in the characteristics of the photoplethysmographic pulse shape at various body sites. Physiological Measurement 24 (2).

Aoyagi, T., Fuse, M., Kobayashi, N., Machida, K., Miyasaka, K., December 2007. Multiwavelength pulse oximetry: theory for the future. Anaesthesia and Analgesia 105 (6), S53−S58.

Apple, 2020. Your heart rate. What it means, and where on Apple Watch you'll find it," Apple support. https://support.apple.com/en-gb/HT204666. (Accessed 31 March 2020).

Baker, B., 2017-06-28. How to Design Stable Transimpedance Amplifiers for Automotive and Medical Systems. Digi-Key. https://www.digikey.ca/en/articles/how-to-design-stable-transimpedance-amplifiers-automotive-medical-systems. (Accessed 21 November 2020).

Bashkatov, A.N., Genina, E.A., Kochubey, V.I., Tuchin, V.V., 2005. Optical properties of human skin, subcutaneous and mucous tissues in the wavelength range from 400 to 2000 nm. Journal of Physics D 38, 2543–2555.

Bent, B., Goldstein, B.A., Kibbe, W.A., Dunn, J.P., Dec. 2020. Investigating sources of inaccuracy in wearable optical heart rate sensors. Springer Nature Digital Medical 3 (1), 1–9.

Cajas, S.A., Landínez, M.A., López, D.M., Jan. 2020. Modeling of motion artifacts on PPG signals for heart-monitoring using wearable devices. In: 15th International Symposium on Medical Information Processing and Analysis, vol 11330, no. 3.

Charlton, P.H., Harana, J.M., Vennin, S., Li, Y., Chowienczyk, P., Alastruey, J., 2019. Modelling arterial pulse waves in healthy ageing: a database for in silico evaluation of haemodynamics and pulse wave indices. American Journal of Physiology-Heart and Circulatory Physiology 317 (5), H1062–H1085.

Chatterjee, S., Phillips, J.P., 2017. Investigating optical path in reflectance pulse oximetry using a multilayer Monte Carlo model. In: Quincy Brown, J., van Leeuwen, T.G. (Eds.), Clinical and Preclinical Optical Diagnostics, vol 10411. Proc. of SPIE-OSA.

Chowdhury, M.H., et al., June 1, 2020. Estimating blood pressure from the photoplethysmogram signal and demographic features using machine learning techniques. Sensors, MDPI 20 (11), 3127.

Chung, H.U., et al., March 1, 2019. Binodal, wireless epidermal electronic systems with in-sensor analytics for neonatal intensive care. Science 363 (Issue, 947).

COVID-19 Treatment Guidelines Panel. Coronavirus Disease 2019 (COVID-19) Treatment Guidelines. National Institutes of Health. Available at: https://www.covid19treatmentguidelines.nih.gov/. Accessed on October 18, 2021.

Davies, C., Takatani, S., Sakakibara, N., Nose, Y., 1989. Application of the Kubelka Munk equation to characterizing a reflectance pulse oximeter. In: Annual International Conference of the IEEE Engineering in Medicine and Biology Society, vol. 4, pp. 1095–1097.

Davies, H.J., Williams, I., Peters, N.S., Mandic, D.P., 2020. In-ear SpO2: a tool for wearable, unobtrusive monitoring of hypoxaemia in COVID-19. Sensors, MDPI 20 (17), 4879.

Dong, J., Xiong, D., 2017. Applications of light emitting diodes in health care. Ann Biomed Eng. 45 (11), 2509–2523.

Elgendi, M., 2012. On the analysis of fingertip photoplethysmogram signals. Current Cardiology Reviews 8 (1), 14–25.

Fung, Y.C., Zweifach, B.W., Intaglietta, M., 1966. Elastic environment of the capillary bed. Circ. Res. 19, 441–461.

Hartmann, V., Liu, H., Chen, F., Qiu, Q., Hughes, S., Zheng, D., 2019. Quantitative comparison of photoplethysmographic waveform characteristics: effect of measurement site. Front Physiol 10, 198.

Hayes, M.J., Smith, P.R., 1998. Artifact reduction in photoplethysmography. Appl. Opt. 37 (31), 7437–7446.

IEC 60601-1-2:2014+AMD1:2020 CSV Consolidated Version: Medical Electrical Equipment - Part 1-2: General Requirements for Basic Safety and Essential Performance—Collateral Standard: Electromagnetic Disturbances—Requirements and Tests, 2020.

ISO 80601-2-61 Medical Electrical Equipment Part 2-61: Particular Requirements for Basic Safety and Essential Performance of Pulse Oximeter Equipment. Geneva, Switzerland, 2011.

Jacques, S., February 2, 2020. Tissue Optics. SPIE Photonics West Short Course SC029, San Francisco.

Jacques, S., Li, T., Prahl, S., 2017. mcxyz.c, a 3D Monte Carlo simulation of heterogeneous tissues. omlc.org/software/mc/mcxyz. (Accessed 16 May 2020).

Karlen, W., Ansermino, J.M., Dumont, G., 2012. Adaptive Pulse Segmentation and Artifact Detection in Photoplethysmography for Mobile Applications. International Conference of the IEEE Engineering in Medicine and Biology Society, San Diego, CA, pp. 3131–3134.

Li, Q., Mark, R.G., Clifford, G.D., 2008. Robust heart rate estimation from multiple asynchronous noisy sources using signal quality indices and a Kalman filter. Physiological Measurement 29, 15–32.

Li, S., Tan, S., Lee, C., Waffenschmidt, E., Hui, S., Tse, C., 2016. A survey, classification, and critical review of light emitting diode drivers. IEEE Transactions Power Electronics 31 (2), 1503–1516.

Liang, Y., et al., 2018. A new, short-recorded photoplethysmogram dataset for blood pressure monitoring in China. Scientific Data 5 (180020).

Lopez, S., November 2012. Pulse oximeter fundamentals and design, freescale semiconductor application note, document number: an4327. Rev 2.

Marti, D., Aasbjerg, R.N., Andersen, P.E., Hansen, A.K., 2018. MCmatlab: an open-source, user-friendly, MATLAB-integrated three-dimensional Monte Carlo light transport solver with heat diffusion and tissue damage. Journal of Biomedical Optics, SPIE 23, 1–6.

Medical Electronic Equipment—Part 1: General Requirements for Basic Safety and Essential Performance", IEC 60601-1 Edition 3.1, 2012-08.

Nilsson, L., Goscinski, T., Kalman, S., Lindberg, L.G., Johansson, A., 2007. Combined photoplethysmographic monitoring of respiration rate and pulse: a comparison between different measurement sites in spontaneously breathing subjects. Acta Anaesthesiologica Scandinavia 51, 1250—1257.

O'Carroll, O., et al., August 2020. Remote monitoring of oxygen saturation in individuals with COVID-19 pneumonia. The European Respiratory Journal 56 (2).

Peñáz, J., 1973. Photoelectric measurements of blood pressure, volume and flow in the finger. Proceedings of the Digest of the 10th International Conference on Medical and Biological Engineering, Dresden, Germany, 13—17 August 1973.

Pollreisz, D., TaheriNejad, N., Aug. 2019. Detection and removal of motion artifacts in PPG signals. Mobile Networks and Applications 1—11.

Pulse Oximeters—Premarket Notification Submissions [510(k)s]: Guidance for Industry and Food and Drug Administration Staff, March 4, 2013. FDA.

Reuss, J.L., 2005. Multilayer modeling of reflectance pulse oximetry. IEEE Transactions on Biomed Eng 52 (2), 153—159.

Reuss, J.L., Siker, D., 2004. The pulse in reflectance pulse oximetry: modeling and experimental studies. Journal of Clinical Monitoring and Computing 18, 289—299.

Saeed, M., et al., 2011. Multiparameter intelligent monitoring in intensive care II (MIMICII): a public-access intensive care unit database. Critical Care Medicine 39 (952).

Sukor, J.A., Redmond, S.J., Lovell, N.H., 2011. Signal quality measures for pulse oximetry through waveform morphology analysis. Physiological Measurement 32, 369—384.

Tamura, T., 2019. Current progress of photoplethysmography and SpO$_2$ for health monitoring. Biomedical Engineering Letters 9 (1), 21—36.

Tang, Q., Chen, Z., Allen, J., Alian, A., Menon, C., Ward, R., Elgendi, M., 2020. PPGSynth: an innovative toolbox for synthesizing regular and irregular photoplethysmography waveforms. Frontiers in Medicine 7.

Teng, X.F., Zhang, Y.T., 2004. The effect of contacting force on photoplethysmographic signals. Physiological Measurement 25, 1323—1335.

Tuchin, V.V., 1997. Light scattering study in tissues. Physics—Uspekhhi 40 (5), 495—515.

Urpalainen, K., 2011. Development of a Fractional Multi-Wavelength Pulse Oximetry Algorithm. Master Thesis, School of Electrical Engineering, Aalto University.

Volkov, M.V., Margaryants, N.B., Potemkin, A.V., Volynsky, M.A., Gurov, I.P., Mamontov, O.V., Kamshilin, A.A., 2017. Video capillaroscopy clarifies mechanism of the photoplethysmographic waveform appearance. Nature Scientific Reports 7 article num. 13298.

Webster, J.G. (Ed.), 1997. Design of Pulse Oximeters. IOP Publishing Ltd.

Webster, J.G. (Ed.), 2010. Medical Instrumentation: Application and Design, fourth ed. John Wiley and Sons.

Wilson, B.C., Adam, G., 1983. A Monte Carlo model for the absorption and flux distributions of light in tissue. Medical Physics 10 (6), 824—830.

Zhang, Z., Pi, Z., Liu, B., February 2015. TROIKA: a general framework for heart rate monitoring using wrist-type photoplethysmographic signals during intensive physical exercise. IEEE Transactions on Biomedical Engineering 62 (2), 522—531.

Zhou, C.C., Wang, J.Y., Qin, L.P. and Ye, X.S., "Model design and system implementation for the study of anti-motion artifacts detection in pulse wave monitoring," In Proceedings of the 13th International Joint Conference on Biomedical Engineering Systems and Technologies (BIOSTEC 2020) - Volume vol. 1: BIODEVICES, pages 102—109.

CHAPTER

7

Devices based on the ECG signal

Acronyms and explanations

AV	atrioventricular
bSQI	beat detection signal quality index
DRL	driven right leg
HF	high frequency
HR	heart rate
HRV	heart rate variability
iSQI	interchannel signal quality index
LF	low frequency
N—N intervals	R—R intervals of normal beats
pNN50	percentage of the number of pairs of the adjacent N—N intervals that differ by more than 50 ms versus the total number of N—N intervals
R—R interval	the time difference between the peaks of the consecutive R waves.
RMSSD	square root of the mean of the sum of square differences
RSA	respiratory sinus arrhythmia
SA	sinoatrial
SD1	standard deviation 1 obtained using the return map when doing HRV analysis
SD2	standard deviation 2 obtained using the return map when doing HRV analysis
SDNN	standard deviation of N—N intervals
TINN	triangular interpolation of N—N interval histogram
TKEO	Teager-Keiser energy tracking operator
TRI	triangular index
VLF	very low frequency
WiFi	wireless fidelity

Variables used in this chapter include:

A_{dIA}	differential gain of the instrumentation amplifier
A_{dLP}	differential gain of the active lowpass filter
E	potential on the electrode
f_{HP}	the cut-off frequency of a highpass filter
f_{LP}	the cut-off frequency of a highpass filter
f_s	sampling rate

i_d	displacement current
T_{RRmean}	the average duration of the R—R intervals
V	voltage (potential difference) between two electrodes; for example, V_I is the voltage between two electrodes placed on the left and right arms/shoulders that make the lead I
V_{cm}	common mode voltage
$x(n)$	filtered ECG signal in the discrete domain where n is the sample number

7.1 Introduction

An *electrocardiogram* (ECG or EKG) represents the graph of the potential difference between two measurement points on the body versus time, where the potential difference is related to the electrical activity of the heart. ECG is measured using specialized devices that contain electrodes placed on the skin, front-end electronics for filtering and amplifying the signals from the electrodes and processing electronics for displaying the ECG signal and computing parameters of interest. The ECG signal can be recorded intermittently for short intervals or continuously for longer periods. *Intermittent recording* can be performed in hospitals (10 s ECG) or at home using wearable devices such as ECG-enabled smartwatches. *Continuous monitoring* is performed in hospitals for critically ill patients or patients under general anesthesia. In addition, continuous monitoring is required to detect infrequently occurring events such as cardiac arrhythmias. In these cases, long-term (24 h or more) continuous monitoring can be done at home using ambulatory monitors. ECG is used for diagnosing cardiovascular diseases such as different arrhythmias or for prognosis (e.g., predicting the occurrence of infarction).

This chapter introduces how ECG is generated and shows how ECG can be measured using multiple leads. An *ECG lead* is a pair of two electrodes used to measure the potential difference between two places on the body and obtain the ECG signal. Then we explain the morphology of the ECG signal. Next, we will discuss noise and interference common in ECG signal acquisition. Then we present several ways to simulate the ECG signal. We also describe different databases of ECG waveforms. These databases are important when developing algorithms for classifying arrhythmias, predicting different events, and so on. Next, we present standards related to ECG.

Segmentation or *delineation* of the ECG signal includes extracting each beat and determining fiducial points (points of interest) on the ECG waveform. These fiducial points include the QRS complex, P and T waves, and others. We show a basic algorithm for QRS complex detection. Also, we show, without going into detail, how one can segment the ECG signal to detect P and T waves and the QRS complex. Next, we show the overall block diagram of ECG systems and details of the front-end, including amplifiers and filters. ECG circuitry is simulated in Simscape. We connect the circuit to the models of dry electrodes introduced in Chapter 2. Then, we show the effects of DC bias and common mode voltage on the output of the ECG front-end circuit and introduce the driven right leg circuit that reduces the common mode voltage. Research directions regarding electrode design, the design of novel wearable devices, and machine learning are presented next.

The time difference between successive QRS intervals is not fixed over time, and it is termed *heart rate variability*. Different methods for analyzing heart rate variability have been established in the last several decades, and they are also reviewed in this chapter. Heart rate variability analysis can be easily done with ECG signals collected using wearable devices pervasively; therefore, it is presented in this book.

In this chapter, several Matlab programs for analysis of the ECG signal have been provided on the book webpage. The code for the simulation of the ECG signal generation and for heart rate variability is adapted from the works of McSharry et al. (2003), Vollmer (2019). A signal quality assessment is also presented. All of this, together with the code presented in Chapters 2 on modeling electrodes, Chapter 3 on filters and instrumentation amplifiers, and Chapter 4 on modeling motion artifacts and signal quality, can be combined to allow for the following analyses:

— Analysis of the effect of noise and motion artifacts on the quality of the ECG signal at the output of the ECG front-end.
— Analysis of the electronic modifications needed when connecting wet, dry, or capacitive ECG electrodes.
— Analysis of the effects of noise and motion artifacts on the robustness of the algorithms for heart rate estimation.
— Importance of different signal quality estimators under different noise and motion artifact conditions.

Please note that these analyses were not performed in this chapter and are left as an exercise.

The chapter mainly focuses on single-lead ECG but also mentions 12-lead ECG. Even though much work in the ECG field is around advanced signal processing and machine learning algorithms, they have just been mentioned in this chapter.

7.2 What is measured using ECG

7.2.1 Physiology

Electrical conduction of the heart is shown in Fig. 7.1, while the ECG signal generated as a result of the electrical conduction is shown in Fig. 7.2. The human heart, as shown in Fig. 7.1, contains four chambers: left and right atrium and left and right ventricle. Fig. 7.1 on the left shows the conduction in the atria. The cardiac cycle starts with an electrical signal generated from the sinoatrial node (SA). The signal spreads across the cells of the heart's right and left atria which causes the atria to contract, pushing the blood from the atria into both ventricles. Contraction of the heart's atria is related to the P wave on the ECG waveform. Please note that we did not cover the action potential in this book and, therefore, will not go into the propagation of the electrical signal—for more details, please refer to Malmivuo and Plonsey (1995), Webster (2010).

The electrical signal arrives at the atrioventricular (AV) node and propagates through the ventricles after a short delay, as shown in Fig. 7.1 on the right. The signal propagates through the conduction system of ventricles, including the Bundle of His, the bundle branches and the

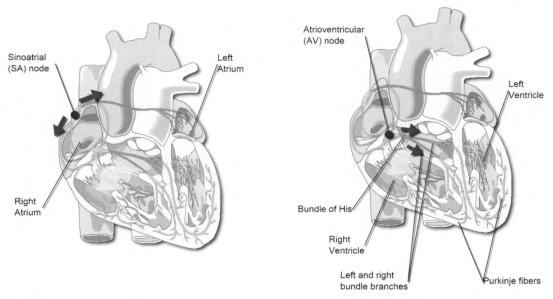

FIGURE 7.1 Electrical conduction system of the heart. *Source: U.S. Department of Health and Human Services, National Institutes of Health, National Heart Lung and Blood Institute, provided by SmartDraw Inc.*

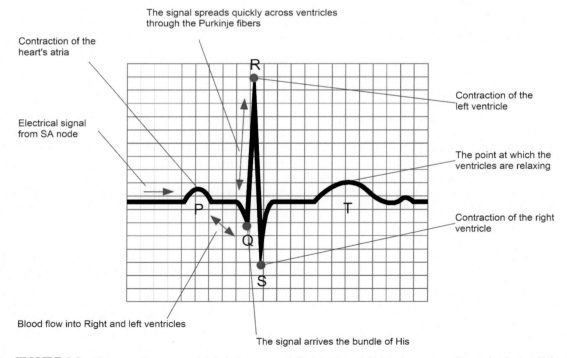

FIGURE 7.2 Electrocardiogram with labeled waves and the heart's activity during one cardiac cycle. *Source: U.S. Department of Health and Human Services, National Institutes of Health, National Heart Lung and Blood Institute, provided by SmartDraw Inc.*

Purkinje fibers. The propagation of the electrical signal through the ventricles corresponds to the QRS interval. As the signal spreads across the cells of the heart's ventricle walls, both ventricles contract pushing the blood to the lungs for the right ventricle and to the aorta for the left ventricle. The T wave corresponds to the point when the ventricles are relaxing.

The characteristic waves are shown in Fig. 7.2 and are called *P wave, QRS complex,* and *T wave.* QRS complex includes *Q, R and S waves.* The time difference between fiducial points on the ECG waveform is of clinical importance. The time difference between the end of one ECG wave to the beginning of the next wave is referred to as a *segment.* The time difference between the fiducial points that include a segment and one or more waves is called an *interval.* The PR interval represents the time difference between the beginning of the P wave and the beginning of the QRS complex (Q wave); therefore, it includes the P wave. The PR interval is important as it is related to the complete excitation of the atria. The PR segment is defined between the end of the P wave to the beginning of the QRS complex. The QT interval is the time difference between the Q wave and the end of the T wave. It is important because it covers the overall ventricular activity. The ST segment is the time difference between the S wave and the beginning of the T wave, while the ST interval is the time between the S wave and the end of the T wave.

R—R interval is the time difference between the peaks of the consecutive R waves. *Heart rate variability (HRV)* is beat-to-beat variation in the duration of the R—R intervals. Normal cardiac rhythm is called *sinus rhythm.* An *arrhythmia* is an event where the rate or rhythm of the heartbeat is abnormal. When a heart beats too fast, the condition is called tachycardia. When a heart beats slower than normal, the condition is called bradycardia. *Atrial fibrillation* is an arrhythmia characterized by high-frequency excitation of the atrium. Atrial fibrillation is recognized in the electrocardiogram (ECG) as an irregular rhythm, with no noticeable P-waves before the QRS complex.

7.2.2 Measurement method: leads

The heart's electrical activity is recorded by placing electrodes on the skin and recording the difference in potential between the electrode pairs. An ECG lead consists of one positive and one negative electrode (bipolar lead) or one positive surface electrode and a reference point. A standard diagnostic ECG system uses *12 leads* formed by 10 electrodes (Malmivuo and Plonsey, 1995). The placement of 10 electrodes is shown in Fig. 7.3.

The three electrodes are placed on the right side, left side, and lower left side. Commonly, they are placed on the right arm (RA), left arm (LA), and left leg (LL), and therefore they form the limb leads. The limb leads are:

Lead I: $V_I = E_{LA} - E_{RA}$
Lead II: $V_{II} = E_{LL} - E_{RA}$
Lead III: $V_{III} = E_{LL} - E_{LA}$

Here, E_{LA}, E_{RA}, and E_{LL} are the potentials of the electrodes placed on the left arm, right arm, and right leg, respectively while V_I, V_{II}, and V_{III} are the potential differences between leads I, II, and III, respectively. Please note that the potential differences of leads I, II, and III are dependent and that $V_I + V_{III} = V_{II}$. Even though the potential differences between the leads are dependent, all three are commonly used in clinical practice. The three limb leads

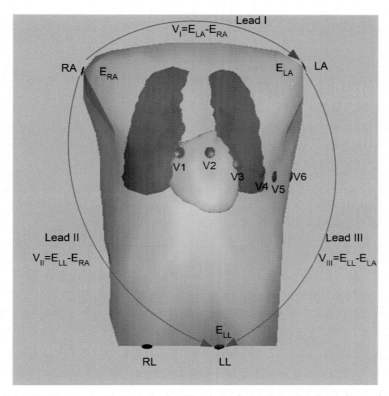

FIGURE 7.3 12 lead ECG system with 10 electrodes. The electrodes are LA—the electrode attached to the left arm, RA—the electrode attached to the right arm, LL—the electrode attached to the left leg and RL—the electrode attached to the right leg that is normally used as a ground electrode, V1, V2, ..., V6 are six chest electrodes. In addition, leads I, II, and III formed using RA, LA, and LL electrodes are shown in the figure. E refers to the potential of the electrode, while V refers to the voltage. The figure is extracted from the simulator ECGSim (Arts et al., 2005) for a default configuration of a normal young male.

are also called Einthoven limb leads, and the triangle made by connecting the limb electrodes using straight lines (not shown in Fig. 7.3) is called the Einthoven triangle.

The other nine leads are unipolar and formed as a difference between the potential of one electrode and the average potential of two or more other electrodes.

Three additional leads can be formed by combining the limb electrodes and they are named augmented limb leads. Their potential differences are labeled as aVL, aVR, and aVF, where aVL is augmented voltage of the left arm, aVR is used for the right arm, and aVF for the left foot. Here, the voltage is computed as a potential of the electrode of interest versus the reference point formed as an average potential of the other two limb electrodes. For example, $aVL = E_{LA} - (E_{RA} + E_{LL})/2$.

Another set of electrodes used in 12-lead ECG consists of six electrodes attached to the chest. These electrodes are unipolar electrodes where the positive electrode is a chest electrode, and the mean of the potentials of the three limb electrodes acts as a virtual negative electrode.

BOX 7.1

What ECG lead configuration is used when ECG is measured using smartwatches?

ECG signal obtained from smartwatches is normally based on the bipolar lead configuration in which the negative electrode is placed on the crown of the watch and normally represents the RA electrode and the positive electrode is on the back of the watch and represents the LA electrode. Therefore, smartwatches are configured in Einthoven's

ECG lead I. It has been shown that by placing the watch on the left lower abdomen region and by touching it with the left or right index finger, it is possible to simulate Lead II and III configurations (Samol et al., 2019). In Samol et al. (2019), it was also shown that it is possible to simulate leads V1, V4, and V6 of a standard 12-lead ECG.

Therefore, the voltage of the electrodes is computed as $V_n = E_n - (E_{LA} + E_{RA} + E_{LL})/3$, where E_n is the potential of the nth electrode and $n = 1, 2, ..., 6$. The six leads formed with these electrodes as positive electrodes and the average of the potential of the limb leads as a negative electrode are called precordial leads. Please note that the symbol V_1 is used for the position of the electrode on the chest (see the position in Fig. 7.3) and for the potential difference between that electrode and the mean of the potentials of the three limb electrodes. The same is for the other precordial leads.

In addition to the nine electrodes mentioned earlier, the 10th electrode (the right leg electrode) acts as a ground or virtual ground, as described in Section 7.6.4.2.

In this book, we are interested in pervasive devices that rarely use 12-lead ECG systems. Instead, it is much more common to encounter *single-lead* ECG devices consisting only of 2 electrodes. Please see Box 7.1 for the comment on using ECG in smartwatches.

7.3 Signal properties, simulators, and databases

7.3.1 How is the ECG signal affected by physiological parameters and noise?

ECG morphology is affected by noise, motion artifacts, muscle movements, and so on. Noise modeling for biomedical signals was introduced in Chapter 4, Section 4.2. The ECG signal can be modulated by respiration both in time and in the amplitude of the peaks.

Changes in the R–R interval or large HRV are normally a sign of good health. Changes in R–R interval over time are synchronous with respiration so that the R–R interval shortens during inspiration and prolongs during expiration and are, therefore, called *respiratory sinus arrhythmia* (RSA). The RSA manifests itself as a peak in the high-frequency spectrum of R–R intervals in the range between 0.15 Hz and 0.4 Hz. In addition to respiration, the R–R intervals are often modulated with a signal that has a period of about 10 s. This signal produces a peak in the low frequency (LF) band between 0.04 Hz and 0.15 Hz, and it is called the *Mayer wave* in arterial pressure signals. There is a debate in the medical research community about the origin of the LF modulations of the R−R intervals (Cohen and Taylor, 2002).

Motion artifacts are caused by the relative motion of electrodes with respect to the skin (Li et al., 2016). This relative motion results in stretching or compressing the skin leading to changes in the thickness of the skin and, therefore, changing the skin-electrode impedance. Variations in skin-electrode impedance will result in alterations in the potential difference between the ECG electrodes and, therefore, will also result in artifacts in the ECG signal. It has been shown that the effect of a motion artifact can be represented by a single equivalent variable voltage source (Zipp and Ahrens, 1979). The electric charge that accumulates on the surfaces of two materials rubbing against each other also contributes to forming a motion artifact. This phenomenon is termed *triboelectricity* and affects dry electrodes (those without gel or sweat) more than wet electrodes (Li et al., 2016).

7.3.2 Signals quality

As pointed out in Orphanidou (2018), the signal quality assessment can be divided into basic and diagnostic signal quality. Basic signal quality indicators give a high score to the signal with clear R waves without being concerned about detecting other waves in the ECG waveform. The basic signal quality indicators are used while estimating the heart rate, doing HRV analysis and detecting some arrhythmias. Diagnostic signal quality indicators give high scores when the ECG signal can be segmented so that the P wave, QRS complex, and T wave are clearly identifiable. In this case, the signal can also be used for clinical diagnosis. Since this book focuses on pervasive devices, we will discuss only basic signal quality assessment methods.

Signal quality assessment for ECG signal was introduced in Chapter 4, Section 4.4.3. Three SQIs for the ECG signals were shown, including:

— Computing kurtosis and determining the quality of a segment of the signal of interest,
— Computing spectral power ratios between the power of the signal and the power of the signal and noise for the segment of the signal of interest,
— Performing template matching and computing the quality index for each beat in the signal segment.

All these quality indices deal with basic signal quality and can be applied to almost any physiological signal considered in this book. In addition to these quality indices, several other basic beat-to-beat signal quality indices for ECG are commonly used, including comparing the results of two beat detection algorithms on a single lead and comparing beat detection using multiple ECG leads.

7.3.2.1 Comparison of multiple beat detection algorithms on a single lead

The algorithm for QRS detection will be introduced in Section 7.5.1 (Pan and Tompkins, 1985). The idea behind this signal quality assessment method is to use two very different QRS detection algorithms and compare their results beat by beat. If both algorithms detected the same beats, we assume that the ECG waveform is of high quality. The requirement is, however, that the algorithms show different sensitivity to different types of noise. The corresponding equation of the signal quality index is

$$bSQI = N_{matched}/N_{all}$$

where $bSQI$ is the beat detection signal quality index that ranges between 0 and 1, $N_{matched}$ is the number of beats that were detected by both algorithms synchronously, N_{all} is the number of beats that were detected by either of the algorithms (Li et al., 2008).

7.3.2.2 Beat detection comparison using multiple ECG leads

If the synchronous signals from L ECG leads are available, it is possible to compare the results of a QRS detection algorithm applied to signals from two or more leads. The SQI is called an interchannel signal quality index ($iSQI$). SQI for lead i ($iSQI_i$) represents the maximum ratio of the number of beats that are matched $N_{matched\,i,j}$ between the signals from the lead i and other leads j to the number of all detected beats N_{all} by the lead i:

$$iSQI_i = max(N_{matched\,i,j}/N_{all\,i}) \quad j = 1,...,L, j \neq i$$

This method is useful when multiple leads are available.

7.3.3 Simulators

ECG models and simulators can be divided based on what type of information is required from the simulator based on the application. They include:

- Physiological simulators that show the action potential and allow for analyzing the effect of different health conditions on ECG morphology.
- White-box mathematical or machine learning-based models and simulators that generate an ECG signal with a focus on its morphology, the effects of noise, breathing, and arrhythmias.
- Multimodal simulators that allow for generating ECG together with other physiological signals such as PPG and blood pressure.
- ECG simulators for different arrhythmias that allow for showing the effects of arrhythmias on ECG morphology or detecting and classifying the arrhythmias.

ECGSim is a freely available physiological simulator that presents ECG waveforms on the thorax based on the modeled electrical activity of the heart for several ECG lead configurations, including 12 lead and single lead I configurations (Arts et al., 2005). It shows the action potential in the different areas of the heart as well as the ECG signal. The waveforms can be stored and processed offline in languages like Matlab and Python. Fig. 7.3 was extracted from ECGSim. ECGSim allows for observing the action potential at the different places on the heart, visualizing the heart vector in 3D over time and relating it with the morphology of the ECG signal, observing the signal for different lead configurations including the single lead.

The morphology of the waveform is modeled by approximating ECG waves with different functions, including polynomials, Gaussian functions, and functions that simulate burst-type signals. Models commonly allow for adding noise, breathing artifacts, and arrhythmias. One

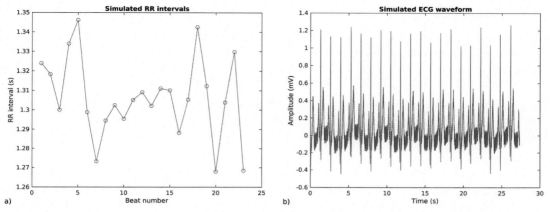

FIGURE 7.4   (A) simulated R—R intervals and (B) simulated ECG waveform based on McSharry et al. (2003).

of the most used models is described in McSharry et al. (2003), representing a dynamic model of coupled oscillators modeled using differential equations. The model includes R—R interval variations and the contribution of respiration and the Mayers wave on the R—R intervals. We describe the idea behind the model and show the simulation results in Fig. 7.4. First, the R—R interval is specified using its mean (based on the desired heart rate) and the standard deviation. In the case shown in Fig. 7.4, the heart rate is 45 beats per minute and the standard deviation of the heart rate is 1 beat per minute. Then, the ratio of the signal powers formed from R—R intervals between low frequency (LF) band 0.015—0.15 Hz and high frequency (HF) band 0.15—0.4 Hz is provided. We will discuss the R—R intervals and the frequency bands in Section 7.8. LF/HF ratio in the power spectrum is selected to be 0.5 in Fig. 7.4. From the power spectrum, using the inverse Fourier transform, the variations of R—R intervals from beat to beat can be simulated, and the beginning of every beat can be determined. R—R intervals are shown in Fig. 7.4A. As can be observed, there is a large variation in the R—R interval even during a short period. After the beginning of each wave in the ECG signal is specified, several other parameters need to be estimated to simulate the ECG waveform using Gaussian functions. We will not go into details in the book on how the differential equations are solved to determine the parameters—the code on the book webpage will allow for experimentation and generation of different waveforms. The simulated signal from the model given by McSharry et al. (2003) is shown in Fig. 7.4B. The signal looks realistic as it provides realistic timing of ECG waves and includes the modulation due to respiration in both amplitude and frequency. In addition, it is easy to add noise and motion artifacts to the signal.

Multimodal simulators have been introduced recently based on a similar concept of coupled oscillators. The one presented in Antink et al. (2017) allows for generating ECG, PPG, respiratory, and BCG signals.

A number of models and simulators are used to generate different arrhythmias. Often, a hidden Markov model is used to switch between normal beats and beats with arrhythmias. Recent work on simulating ECG signals with arrhythmias and different noise types is Awal et al. (2021). This work simulates ECG using a simplified model with two Gaussian functions and allows for generating several types of arrhythmias and compressing the signal.

7.3.4 Databases

A large number of ECG signal databases exist compared to the number of databases for other physiological signals. Here, we present a short list of the databases—for more information, please consult (Flores et al., 2018). Several of these databases are located on PhysioNet (Goldberger et al., 2000). If the databases contain only ECG signals, then they are used for:

(a) classifying *beats and rhythm* as normal or with arrhythmias.
(b) classifying the patient conditions based on the analysis of multiple beats, and
(c) biometric/security applications.

Databases containing other physiological signals besides ECG are used for other applications depending on the physiological signals provided. For example, ECG measured in combination with PPG that uses blood pressure signal as a reference can be used for estimating blood pressure based on the pulse arrival time—for more information, please refer to Chapter 8.

The (Physikalisch-Technische Bundesanstalt (PTB) database contains 12-lead ECG signals sampled at 1000 samples per second with 16-bit resolution. There are 52 healthy controls and 238 patients. Information about patients is provided; therefore, it can be used for the application (b) (see the previous paragraph): developing algorithms to classify patient conditions. The database is available at https://www.physionet.org/physiobank/database/ptbdb/.

A commonly used database that has each beat annotated is MIT-BIH arrhythmia, and as such, it can be used for the application (a): developing models and algorithms for detecting and classifying arrhythmias. The database contains 48 30-min segments from 47 patients (Moody and Mark, 2001). The waveforms from two ECG leads are provided. The signal was sampled at 360 samples per second per channel with an 11-bit resolution. The annotations are divided into beat and rhythm annotations. *Beats* are annotated as normal, atrial premature, ventricular fusion, ventricular premature beat, and so on. *Rhythm* annotations cover a sequence of beats, and rhythms are classified as normal sinus rhythm, ventricular tachycardia, atrial flutter, atrial fibrillation, and so on. ECG waveforms from 2 leads, as well as beat and rhythm annotations, are shown in Fig. 7.5. Beat annotations can be seen next to each pulse—dot means normal beat, *a* means the aberrated atrial premature beat, while *A* means the atrial premature beat. Rhythm annotations start with left parenthesis followed by the type of the rhythm—for example, "AFIB" in Fig. 7.5 means that the detected rhythm is atrial fibrillation. The database is available at: https://www.physionet.org/content/mitdb/1.0.0/.

Besides using ECG for diagnostics and health monitoring, it is also used for evaluating stress and emotions, as well as for biometrics. A database (Pouryayevali et al., 2014) contains short ECG recordings from 1012 subjects using Lead I configuration. For 43 patients, their recording was repeated over 6 months six times. For some subjects, waveforms are collected at different body postures and pre- or postexercise. This database is intended for biometric research and can be obtained at: http://www.comm.utoronto.ca/~biometrics/databases. html. Several large databases suitable for developing machine learning models were described in Siontis et al. (2021).

Multiparameter physiological signal databases exist as well. A comprehensive database of more than 3300 surgical patients with anesthesia containing many physiological signals acquired synchronously, including ECG signals, is described in Lee and Jung (2018). Data

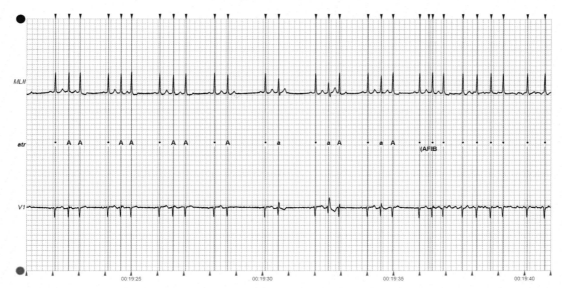

FIGURE 7.5 Annotated ECG signal for subject 201 from ECG. Waveforms from two leads are shown for a time interval of about 20 s. Beat annotations are: *a*: aberrated atrial premature beat, *A*: atrial premature beat, and · : normal. Rhythm annotation is AFIB: atrial fibrillation.

are collected from patient monitors that provide some or all of the following signals: ECG, capnogram for monitoring concentration of carbon dioxide, plethysmogram, respiration signal, blood pressures, as well as signals/outputs from anesthesia machines and cardiac output monitors. ECG signal is sampled at 500 samples per second. The database is available at https://vitaldb.net/.

7.4 Types of devices and standards

7.4.1 What does the ECG device display?

Ambulatory ECG monitoring systems record an ECG signal for prolonged periods during which the patient does not need to be still. As such, they are useful for detecting short, intermittent events such as arrhythmias. The definition of the ambulatory ECG monitor based on Steinberg et al. (2017) is: "External ambulatory ECG serves to detect, document, and characterize abnormal cardiac activity during ordinary daily activities, extending the role of ECG recording beyond the bedside 10 s standard 12-lead resting ECG." An example is a *Holter monitor*, a battery-operated device with a minimum of three leads that commonly records ECG for 24–48 h. Processing of the ECG waveform is commonly performed offline. An alternative to traditional Holter monitors is a single-lead or two-lead *wireless patch* that is attached to the chest and can record the ECG signal for up to 28 days. These patches are comfortable to wear and are water-proof. As such, they do not interfere with patients' activities. Newer patch ECG monitors can also measure other vital signs (body temperature, respiration) and monitor patient activities. Both patches and Holter monitors often do not have an interface to display the recorded waveform. The waveforms are first transferred to a computer or a smartphone, stored and then displayed on the computer or smartphone screen. Holter monitors and patches are able to detect and sometimes classify arrhythmias and estimate the heart rate and heart rate variability indices.

In contrast to the continuous ECG monitors that always record the ECG signal, *event recorders/monitors* record the ECG signal for a predefined period after the patient detects the symptoms and activates the recorder. The symptoms include bradycardia and supraventricular and ventricular arrhythmias. *External loop recorders* are worn continuously and activated to record the ECG signal for a predefined period (Steinberg et al., 2017). The recorder can be activated by the patient or automatically by its software for arrhythmia detection. *Mobile cardiac telemetry devices* record the ECG signal, perform real-time processing to detect arrhythmias, and stream data wirelessly. They combine the benefits of Holter monitors (or patches) with external loop recorders. Event monitors and patches are mainly implemented using a single lead. The placement of the patches on the body is discussed in Box 7.2.

Medical-grade diagnostic ECG monitors normally record 12 lead ECG for a short time, such as 10 s.

7.4.2 Standards

At this point, a new IEC 80601−2-86 standard that represents a set of general basic safety and performance requirements for all types of ECG devices is still in the process of

---

BOX 7.2

Position of the wireless patch on the body

Wireless patches used for long-term monitoring are very compact devices that measure up to 10 cm in length. As such, the electrodes are placed close to each other, which is not common in traditional ECG devices. By decreasing the distance between the electrodes placed on the chest, the amplitude of the ECG signal becomes smaller while the noise remains at the same level as if the electrodes are placed at their "traditional" positions. Therefore, the patches need to be placed on the body at the specific location they are designed for. Two positions are most often recommended by the manufacturers: horizontal in which the electrodes are placed

on the upper left side of the thorax or vertical where they are placed in the middle of the thorax. In Rashkovska et al. (2020), the patch has been placed horizontally in nine different positions across the thorax and vertically in four different positions in the middle of the thorax. Almost all of them gave very clean ECG signal with clearly discernible P waves—however the amplitude of the QRS complex varied a lot among these different positions. The signal obtained from the patches might not correspond to any of the signals obtained using traditional 12-lead ECG systems (Rashkovska et al., 2020).

development (ISO/IEC 80601, 2018). It combines six existing standards that are commonly used when evaluating ECG devices, including (Young and Schmid, 2021):

- IEC 60601-2—25: Particular requirements for the basic safety and essential performance of electrocardiograph.
- IEC 60601-2—27: Particular requirements for the basic safety and essential performance of electrocardiographic monitoring equipment.
- IEC 60601-2—47: Particular requirements for the basic safety and essential performance of ambulatory electrocardiographic systems.
- ANSI/AAMI EC12: Requirements for disposable ECG electrodes.
- ANSI/AAMI EC53: Requirements for ECG cables and lead wires.
- ANSI/AAMI EC57: Requirements for testing and reporting performance results of the algorithms for detecting cardiac rhythm and the ST segment.

IEC 80601—2-86 standard is supposed to unify standards for diagnostic electrocardiographs, ECG patient monitors used in hospitals and ambulances, and ambulatory ECG equipment.

The minimum frequency response for ECG patient monitors (0.67—40 Hz) now applies to all ECG equipment with the rationale that this is the minimum bandwidth needed to allow accurate ECG reproduction for rhythm interpretation. In addition, if a device is intended for the ST segment measurements, it must meet the additional bandwidth requirement of a low-frequency cut-off of 0.05 Hz (ISO/IEC 80601, 2018).

The other relevant standards are related to biocompatibility. Any devices in direct contact with the skin of the human body should comply with the requirements of biocompatibility standards. The device should not cause any toxic, injurious, or physiologically reactive reaction or immunological rejection when in contact with living tissue or a living system. ANSI/AAMI/ISO 10,993 standard is used to evaluate and test biocompatibility. ECG electrodes should be biocompatible. For applications where the electrodes are in direct contact with the skin, biocompatibility requires evaluating cytotoxicity, skin irritation, and skin sensitization. According to the FDA regulations, complying with the requirements of the following standards is essential.

- Dermal irritation and delayed-type sensitivity (See ISO 10,993-10 Biological evaluation of medical devices—Part 10: Tests for irritation and delayed-type sensitivity),
- Cytotoxicity (see ISO 10,993-5 Biological evaluation of medical devices—Part 5: Tests for cytotoxicity).

7.5 Processing the ECG signal

To discuss algorithms for processing ECG signals, we need to discuss the applications and the objectives of processing. With ECG, we might be interested in the following applications:

- Presenting filtered ECG signal on the screen.
- Computing the heart rate.
- Computing the heart rate variability indices.

- Detecting the fiducial points and computing the durations of different ECG segments.
- Classifying beats to detect irregular beats or arrhythmias.
- Classifying patients' conditions, including congestive heart failure and episodes of atrial fibrillations.

Therefore, many different signal processing algorithms and machine learning models have been applied in this field. A survey of machine learning models is given in Lyon et al. (2017). We will show a basic algorithm for estimating the heart rate. The algorithm for detecting the fiducial points on the ECG waveform is shown in Appendix 7.10.

7.5.1 Heart rate estimation

One of the important requirements when processing ECG signals is that the algorithms should be able to detect the QRS complex accurately. Accurate detection of QRS complex is needed for almost all types of applications presented earlier, as well as some other applications such as blood pressure estimation based on the time difference between the fiducial points of different signals, which will be discussed in Chapter 8. The algorithm presented here is based on the classical Pan and Thompson algorithm for QRS detection and heart rate estimation (Pan and Tompkins, 1985).

Fig. 7.6 shows the processing steps of the algorithm, while Fig. 7.7 shows the waveforms after each processing step. A noisy ECG signal is the input to the algorithm, and it is shown in Fig. 7.7A. The algorithm has six processing steps, and it outputs the clean ECG signal, time instants of the QRS complex, heart rate estimates over time, and R—R intervals. In our example, the signal is sampled at 250 Hz, and the amplitude of the QRS complex is about 1 mV. Two types of noise have been added to the original signal using the methods described in Chapter 5: white Gaussian noise with a standard deviation of 0.1 mV and 60 Hz interference with an amplitude of 0.1 mV. The first step is bandpass filtering, and we selected a bandpass filter with cut-off frequencies of 1 Hz and 20 Hz. Bandpass filtered signal is shown in Fig. 7.7B. It is clear that the SNR is improved after bandpass filtering. The slope of the R wave is an important feature in detecting the QRS complex. Therefore, the next processing step (step 2) in the pipeline is the differentiator, whose output is shown in Fig. 7.7C. The signal has a sharp positive peak where the rising edge of the R wave is the steepest and a sharp negative peak where the falling edge of the R wave is the steepest. In step 3, each sample of the waveform is squared. This emphasizes the higher frequencies in the ECG signal and amplifies the signal after differentiation. The squared waveform scaled in the range between 0 and 1 is shown in Fig. 7.7D. The squared signal in Fig. 7.7D has two very close peaks in each cardiac cycle per each QRS complex. These two peaks are due to the steep rising and falling edges of the QRS complex. Step 4 is to perform averaging and convert these two peaks into a single peak. The number of samples of the sliding window needs to be chosen so that the filter output is the single peak per beat. In our case, we used the moving average filter with a sliding window of 30 samples. The output is shown in Fig. 7.7E. This figure shows pulses at the QRS complex positions shifted by the filters' processing delay. To correctly determine the exact position of the QRS complex, the original ECG signal should be shifted by the number of samples corresponding to the processing delay of the filters. Step 5 includes the conversion of the averaged pulses into binary pulses with amplitudes 0 and 1. In our case, this is done by setting a threshold corresponding to a

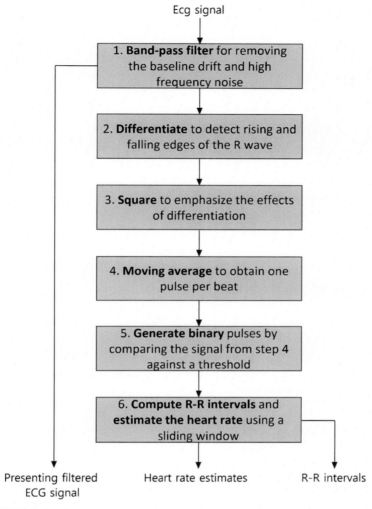

FIGURE 7.6 QRS complex detection algorithm based on Pan and Tompkins (1985).

percentage of the maximum amplitude of the squared waveform—for example, 50% of the maximum amplitude. The digital pulse train is shown in Fig. 7.7F.

Ultimately, it is possible to estimate the heart rate based on the pulse train. It is normally expressed as the number of beats per minute. The heart rate can be estimated as the number of pulses detected over the period or as an inverse of the average R—R intervals. R—R intervals are computed as the time difference between each rising edge of the pulses in the pulse train. Then, one can compute the average R—R interval T_{RRmean} over the time window T_w. The heart rate in beats per minute is computed as

$$HR = 60/T_{RRmean}$$

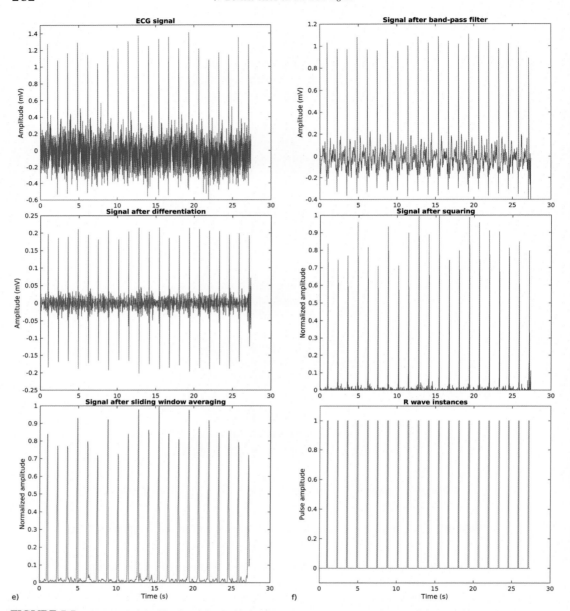

FIGURE 7.7   Outputs of the blocks of the QRS detection algorithm shown in Fig. 7.6. Figures (B) to (F) are related to steps 1—5 in Fig. 7.6. The figures are (A) original noisy ECG signal, (B) output of the bandpass filter, (C) output of the differentiator, (D) output of the squaring function, (E) output of the averaging block, and (F) the train of pulses with logic one at the position of QRS complex.

7.5.1.1 *Improvements to the heart rate estimation algorithm*

It is possible to detect false QRS complex or to miss some of the QRS complexes either due to noise or dominant T waves, or other reasons. Therefore, several physiological and other measures are commonly used to determine whether to select the peak as the QRS complex or not. For example, it is proposed in Pan and Tompkins (1985) to select only those peaks whose R—R intervals are greater than 360 ms. If the peaks are too close, they can represent noise or T waves. If the time difference between the peaks is larger than $1.5T_{RRmean}$, then the maximum peak between the two R—R intervals can be selected as the QRS complex.

Quality indicators can be used to select beats of satisfactory quality. Therefore, some beats will be rejected as low-quality beats and should not be considered in heart rate estimation.

In the algorithm presented in Section 7.5.1, the threshold was selected as a percentage of the maximum amplitude in the time window T_w. It was proposed in Pan and Tompkins (1985) to compute the threshold based on the weighted average obtained from the maximum amplitude of the selected QRS complex and the maximum amplitudes of nonselected pulses that correspond to noise.

Heart rate estimation algorithms normally operate in real time, and therefore, they compute the heart rate continuously in a given window T_w. In addition, these monitors adjust the threshold adaptively based on the signal amplitude and the noise level. Implementation of a similar algorithm that processes the ECG signal continuously is presented in Real-Time ECG QRS Detection.

7.6 Implementation of a single-lead ECG system

7.6.1 An overall ECG system

We will start with the description of the overall ECG device based on a single lead configuration. The device has two electrodes connected to an instrumentation amplifier that is further connected to a bandpass filter, additional amplifiers, an A/D converter, and processing logic. As we will explain later, amplification is often done in two stages to avoid amplifying the noise and the DC signal in case of the baseline drift. In addition, the third electrode is often added, which acts as a ground electrode. The third electrode can be connected with the rest of the circuitry through an amplifier in a negative feedback loop instead of being connected to the ground. That circuit is called the driven right leg circuit, and its goal is to reduce the common mode voltage that appears on the other two electrodes. The block diagram of the ECG device is shown in Fig. 7.8. Blocks specific to ECG processing are shown in blue in Fig. 7.8. Nowadays, the sampling rates of A/D converters used in ECG devices vary between around 100 Hz and 8000 Hz. The resolution of the A/D converters is between 16 and 24 bits. The analog front end includes an instrumentation amplifier with a common mode rejection ratio (CMRR) better than 100 dB to 1 kHz frequency band, about 50 μV offset voltage, and low input bias current in the order of nA.

ECG systems also have several general-purpose components shown in gray. They include battery and power management circuits and a wireless communication unit (Bluetooth, WiFi) with an antenna. ECG circuits normally contain the lead-off detector circuit to determine whether the electrodes are properly attached. The radio frequency protection circuit and

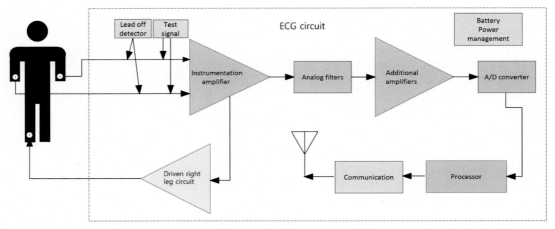

FIGURE 7.8   Block diagram of an ECG device. White circles on the subject represent electrodes. Blue blocks are related to processing the ECG signal. The yellow block represents the feedback, while the gray blocks present general-purpose hardware or other elements not directly related to processing the signal.

protection against electrostatic discharge (ESD) are often included at the input of the circuits and are not shown in the figure. The test circuit generates a pulse for use in subsystem verification at power-up.

7.6.2 Amplifiers and filters

The amplifier and filter circuit blocks are presented in Fig. 7.9. They include an instrumentation amplifier, a passive highpass filter, and an active lowpass filter. The design is based mainly on Webster (2010). The instrumentation amplifier provides high-impedance input that connects electrodes with the rest of the system. Amplification in the circuit is divided between the instrumentation amplifier and the active lowpass filter. The reason for that is the existence of DC bias at the input that would cause saturation of the instrumentation amplifier if the gain of the instrumentation amplifier is too high. The DC component is caused by the baseline drift due to motion artifacts or external interference. Therefore, the idea is to amplify the signal in two stages with the highpass filter for removing the DC offset between them.

The differential gain of the instrumentation amplifier (IA) A_{dIA} is computed as

$$A_{dIA} = -\left(1 + \frac{R_1 + R_2}{R_3}\right) \times \frac{R_6}{R_5} = -\left(1 + \frac{25 \text{ k}\Omega + 25 \text{ k}\Omega}{5.5 \text{ k}\Omega}\right) \times \frac{50 \text{ k}\Omega}{25 \text{ k}\Omega} = -20.2$$

The gain of the active lowpass filter A_{dLP} is

$$A_{dLP} = -\frac{R_{10}}{R_8} = -31.9$$

Therefore, the differential gain of the whole circuit is $A_{dIA} \cdot A_{dLP} = 644.4$

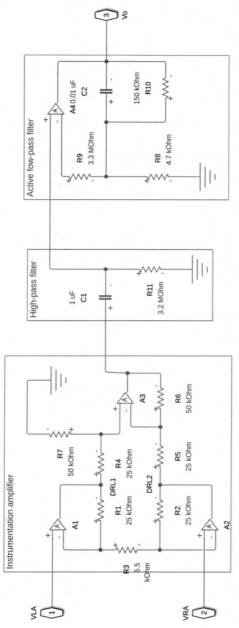

FIGURE 7.9 ECG amplification and filtering stages.

The cut-off frequency of the highpass filter f_{HP} is given by the following equation:

$$f_{HP} = \frac{1}{2\pi R_{11}C_1} = \frac{1}{2\pi 3.2 \text{ M}\Omega \times 1 \text{ }\mu\text{F}} = 0.05 \text{ Hz}$$

The cut-off frequency of the active lowpass filter is

$$f_{LP} = \frac{1}{2\pi R_{10}C_2} = \frac{1}{2\pi 150 \text{ k}\Omega \times 0.01 \text{ }\mu\text{F}} = 106.1 \text{ Hz}$$

Therefore, the circuit filters out the frequency components outside the band [0.01 Hz, 106.1 Hz].

The input and output signals of the front-end circuit are shown in Fig. 7.10. Fig. 7.10A shows the input signal, which is in the order of mV. Fig. 7.10B shows the output signal that is amplified around 644.4 times. To compare the performance of this circuit with some other circuits introduced later in this chapter, we will compute the spectral distribution ratio of the signal at the output of the front-end circuit. The spectral distribution ratio corresponds to the signal-to-noise ratio, and it was introduced when we discussed signal quality assessment in Section 4.4.3. Here, we are looking at the ratio of the power of the signal in the range of 5–14 Hz versus the power of the signal and noise in the range between 5 and 70 Hz. The spectral distribution ratio is 0.97.

7.6.3 Simulation of a single lead ECG system

A block diagram of a model of the idealized ECG system is shown in Fig. 7.11. It consists of the bioelectric sources, electrode-skin interface models for both electrodes, and circuits, including amplifiers and filters.

The bioelectric signal at the input of the electrodes is presented using the following approach. Here, we first assume that the voltage at the electrodes includes biopotential inside the body as well as common mode voltage. Common model voltage is generated by

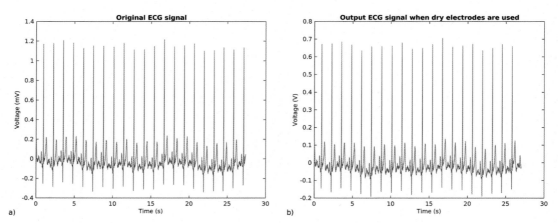

FIGURE 7.10   ECG signal (A) at the input and (B) at the output of the front-end circuit.

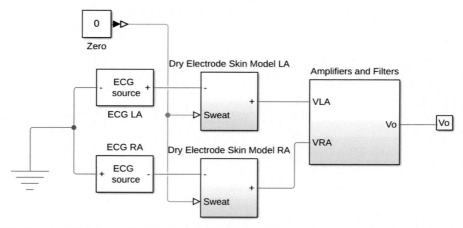

FIGURE 7.11 A high-level model of the lead I ECG system without a common mode voltage source. It contains the ECG signal source(s), electrode and electrode–skin interface models, and the circuit that contains amplifiers and filters.

displacement current flowing through the body due to 60 Hz interference. Therefore, we can present the voltage at the input of the electrode model for the left arm V_{LA} and the right arm V_{RA} as

$$V_{LA} = V_{cm} + V_I/2$$

$$V_{RA} = V_{cm} - V_I/2$$

where V_{cm} is common mode voltage and V_I is the differential voltage over the lead I. Voltages V_{cm} and V_I can also be expressed as

$$V_I = V_{LA} - V_{RA}$$

$$V_{cm} = (V_{LA} + V_{RA})/2$$

In Fig. 7.11, we do not consider the common mode voltage ($V_{cm} = 0$), and therefore, the input is modeled as a positive Lead I ECG signal connected to the LA electrode and a negative Lead I ECG signal connected to the RA electrode. Electrode models were introduced in Chapter 2. In this example, we will assume we are using a Lead I configuration with two dry electrodes.

Example 7.1. Analyze the effects of DC bias on the output of the ECG amplification and filtering circuit in case there is no DC bias, where the DC bias is 100 mV and when the DC bias is 300 mV.

Solution. Even though the amplification is divided into two stages, the DC bias affects the output. The DC signal of 100 mV is added to the ECG signal that is in the range of 1 mV and fed to the input of the circuit in Fig. 7.9. The output is shown in Fig. 7.12. It can be seen that it takes about 15 s for a signal to stabilize.

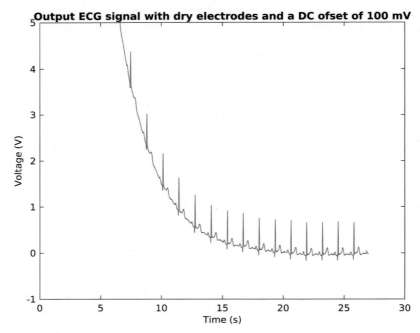

FIGURE 7.12   The waveform at the output of the ECG amplifier and filter circuit in case the DC offset is 100 mV.

The bias of 200 mV is causing the signal to saturate at the output, and the ECG signal is not visible anymore.

One way to prevent the amplifier's saturation in the presence of DC bias and common mode signals is to boost the supply voltage of the instrumentation amplifier internally. Several industrial solutions integrate the charge pump that increases the supply voltage of the instrumentation amplifier in the chip. An example is a single lead ECG front-end chip 8233 CE from Analog Devices that integrates a variation of the circuit presented in Fig. 7.9 on a single chip. Besides amplification and filtering stages, it also integrates the driven right leg circuit explained in the next section. The chip has a low supply current of 50 μA and a CMRR of 80 dB, and it can be used in two and three-electrode configurations.

7.6.4 Dealing with interference in ECG circuits

Common mode voltage can appear on the electrodes due to interference from other electronic devices and power lines. The current in the body is induced from the power lines, and it flows through the body to the ground impedance. We call this current the *displacement current* i_d. If the ground impedance is R_G, then common mode voltage is $V_{cm} = i_d R_G$. As pointed out in Chapter 3, instrumentation and operational amplifiers for biomedical applications normally have a large CMRR. However, besides using amplifiers with large CMRR, it is also possible to reduce V_{cm} through the feedback loop. Therefore, in many ECG systems, the electrode attached to the right leg is not connected to the ground but to the feedback circuit called the driven right leg (DRL) circuit. The goal of the DRL circuit is to reduce V_{cm}.

Fig. 7.13 shows the model of the system similar to the one presented in Fig. 7.11. The main differences include the displacement current source, the right leg electrode of the same type as the other two electrodes and the DRL circuit. While the signal in the left arm and right arm electrodes propagates from left to right in the figure, the signal in the feedback loop is directed from right to left. In the amplifier and filter circuit (Fig. 7.9), signals DRL1 and DRL2 are the outputs of the amplifiers A1 and A2. These signals are then inputs to the DRL circuit.

7.6.4.1 System in which the right leg is grounded

Fig. 7.14 shows the output of the ECG system in case the DRL circuit is not used (the DRL_Feedback circuit in Fig. 7.13 is connected to the ground) and the right leg electrode is connected to the ground. To reduce the CMRR of the instrumentation amplifier, the resistors R_4, R_5, R_6, and R_7 have values modified by 1% of their nominal values. The displacement current is a 60 Hz sine wave signal with a DC value of 0.2 μA and an amplitude of 0.2 μA. We can see the 60 Hz interference in the ECG signal in Fig. 7.14A. The spectral distribution ratio drops to 0.28. Fig. 7.14B shows a combined common mode and ECG signals at the point

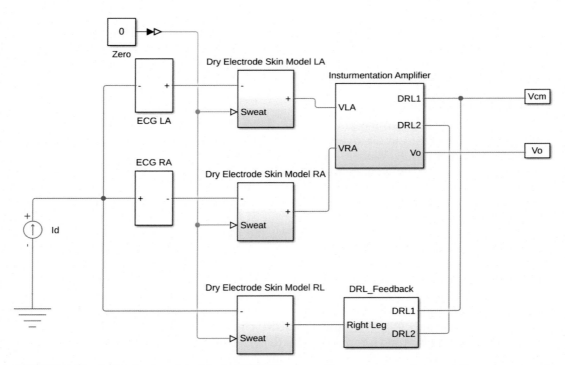

FIGURE 7.13 A high-level model of the lead I ECG system with a common mode source. The common mode signal is simulated by a current source i_d which represents the displacement current through the human body. Besides the current source, the model contains ECG signal source(s), left arm, right arm, and right leg electrode and electrode–skin interface models, the circuit that contains amplifiers and filters, and the driven right leg circuit. Please note that if we want to consider the effects of the common mode voltage on the output of the circuit without the driven right leg circuit, then we need to connect the output of the driven right leg feedback circuit to the ground.

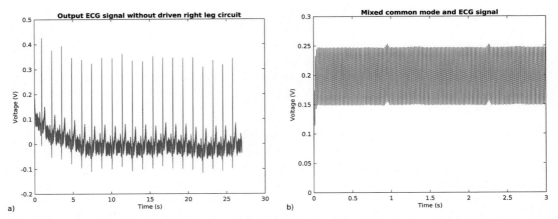

FIGURE 7.14 (A) ECG signal V_o in Fig. 7.13 in case when the driven right leg circuit was not used, and the right leg electrode was connected to the ground. (B) Common mode voltage V_{cm} in Fig. 7.13 that corresponds to the point DRL1 in Fig. 7.9.

DRL1 in Fig. 7.9 in 3 s. It is clear that the common mode signal is dominant (the amplitude is 0.1 V), and only the ECG R peak is visible at around 1 s and 2.3. This common mode signal is not completely suppressed by the instrumentation amplifier because of the large mismatch between the resistor values.

7.6.4.2 Driven right leg circuit

The RL electrode is not connected to the ground but to a feedback circuit called the *driven right leg circuit*. The DRL circuit inverts the common-mode signal that is present at the instrumentation amplifier inputs. When the current at the output of the DRL circuit is injected into the subject, it generates the voltage that cancels out the common-mode voltage, reducing the amplitude of the common-mode signal. We have already seen in Fig. 7.14B that the common mode voltage at the point DRL1 in Fig. 7.9 is very large. So, instead of connecting the RL electrode to the ground, we connect it to the output of the feedback circuit shown at the point called "Right Leg" in Fig. 7.15A. In Fig. 7.15A, the inputs are the signals DRL1 and DRL2 from the instrumentation amplifier, and the output is connected to the right leg electrode. The circuit combines the circuits presented in Hann (2012), Webster (2010). If we assume that all the electrodes are the same, the electrode-skin interface has the same parameters, and that the operational amplifiers are ideal with the zero input current, then the potential of the points DRL1 and DRL2 in Fig. 7.9 will be equal to V_{cm}. Therefore, the voltage at the point V_{b_in+} will be also V_{cm}. The equivalent DRL circuit is shown in Fig. 7.15B. The impedance of the right leg electrode is replaced with R_{RL}. The voltage $V_{in-} = 0\ V$, and the displacement current is i_d. Therefore, the common mode voltage can be computed as

$$V_{cm} = \frac{R_{RL}}{1 + \dfrac{R_f}{R_a}} \times i_d$$

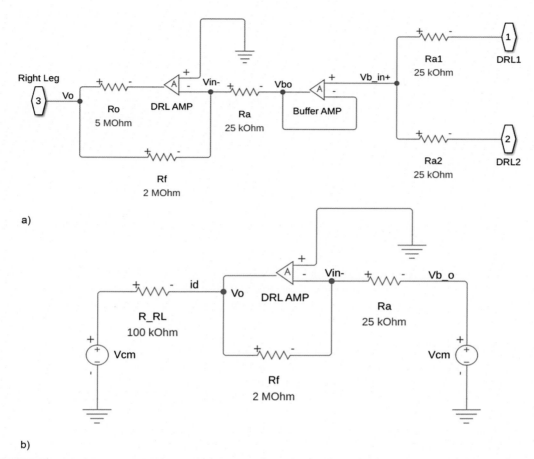

a)

b)

FIGURE 7.15   (A) Driven right leg circuit. (B) The equivalent circuit for computations of the effect of the driven right leg circuit to the common mode voltage.

In this example, the common mode voltage is reduced $1 + R_f/R_a = 81$ times. Increasing the ratio of R_f/R_a would further reduce V_{cm} up to a point (please run the simulations by increasing R_f up to 5 MΩ while keeping $R_a = 25$ kΩ and observe the amplitude of the common mode voltage at the point DRL1).

R_o is used to reduce the output current of the DRL circuit that is injected into the human body to less than 10 μA for safety reasons. In this case, the resistor is selected to be $R_o = 5$ MΩ which means that for the supply voltage of ±5 V, the maximum current will be 1 μA.

Fig. 7.16A shows the output of the ECG device with the DRL circuit. Also, to reduce the CMRR of the instrumentation amplifier, the resistors R_4, R_5, R_6, and R_7 have values modified by 1% of their nominal values. The displacement current is a 60 Hz sine wave signal with a DC value of 0.2 μA and the amplitude of 0.2 μA. We can see that the ECG signal at the output does not contain a significant 60 Hz component and that the common mode signal at DRL1 is significantly

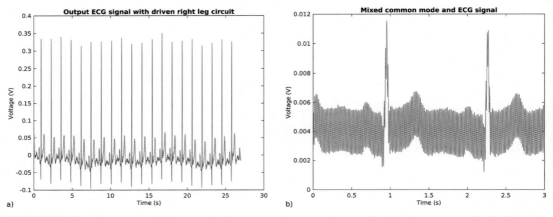

FIGURE 7.16 (A) ECG signal V_o in Fig. 7.13 in case when the driven right leg circuit was used. (B) Common mode voltage V_{cm} in Fig. 7.13 that corresponds to the point DRL1 in Fig. 7.9.

reduced (Fig. 7.16B) even though it is still high. The spectral distribution ratio is again 0.97, which corresponds to improved SNR in comparison to the SNR of the circuit without DRL feedback. Therefore, the DRL circuit is useful in reducing the effect of the common mode voltage.

7.6.5 Lead-off detector

A lead on/off detection circuit can be implemented in different ways. AC lead-off detection mode requires that a very small amplitude current at a high frequency (e.g., 100 kHz) is injected into the body between two electrodes and the voltage is measured between them. If the electrodes are properly attached to the body, we expect to observe low voltage amplitude. If one or both electrodes are detached or do not have good contact, the voltage measured between them will be close to or equal to the supply voltage. DC lead-off detection mode would require a pull-up resistor connected between the supply voltage and the positive input of the instrumentation amplifier and a pull-down resistor connected between the negative input of the instrumentation amplifier and the ground (or negative supply voltage). When the electrodes are connected, the resistors act like a voltage divider that sets the common mode voltage to the middle between the positive and negative (or ground) supply voltages. When the electrodes are disconnected, the amplifier saturates. Sensing the response can be done by monitoring the voltage on the instrumentation amplifier with an on-chip comparator. Some more advanced industrial solutions have low-resolution A/D converters that can provide information about lead-off detection—for example, ADS129x by Texas Instruments. The advantage of AC lead-off detection is that the detection can be achieved during ECG signal acquisition because the AC excitation signals occupy the frequency band outside of the ECG signal band.

7.7 Current trends and research directions

We will consider several research directions, including the research on electrode placements for wearable devices, design of the electrodes, design of wearable ECG devices, and machine learning for arrhythmia detection and classification.

7.7.1 Electrode placement

With the development of new wearable devices, research has been performed on the proper placement of ECG electrodes. Ideally, one would obtain an ECG signal of satisfactory quality when ECG electrodes are close to one another, and a single lead configuration is used. Wireless patches with electrodes in close proximity to one another were successfully commercialized. An effort has been made to obtain the ECG signal from electrodes attached next to one another on body sites further away from the heart, such as the left or right arm. Research on body surface ECG mapping of the whole left arm and the left hand for 37 subjects was performed using electrodes placed close to one another (Lynn et al., 2013). Overall, the single-arm ECG signals are weak and smaller than 80 μV even when wet Ag/AgCl electrodes are used. The signals obtained when the electrodes are placed near the shoulder deltoid muscle and the armpit are the strongest. However, the amplitude of the signals reduces rapidly as the distance from the shoulder increases. The signal amplitude near the elbow is four times smaller than that on the shoulder. Motion artifacts are a significant problem when we use dry electrodes placed close to one another. An ensemble averaging of the pulses is normally used to improve the signal-to-noise ratio of single-arm ECG signals. The systems that use dry ECG electrodes close to one another but far from the heart require special attention during the design to extract an ECG signal with high-enough SNR so that R peaks can be extracted from it. The design needs to include an active shielding circuit, the high input impedance of the order of Gigaohms, and so on. However, the signal will always be very noisy, and only R waves can be extracted.

7.7.2 Academic solutions for electrode design

Patch-based devices are normally rigid and do not allow for long-term comfortable measurements. Several solutions have been proposed to address these issues that contain sensors and associated circuits embedded in thin layers that are soft and flexible and, therefore, can achieve intimate contact with skin. It was pointed out in Wu et al. (2020) that on-skin electrodes should have the following features: conductivity, stretchability, softness, adhesion to the skin, breathability, and biocompatibility. Compared with Ag/AgCl electrodes, on-skin electrodes have much lower conductivity. The materials used to build on-skin electrodes include metal thin film, carbon materials, metallic nanomaterials, and other materials. A smart-skin solution for neonatal intensive care was developed and presented in Chung et al. (2019). It allows for stretching electrodes and electronics by 16%. Another flexible ECG patch is shown in Lee et al. (2018). The device's thickness is about 1 mm, and the length is 55 mm. The electrodes are placed on the opposite ends of the patch, and the patch is attached on the left upper side of the chest. The distance of less than 5 cm between the centers of the electrodes is considered enough for collecting an ECG signal of satisfactory quality. In these solutions, the electrodes are flexible, while the electronic components are rigid integrated circuits. Besides these solutions, there has been a surge of works in which flexible electronics is developed as well—for more details, please refer to Zulqarnain et al. (2020).

Capacitive electrode-based ECG systems have advantages because they do not require direct contact between the electrodes and the skin. Therefore, electrodes can be embedded in chairs, mattresses, and T-shirts. In these solutions, significant emphasis is on active electrode design to reduce the effect of common mode voltage and motion artifacts.

7.7.3 Devices

Wearable devices for ECG monitoring include patches, clothing-based monitors, chest straps, upper arm bands, devices that one can touch with fingers from different arms, and wristbands. Patches are popular for long-term ambulatory monitoring. Several recent solutions include patches attached on the left side of the chest that can monitor for several weeks (e.g., Vital Connect Patch). Others include a small device attached to a strip or two electrodes on the chest (e.g., Body Guardian Heart). Devices often measure temperature and impedance and have accelerometers to detect motion.

AliveCor by KardiaMobile used two dry metal electrodes that need to be touched by fingers. The device received FDA clearance to detect atrial fibrillation, bradycardia, tachycardia, and normal heart rhythm in 30 s. A similar device is ECG Check by Cardiac Design, which records ECG for 30 s and classifies the rhythm as normal or irregular.

7.7.4 Machine learning

Several applications have used machine and deep learning models on ECG data. The development of machine learning models for ECG is facilitated by the existence of comprehensive databases with labeled data, as described in Section 7.3.4. Many researchers worked on arrhythmia detection and classification. Arrhythmia detection and classification have already been included in several commercial devices mentioned earlier. Arrhythmia classification includes several processing stages: preprocessing/denoising, segmentation of the waveforms, feature extraction, feature selection, and machine learning classification. Atrial fibrillation is traditionally detected using algorithms based on 12-lead ECG. However, machine learning and deep learning approaches allowed the detection and classification of atrial fibrillation using single lead ECG. Databases used for classifying atrial fibrillation based on single lead ECG are described in Liu et al. (2021). The review of deep learning models for automated ECG interpretation, stroke risk assessment, and other applications is given in Siontis et al. (2021).

7.8 Heart rate variability

7.8.1 Introduction

A time interval of a heartbeat is called an *interbeat interval (IBI)*. *Heart rate variability* (HRV) is the fluctuation of IBI between adjacent heartbeats. As pointed out in Pichot et al. (2016), HRV is used for examining noninvasively cardiac autonomic functions, cardiac and respiratory rehabilitation, sleep-disordered breathing, stress, and emotion detection/classification. Even though HRV is easy to extract from the ECG waveform and has the potential to be used in many applications, its use is still limited due to several reasons, such as a lack of understanding of metrics that are derived from HRV and their physiological interpretation. In addition, HRV analysis results in a number of metrics or indices. This makes it much more complicated to analyze in comparison with the analysis of some other physiological signals/variables, such as blood pressure, that result in one or two most important parameters being interpreted. This has been recognized as a problem and partially addressed by the European Society of Cardiology and the North American Society of Pacing and Electrophysiology in 1996 (European Society of Cardiology, 1996). As HRV is easily obtained whenever ECG is collected for more than 5 min, there is a potential for the pervasive use of HRV. Therefore, in this section, we will introduce metrics used to analyze HRV, show how they are applied, and present their ranges for normal subjects.

The autonomic nervous system has continuous control over the SA node firing. The autonomic nervous system consists of sympathetic and parasympathetic branches. An increase in sympathetic stimulation increases the heart rate and stroke volume. On the other hand, increased parasympathetic simulation is associated with decreased heart rate and stroke volume. HRV analysis is important as it reveals the state of the sympathetic and parasympathetic branches of the autonomous nervous system and the overall cardiac health. Sympathetic stimulation occurs in response to stress, exercise, and heart disease. Parasympathetic stimulation occurs in response to trauma, allergic reactions, and so on (Acharya et al., 2010). A detailed description of the relationship between HRV and myocardial infarction, depression, arrhythmias, diabetes, and renal failure, as well as the relationship with age, gender, fatigue, drugs, smoking, and alcohol intake, are given in Acharya et al. (2010).

7.8.2 Processing steps

Processing steps in HRV analysis are shown in Fig. 7.17. As an input ECG signal, we used simulated data with the same parameters explained in Section 7.3.3, except that we generated 500 beats here. One short segment of the ECG signal is shown in Fig. 7.4B. HRV analysis is done in the time or frequency domain or using geometric methods that will be explained later. All three methods result in different metrics. The long-term analysis can take up to 24 h, and the short-term analysis usually takes 5 min, and they are both performed if long-term data are available. A minimum of the ECG segment is 2 min when computing the power in the LF band. Global analysis is performed on the overall available ECG sequence, while local analysis can be performed on any specified interval.

The input data to the HRV analysis algorithms is either an ECG signal or an already extracted sequence of pulses that correspond to R wave positions (Fig. 7.7F). If the input is an ECG signal, as in Fig. 7.17, then the R peaks need to be detected, and the beats need to be extracted (Step 1 of the algorithm in Fig. 7.17. Next, in Step 2, abnormal or noise-corrupted beats are removed either automatically by signal quality assessment methods or manually by selecting the peaks to be removed. In the end, the sequence of R−R intervals of normal beats was kept for further analysis. These intervals are called normal-to-normal or N−N intervals. Therefore, in HRV analysis terminology, N−N intervals are often used to indicate the sequence of cleaned R−R intervals. After the N−N sequence is formed, time domain analysis (Step 5a) and analysis based on geometric methods (Step 5b) can be performed. Also, since R waves are detected at intervals that are not fixed, interpolation of the N−N sequence need to be performed before the Fast Fourier transform or other frequency domain methods can be applied. This is done in Step 4 of the algorithm. Finally, frequency domain metrics are extracted in Step 5c.

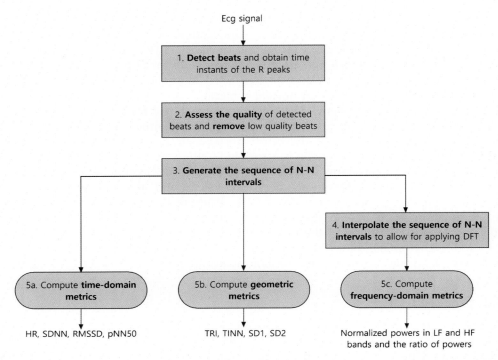

FIGURE 7.17   Processing steps in HRV analysis—modified from (European Society of Cardiology, 1996). The abbreviations of HRV metrics are explained in Section 7.8.3.

7.8.3 HRV metrics

The examples of time-domain, frequency-domain, and geometric method graphs that can be used to compute the HRV metrics are shown in Fig. 7.18. The graphical user interface and the results are obtained by running the code HRVTool version 1.07, explained in Vollmer (2019). A detailed description of the metrics, their uses, their typical values, and ranges are given in Table 7.1.

7.8.3.1 Time domain metrics

R—R intervals are presented in the time domain using an R—R tachogram graph at the top of Fig. 7.18A. The tachogram shows the R—R intervals versus time. It is shown for 120 s so that it can be used for short-term analysis. Please note that all the beats are of high quality; therefore, the R—R intervals are the same as N—N intervals.

Time domain metrics are computed from the R—R tachogram, and commonly used ones include:

- Heart rate (HR) given in beats per min.
- Mean R—R intervals given in ms.
- The standard deviation of N—N intervals (SDNN) given in ms.

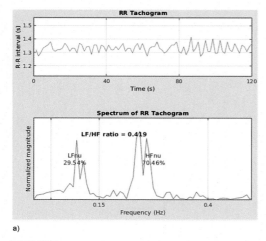
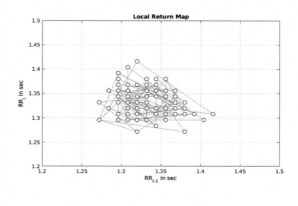

a)   b)

FIGURE 7.18 (A) Top figure: R—R intervals over time; Bottom figure: The spectrum of RR intervals. (B) Return map of the R—R intervals. The figure shows examples of time domain, frequency domain, and geometric method graphs that can be used to compute the HRV metrics. The program used to perform the analysis is the open-source HRVTool (Vollmer, 2019). The ECG signal is generated using the same parameters as in Fig. 7.4 based on the simulator described in McSharry et al. (2003).

- The square root of the mean of the sum of square differences (RMSSD) between the adjacent N—N intervals given in ms. It reflects beat-to-beat standard deviation in heart rate.
- Percentage of the number of pairs of the adjacent N—N intervals that differ by more than 50 ms versus the total number of N—N intervals (pNN50).

These measures are easy to compute. The first two metrics are based on N—N intervals, while the second two measures are based on the differences between the adjacent N—N intervals. They directly depend on the length of the interval selected, and therefore, the length should always be specified. SDNN is a commonly used metric for long-term cardiac monitoring. It is determined that SDNN<50 ms is normally a sign of cardiac problems—see Table 7.1 for the normal values for long and short-term monitoring. The RMSSD is used for emotion regulation.

7.8.3.2 Frequency domain metrics

The spectrum of R—R intervals is divided into

- very low frequency (VLF) band that occupies the frequency range below 0.04 Hz.
- low frequency (LF) band that corresponds to frequencies between 0.04 and 0.15 Hz.
- high frequency (HF) band that represents frequencies between 0.15 and 0.4 Hz.

Parasympathetic activity is associated with the HF band. Sympathetic activity occurs in the LF band together with other factors (Shaffer and Ginsberg, 2017); therefore, the power in the LF band spectrum cannot be used as a single measure of sympathetic activity.

TABLE 7.1 HRV metrics and their major applications (Shaffer and Ginsberg, 2017), as well as the range of metrics for normal subjects (Nunan et al., 2010).

Metrics	Type	Major use	Description	Typical values and ranges in healthy adults for short-term HRV analysis (Nunan et al., 2010).	The values for the 120 s ECG segment for the example in Fig. 7.18.
SDNN	Time domain	Cardiac risk based on 24-h monitoring	Long-term analysis: SDNN <50 ms unhealthy, 50 ≤ SDNN <100 ms have compromised health, SDNN≥100 ms healthy.	Mean SDNN = 50 ms Range of SDNN = [32 ms, 93 ms]	30.8 ms
RMSSD		Emotion regulation, short-term HRV	RMSSD has the potential to predict different patterns of neural activity during explicit emotional regulation (Steinfurth et al., 2018)	Mean RMSSD = 42 ms Range of RMSSD = [19 ms, 75 ms]	47.4 ms
pNN50		Parasympathetic activity	Correlated with RMSSD and HF	—	25.6%
Normalized power in LF	Frequency domain	Sympathetic activity	Measuring sympathetic activity is questionable due to the contributions of other factors in the LF frequency band	Mean LF = 52 Range of LF = [30, 65]	29.5%
Normalized power in HF		Stress	Lower HF power is correlated with stress, panic, anxiety, or worry.	Mean HF = 40 Range of HF = [16, 60]	70.5%
LF/HF ratio		It was originally used to determine the ratio between sympathetic and parasympathetic activities.	The LF/HF ratio should be used with caution, especially when explaining the balance between sympathetic and parasympathetic activities	Mean LF/HF ratio = 2.8 Range of RMSSD = [1.1, 11]	0.41
TRI	Geometric	Arrhythmia detection	When long-term TRI ≤20.42 and RMSSD ≤68 ms, the heart rhythm is normal. When long-term TRI >20.42, the pattern is arrhythmic	—	8.2
SD1		Short term HRV	Correlates with the change in IBI duration per unit change in blood pressure, and HF power	—	33.5 ms
SD2		Short- and long-term HRV	Correlates with LF power	—	28.1 ms

Frequency domain metrics include normalized power in LF and HF frequency ranges, as well as the ratio of normalized power in these ranges called LF/HF. Normalized power is computed as a power in the particular band of interest (LF or HF) divided by the power in these two bands together (without including the power in the VLF band). Normalized power is presented in percentages and is shown in the bottom figure of Fig. 7.18A. Some metrics are outside of the normal range because we used a simulated ECG signal. This is the case with the ratio LF/HF—we expect the larger power in the LF range, which was not the case with our simulated signal, where LF/HF ratio was set to 0.5 in the simulation.

7.8.3.3 Geometric metrics

Several *geometric methods* and their metrics are introduced next. We will first discuss metrics based on the shape of the histogram of N—N values. The histogram shows the number of pulses in each histogram bin, while each bin has a length (duration) of 1/128 s typically. The histogram is often interpolated using a triangle with the maximum at the maximum height of the histogram. Features of interest are the maximum of the triangle and the length of the triangle's base. The following metrics are extracted from the histogram.

- Triangular index (TRI), which is the number of all N—N intervals divided by the maximum height of the histogram, and it can be used for arrhythmia detection
- Triangular interpolation of N—N interval histogram (TINN), which represents the baseline width of the triangle and is given in ms. The TINN describes the variability of N—N intervals that is much less susceptible to outliers than, for example, the standard deviation of N—N intervals

The Poincare plot or Return Map presents R—R (or N—N) interval versus the next adjacent R—R (or N—N) interval. The return map is shown in Fig. 7.18B. The plot's shape and orientation can be used to categorize different conditions, such as the degree of heart failure (Woo et al., 1992). The plot should be fitted using an ellipse that is rotated by 45 degrees. Its main axis is in a 45 degrees direction, and the length of the cloud along the main axis represents the long-term variability called standard deviation 2 (SD2). SD1 is defined as a standard deviation of the difference between the adjacent R—R intervals, and it is related to the smaller axis of the ellipse. It shows practically the same information as RMSSD. Instead of using R—R intervals, the return map can be presented using relative R—R intervals computed as follows: $rr_i = 2(RR_i - RR_{i-1})/(RR_i + RR_{i-1})$.

Please note that the signal generated in Fig. 7.4 is a simulated signal, and therefore, some of the metrics that are obtained using the HRV analysis do not correspond to the physiological values. It is expected that SD2 is larger than SD1, which is not the case in our simulation.

7.8.3.4 Summary

As HRV analysis uses a large number of metrics, it is not clear when to use each of them and what values one should expect for normal subjects. The major applications of each metric are given in Table 7.1 based on the analysis provided in Shaffer and Ginsberg (2017). The mean of the metrics and their ranges were provided in Nunan et al. (2010) based on the short-term analysis of HRV of 44 healthy subjects with a minimum age of 40. It was determined that time domain metrics are more consistent throughout the studies than frequency domain metrics.

A number of open-source simulators and tools for processing ECG as well as other signals to extract HRV indices exist. PhysioNet Cardiovascular Signal Toolbox contains Matlab routines for ECG signal analysis, removing arrhythmias, and HRV analysis (Vest et al., 2018). A comprehensive HRV analysis software, without the source code, is presented in Pichot et al. (2016).

7.9 Summary

In this chapter, we showed the basic aspects of signal acquisition and processing of ECG signals. We focused on single-lead ECG systems because they are used pervasively. After an ECG signal is acquired, a number of different processing and machine-learning techniques can be applied to process different information depending on the need. The majority of the applications require accurate detection of QRS complex and R wave. Detected QRS complex can then be used for heart rate estimation, heart rate variability analysis, computation of the pulse arrival time for blood pressure estimation (discussed in Chapter 8), and classification of normal beats versus arrhythmia. To do additional analysis, ECG segmentation is required that provides the position and amplitudes of all the waves of interest. After the segmentation, different timing information can be extracted from the ECG signal and presented to a physician, or the classification can be performed to determine the type of arrhythmias.

This chapter did not cover medical-grade 12 lead ECG devices because they are not used pervasively outside the hospital. Electronics for these devices often includes multiplexers and control logic that allow selecting pairs of electrodes connected to the instrumentation amplifiers. In addition, significant attention in designing these devices is given to safety. Therefore, they contain isolation amplifiers that provide physical isolation between the part of the device that is in touch with the patient and the rest of the electronics connected to the AC power.

7.10 Appendix: segmentation of the ECG signal

Besides detecting the R wave or QRS complex, we are interested in detecting other fiducial points on the ECG waveform. In this section, we will show at a very high level an approach that can detect Q, R, S, T, and P waves. The input to the algorithm is the same noisy signal shown in Fig. 7.7A. The output of the algorithm (Fig. 7.19) shows the bandpass-filtered ECG signal with the detected waves of interest. After Q, R, S, T, and P waves are detected, the segments of the ECG waveform, as well as ST and QT intervals, can be computed.

The algorithm starts with bandpass filtering of the signal. Then, instead of differencing and squaring the signal, the Teager–Keiser energy tracking operator (TKEO) is applied (Sedghamiz, 2018). This operator is used when it is necessary to estimate the energy of the signal. Here, we assume that the QRS complex will have higher energy than the rest of the signal. The operator is defined as $x_{TKEO}(n) = x^2(n) - x(n-1)x(n+1)$ where x represents filtered ECG signal, $x(n-1)$ and $x(n+1)$ are the ECG samples before and after the sample of interest. The TKEO is then applied to detect the QRS complex. After the R peaks are detected based on the adaptive threshold introduced by Pan and Tompkins (1985), all other waves are detected as minimums or maximums with reference to the R wave in the interval where we expect that

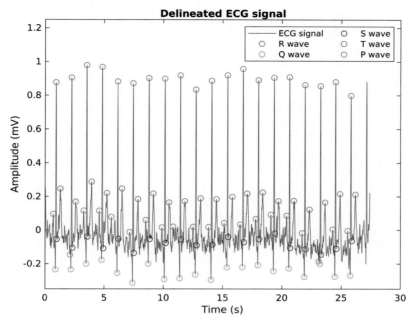

FIGURE 7.19   Segmentation of the ECG signal based on the work of Sedghamiz (2018).

particular wave to appear. The algorithm is implemented as a state machine that identifies waves in the following order: R, Q, S, T, and P. For the implementation details, please refer to MTEO_qrst.m algorithm in BioSigKit (Sedghamiz, 2018).

7.11 Problems

7.11.1 Simple questions

7.1. **(a)** Explain the origin of the ECG signal and the relationship between the waveform and cardiac activity.
   (b) Identify different segments of ECG waveform.
   (c) Derive the formula for voltage at each of the 12 leads.
7.2. **(a)** What are the ideal potential and impedance for electrode-skin contact?
   (b) What is the consequence of the fact that electrodes do not have ideal impedance and potential?
7.3. *AV-node delay is seen in the ECG signals as
   (a) the T wave
   (b) the delay between the T wave and the QRS complex
   (c) the P wave
   (d) the delay between the P wave and the QRS complex

7.4. How many electrodes are used in a 12-lead ECG system?

7.5. How many electrodes and how many independent leads are in the ECG system with three leads (Lead I, II, and III)?

7.6. ECG lead-off detection circuit injects current into the human body.

   (a) True

   (b) False

7.8. R waves occur at the following instances in a given ECG segment: 10.05 s, 11.0 s, 11.9 s, 12.95 s, and 13.95s. Estimate the heart rate in beats/min.

7.9. Unrelated questions related to interference:

   (a) List common sources of electronic/capacitive noise and interference affecting biopotential measurements.

   (b) Describe the effect of electronic coupling on the common mode voltage. How does that voltage affect the amplifier's output? Assume that the amplifier has a finite input impedance, and the electrode-skin contact impedance differs for each electrode.

   (c) *Analyze methods to reduce noise and interference, including shielding.

   (d) *Describe interference from the magnetic field.

7.11.2 Problems

7.10. Answer the following questions (please note that they are unrelated)

   (a) Explain why a high-input impedance amplifier is required in an ECG system, whereas the skin-electrode impedance should be as low as possible.

   (b) Why do we need amplifiers with very high input impedance when dry electrodes are used in an ECG system?

   (c) If the raw ECG signal is 3 mV and the common-mode signal is 1.5 V, how large does the CMRR have to be to reduce the common-mode signal to less than ECG signal?

   (d) Two electrodes are attached to the body and used to measure the biopotential signal. Each electrode and the interface between the electrode and the skin is modeled as one 4 kΩ resistor. The electrodes are attached to a differential amplifier having an input impedance of 1 MΩ. What will be the percentage reduction in the amplitude of the biopotential signal?

7.11. Circuit design:

   (a) Explain the functional blocks of the ECG device and present a block diagram.

   (b) Explain in more detail the preamplifier, lowpass and highpass filters, and the amplifier. Show implementation of a heart rate monitor.

7.12. Fig. 7.20 shows an ECG amplifier with a high gain to boost the 1 mV ECG to a large voltage.

   (a) Describe the function of each part of this circuit. Please note that LT1920 is an instrumentation amplifier and TL032 is a low-power precision operational amplifier. Obtain their datasheets and extract the required parameters.

   (b) What is the range of skin resistance? Based on that range, what would be the amplifier's input impedance that you would select?

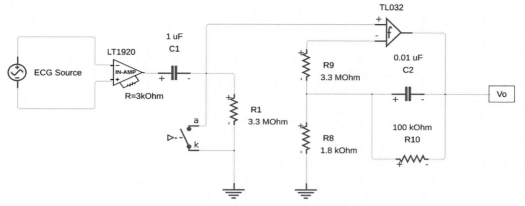

FIGURE 7.20   An ECG amplifier circuit.

    (c) This ECG amplifier contains a highpass and a lowpass filter. Calculate low and high cut-off frequencies of the filters.

    (d) Calculate the ECG amplifier gain.

7.13. A cardiotachograph is a device whose output (in Volts) is proportional to the heart rate.

    (a) Implement the cardiotachograph circuit, assuming that the ECG signal is already available. Assume that the minimum heart rate is 0 and the maximum is 200 beats per minute. 0 beats per minute should produce 0 V, while 200 beats per minute should produce a voltage equal to the supply voltage Vcc.

    (b) Next, the implementation needs to be extended to include the lower and upper limits corresponding to the normal heart rate. This is done using two comparators. The heart rate is compared at the first comparator with the threshold E1, which corresponds to 45 beats per minute, and at the second comparator with the threshold E2, which corresponds to 120 beats per minute. The alarm needs to be generated if the heart rate is below E1 or above E2.

7.11.3 Simulations

7.14. Simulate ECG signal based on the Matlab code provided by McSharry et al. (2003). Analyze the effects of the ratio of LF/HF components on R–R intervals and on the amplitude changes of R waves.

7.15. The notebook for ECG amplifiers is implemented using a dry electrode model. Replace the dry electrodes with the model of the wet electrodes described in Chapter 2. What is the difference in the output result?

7.16. Modify the design in Fig. 7.9 to include a programmable gain amplifier instead of using amplifiers with fixed gains. Also, include the notch filter that can filter 60 Hz frequency.

7.17. For the model presented in Fig. 7.13, simulate cases with and without the driven right leg circuit when all the electrode-skin interface models are replaced with wet electrode—skin interface models presented and simulated in Chapter 2. Observe and comment on the shape of the ECG signal and the amplitude of the common mode voltage.

7.18. Analyze the effect of the mismatch of the parameters of the electrodes (R_{ep}, C_{ep}, V_h) on the output ECG waveform as well as on the Power Ratio value in cases where there is no common mode voltage (Fig. 7.11) and in cases with the common mode voltage (Fig. 7.13).

7.19. Consider textile cotton electrodes that have very large impedances (assume 300 MΩ) and very low capacitance (assume 30 pF) (Chi et al., 2010). How would the behavior of the system change in this case?

7.20. Make the ECG signal zero and consider only the reduction of the common mode signal for the model presented in Fig. 7.13. How is the common mode gain changed when the resistor R_f in the driven right leg circuit shown in Fig. 7.15A takes these values: 1 MΩ, 2 MΩ, or 5 MΩ?

7.21. Tolerances of the resistors R_4 to R_7 or the instrumentation amplifier shown in Fig. 7.13 were set to the worst case values of 1%. Set tolerance of resistors R_4 to R_7 to zero. In this case, all the resistors will have the same values. Analyze if the driven right leg circuit improves the result at the output. Comment on your results.

7.22. Analyze the effects of motion artifacts on the ECG signal. Simulate motion artifacts using the techniques presented in Chapter 4. For the electrode models, use (a) wet, (b) dry, and (c) capacitive electrodes.

7.12 Further reading

Please note that the following material is selected based on availability and the author's preferences and that many good resources might be omitted unintentionally.

An excellent book that describes the origin of the ECG signal and the heart's electrical activation is Malmivuo and Plonsey (1995).

Several toolboxes for processing ECG signals are available. One of them is a Matlab-based set of routines for processing ECG and several other Physiological signals called ecg-kit (Demski and Llamedo, 2016), available at http://marianux.github.io/ecg-kit/. PhysioNet Cardiovascular Signal Toolbox for heart rate variability analysis also includes a number of routines for processing ECG signals (Vest et al., 2018). A Matlab library ECGdeli for ECG segmentation is provided at https://zenodo.org/record/3977971#.YU5d37hKguU.

A very detailed introduction to processing ECG signals and classifying atrial fibrillation is given in Mainardi et al. (2008).

References

Acharya, R.U., Joseph, P.K., Kannathal, N., Choo, L.M., Suri, J.S., 2010. Heart rate variability. In: Acharya, U.R., et al. (Eds.), Advances in Cardiac Signal Processing. Springer, New York.

Antink, C.H., Leonhardt, S., Walter, M., 2017. A synthesizer framework for multimodal cardiorespiratory signals. Biomedical Physics & Engineering Express 3 (3).

Arts, T., Delhaas, T., Bovendeerd, P., Verbeek, X., Prinzen, F.W., 2005. Adaptation to mechanical load determines shape and properties of heart and circulation: the circadapt model. American Journal of Physiology-Heart Circulatory Physiology 288, H1943—H1954.

Awal, M.A., et al., 2021. Design and optimization of ECG modeling for generating different cardiac dysrhythmias. MDPI Sensors 21 (5), 1638.

Chi, Y.M., Jung, T.P., Cauwenberghs, G., 2010. Dry-contact and noncontact biopotential electrodes: methodological review. IEEE Reviews in Biomedical Engineering 3, 106—119.

Chung, H.U., et al., 2019. Binodal, wireless epidermal electronic systems with in-sensor analytics for neonatal intensive care. Science 363 (issue 947).

Cohen, M.A., Taylor, J.A., 2002. Short-term cardiovascular oscillations in men: measuring and modelling the physiologies. Journal of Physiology 542, 669—683.

Demski, A.J., Llamedo, S.M., 2016. Ecg-kit a Matlab Toolbox for cardiovascular signal processing. Journal of Open Research Software 4 (issue 1).

European Society of Cardiology and the North American Society of Pacing and Electrophysiology, 1996. Task force of the European society of Cardiology and the North American society of pacing and Electrophysiology, "heart rate variability. Standards of measurement, physiological interpretation, and clinical use. Circulation 93 (5), 1043—1065.

Flores, N., Avitia, R.L., Reyna, M.A., García, C., 2018. Readily available ECG databases. Journal of Electrocardiology 51 (issue 6), 1095—1097.

Goldberger, A., et al., 2000. PhysioBank, PhysioToolkit, and PhysioNet: components of a new research resource for complex physiologic signals. Circulation 101 (issue 23), e215—e220.

Hann, M., 2012. Use Spice to Analyze DRL in an ECG Front End. EDN News.

ISO/IEC 80601, Part 2—86: Particular Requirements for the Basic Safety and Essential Performance of Electrocardiographs, Including Diagnostic Equipment, Monitoring Equipment, Ambulatory Equipment, Electrodes, Cables, and Leadwires, 2018. Committee Draft.

Lee, H.C., Jung, C.W., 2018. Vital Recorder-a free research tool for automatic recording of high-resolution time-synchronized physiological data from multiple anesthesia devices. Scientific Reports 8 (1).

Lee, S.P., Ha, G., Wright, D.E., et al., 2018. Highly flexible, wearable, and disposable cardiac biosensors for remote and ambulatory monitoring. NPJ Digital Medicine 1 (2).

Li, Q., Mark, R.G., Clifford, G.D., 2008. Robust heart rate estimation from multiple asynchronous noisy sources using signal quality indices and a Kalman filter. Physiological Measurement 29, 15—32.

Li, X., Hui, H., Sun, Y., 2016. Investigation of motion artifacts for biopotential measurement in wearable devices. In: IEEE 13th International Conference on Wearable and Implantable Body Sensor Networks (BSN), pp. 218—223.

Liu, Y., Chen, J., Bao, N., Gupta, B.B., Lv, Z., 2021. Survey on atrial fibrillation detection from a single-lead ECG wave for Internet of Medical Things. Computer Communications 178. Elsevier.

Lynn, W.D., Escalona, O.J., McEneaney, D.J., 2013. Arm and wrist surface potential mapping for wearable ECG rhythm recording devices: a pilot clinical study. Journal of Physics: Conference Series 450.

Lyon, A., Minchole, A., Martinez, J.P., Laguna, P., Rodriguez, B., 2017. Computational techniques for ECG analysis and interpretation in light of their contribution to medical advances. Journal of Royal Society Interface 15.

Mainardi, L., Sörnmo, L., Cerutti, S., 2008. *Understanding Atrial Fibrillation: The Signal Processing Contribution*, Part I and Part II, Synthesis Lectures on Biomedical Engineering No. 24. Morgan &Claypool Publishers.

Malmivuo, J., Plonsey, R., 1995. Bioelectromagnetism - Principles and Applications of Bioelectric and Biomagnetic Fields. Oxford University Press, New York.

McSharry, P.E., Clifford, G.D., Tarassenko, L., Smith, L.A., 2003. A dynamical model for generating synthetic electrocardiogram signals. IEEE Transaction Biomedical Engineering. 50, 289—294.

Moody, G.B., Mark, R.G., 2001. The impact of the MIT-BIH arrhythmia database. IEEE Engineering in Medicine and Biology 20 (issue 3), 45—50.

Nunan, D., Sandercock, G.R., Brodie, D.A., 2010. A quantitative systematic review of normal values for short-term heart rate variability in healthy adults. Pacing Clinical Electrophysiology 33 (issue 11), 1407—1417.

Orphanidou, C., 2018. Signal Quality Assessment in Physiological Monitoring: State of the Art and Practical Considerations. Springer Briefs in Bioengineering.

Pan, J., Tompkins, W.J., 1985. A real-time QRS detection algorithm. IEEE Transactions on Biomedical Engineering (3), 230—236. BME-32.

Pichot, V., Roche, F., Celle, S., Barthélémy, J.C., Chouchou, F., 2016. HRVanalysis: a free software to analysis cardiac autonomic activity. Frontiers Physiology 7 (issue 557).

Pouryayevali, S., Wahabi, S., Siddarth, H., Hatzinakos, D., 2014. On Establishing Evaluation Standards for ECG Biometrics. ICASSP, pp. 3774—3778.

Rashkovska, A., Depolli, M., Tomašić, I., Avbelj, V., Trobec, R., 2020. Medical-grade ECG sensor for long-term monitoring. Sensors 20 (Issue 6), 1695.

Real-Time ECG QRS Detection, The MathWorks, Inc., https://www.mathworks.com/help/dsp/ug/real-time-ecg-qrs-detection.html, last accessed on September 8, 2021.

Samol, A., Bischof, K., Luani, B., Pascut, D., Wiemer, M., Kaese, S., 2019. Single-lead ECG recordings including Einthoven and Wilson leads by a smartwatch: a new Era of patient directed Early ECG differential diagnosis of cardiac diseases? Sensors 19 (Issue 20), 4377.

Sedghamiz, H., 2018. BioSigKit: a Matlab Toolbox and interface for analysis of BioSignals. Journal of Open Source Software 3 (issue 30).

Shaffer, F., Ginsberg, J.P., 2017. An overview of heart rate variability metrics and norms. Frontiers in Public Health 5, 258.

Siontis, K.C., Noseworthy, P.A., Attia, Z.I., et al., 2021. Artificial intelligence-enhanced electrocardiography in cardiovascular disease management. Nature Reviews Cardiology 18, 465—478.

Steinberg, J.S., et al., 2017. ISHNE-HRS expert consensus statement on ambulatory ECG and external cardiac monitoring/telemetry. Annals of Noninvasive Cardiology, Wiley 22 (Issue 3).

Steinfurth, E.C.K., Wendt, J., Geisler, F., Hamm, A.O., Thayer, J.F., Koenig, J., 2018. Resting state vagally-mediated heart rate variability is associated with neural activity during explicit emotion regulation. Frontiers in Neuroscience 12.

Vest, A., Da Poian, G., Li, Q., Liu, C., Nemati, S., Shah, A., Clifford, G.D., 2018. An open source benchmarked Toolbox for cardiovascular waveform and interval analysis. Physiological Measurement 39 (issue 10).

Vollmer, M., 2019. HRVTool - an open-source Matlab Toolbox for analyzing heart rate variability. In: Computing in Cardiology Conference.

Webster, J.G. (Ed.), 2010. Medical Instrumentation: Application and Design, fourth ed. John Wiley & Sons.

Woo, M.A., Stevenson, W.G., Moser, D.K., Trelease, R.B., Harper, R.H., 1992. Patterns of beat-to-beat heart rate variability in advanced heart failure. American Heart Journal 123, 704—710.

Wu, H., Yang, G., Zhu, K., Liu, S., Guo, W., Jiang, Z., Li, Z., 2020. Materials, devices, and systems of on-skin electrodes for Electrophysiological monitoring and human—machine interfaces. Advanced Science, Wiley 8 (2).

Young, B., Schmid, J.-J., 2021. The new ISO/IEC standard for automated ECG interpretation," *hearts*. MDPI 2, 410—418.

Zipp, P., Ahrens, H., 1979. A model of bioelectrode motion artefact and reduction of artefact by amplifier input stage design. Journal of Biomedical Engineering 1 (4), 273—276.

Zulqarnain, M., Stanzione, S., Rathinavel, G., et al., 2020. A flexible ECG patch compatible with NFC RF communication. NPJ Flexible Electronics 4 (13).

Devices based on the time difference between signals: continuous blood pressure measurement

Acronyms and explanations

ICU   intensive care unit
MAD   mean absolute difference
MAPD   mean absolute percentage difference
MAPD   mean absolute percentage difference
MEMS   micro-electromechanical systems
PAT   pulse arrival time
$\textbf{PAT}_{\textbf{D}}$ PAT between the peak of the ECG R wave and the maximum slope of the blood pressure (or PPG) pulse
PAT_{di} PAT between ECG and blood pressure signal at a distal site
$\textbf{PAT}_{\textbf{f}}$ PAT between the peak of the ECG R wave and the foot of the blood pressure (or PPG) pulse
$\textbf{PAT}_{\textbf{p}}$ PAT between the peak of the ECG R wave and the peak of the blood pressure (or PPG) pulse
PAT_{pr} PAT between ECG and blood pressure signal at a proximal site
PTT   pulse transit time
PWV   pulse wave velocity

Variables used in this chapter include:

A lumen area
C_a arterial compliance
$C_{\underline{A}}$ compliance with respect to the area
$\overline{DBP}$ diastolic blood pressure estimate
Y Young's modulus
h arterial wall thickness

H	vertical height relative to the heart level
K	distensibility
l	length of the artery
L	inertance
$p_{dut,i}$	ith measurement of pressure using the device under test
$p_{ref,i}$	ith measurement of pressure using the reference device
$p(t)$	pressure in the artery
Q	blood flow
r	radius of the artery
$\widehat{SBP}$	systolic blood pressure estimate
V	volume of the artery
μ	Poisson ratio
ρ	blood density

8.1 Introduction

The main theme of this chapter is the fusion of different signals based on time or temporal information. The time difference between signals can be measured at the same physiological site, like PPG and ECG signals from a smartwatch at the wrist, or at different sites, such as PPG signals from the wrist and the neck. To determine the time difference among different signals, we detect the fiducial points on the signals and then determine the time difference among these different points.

Pulse analysis and analysis of the time difference between different signals are of great importance leading to various applications, including arterial stiffness measurement, continuous blood pressure estimation, and heart rate variability analysis. A less common application is the measurement of oscillometric blood pressure with the assistance of ECG.

This chapter starts by presenting ECG, PPG, SCG, and blood pressure signals relative to each other and defining the temporal metrics such as pulse transit time and pulse arrival time. Next, we describe models that relate time delays between fiducial points of the pulses collected at different points on the body during the same cardiac cycle and the mechanical and geometric properties of the arteries. After that, we present the application of these models for cuffless estimation of blood pressure, which allows for continuous measurement of blood pressure in each cardiac cycle without using the cuff for occluding the artery. These monitors are also called *beat-to-beat monitors*. It should be noted that this chapter does not include a discussion about continuous blood pressure monitors based on volume clamp technologies because they are expensive for home use and uncomfortable for long-term monitoring. Instead, it focuses only on monitors based on signals and methods described in Chapters 5–7. At the end of the chapter, we discuss noninvasive ways to measure arterial stiffness based on pulse wave velocity.

8.2 Temporal relationship between signals

8.2.1 Wiggers diagram

The temporal relationship among different signals is shown in Fig. 8.1. The figure shows the pressure signal at the top, ventricular volume in the middle, and ECG and PCG signals. PCG signal was mentioned in Overview of Biomedical Instrumentation and Devices and will

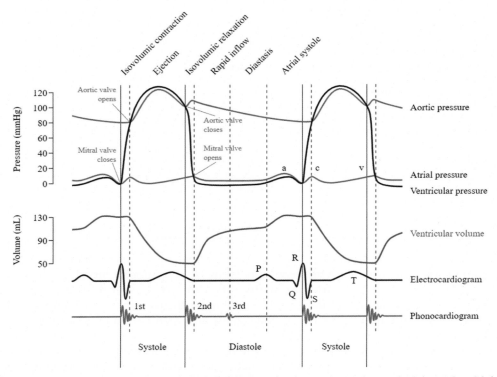

FIGURE 8.1   Wiggers diagram showing the following signals during the cardiac cycle: left atrial and left ventricular pressure, left ventricular volume, aortic pressure, ECG, and PPG. *This figure was drawn by Daniel Chang. It can be downloaded from https://commons.wikimedia.org/wiki/File:Wiggers_Diagram.png and used under the Creative Commons Attribution-Share Alike 2.5 Generic license.*

be discussed in Chapter 10. This way of presenting signals is called the Wiggers diagram. The waveforms shown include left atrial and left ventricular pressure, left ventricular volume, aortic pressure, ECG, and PCG. The cardiac cycle is divided into systole and diastole. The aortic valve opens when the pressure inside the ventricle becomes higher than the pressure in the aorta. Then, the blood flows to the aorta due to left ventricle contraction, and the blood volume in the left ventricle reduces. The aortic pressure during that time follows the ventricular pressure. The aortic valve closes as the ventricular pressure becomes smaller than the aortic pressure. The ventricular pressure quickly drops below the atrial pressure causing the mitral valve to open, which again causes the filling of the left ventricle with blood. Aortic pressure drops during the diastole until it drops below the ventricular pressure in the next cardiac cycle. The figure also shows ECG and PCG signals and their temporal relationship with the events in the cardiac cycle.

8.2.2 PWV, PTT, and PAT

An arterial pressure wave is a pressure impulse wave generated at heart during the aortic valve opening that propagates along the entire arterial tree, widening the walls of each artery (Solà and Delgado-Gonzalo, 2019). *Pulse wave velocity (PWV)* is defined as the velocity of a pressure wave that propagates between two points in the arterial system, which varies from 5 to 15 m/s in the central artery to higher values in peripheral arteries. PWV is measured as the distance traveled by the pressure pulse along the arterial tree divided by the elapsed time or *pulse transit time* (PTT). Therefore, PTT is the time that an arterial pressure wave requires to propagate along the walls of a given segment of the arterial tree. It should be remembered that pressure pulse wave velocity is not blood velocity but rather the velocity of the pressure wave propagation along the wall of the arterial tree. While the blood velocity is between 0.2 m/s and 0.4 m/s, pulse wave velocity is from 5 m/s to more than 20 m/s in different arteries. Higher values of PWV normally mean higher arterial stiffness. The carotid-femoral pulse wave velocity (cfPWV) is the velocity at which the pressure wave propagates between the carotid and the femoral arteries. It can be regarded as an indicator of high cardiovascular risk for cfPWV values above 10 m/s. It is commonly used as an index for arterial stiffness (Townsend et al., 2015).

The arterial blood pressure (ABP) signal is measured at two sites to determine the PTT. This is shown in Fig. 8.2, where the ECG signal, central arterial blood pressure measured invasively, and peripheral blood pressure on the arm are illustrated from top to bottom, respectively. In Fig. 8.2, PTT is measured as a time difference between the instances corresponding to the minima of the arterial and peripheral blood pressure pulses (foot-to-foot delay). Besides the minima of pulses, other fiducial points can also be selected to determine PTT. However, it is not recommended to determine PTT between the time instances that correspond to the maxima of the pulses for several reasons, including (Westerhof et al., 2010): wave reflections contribute differently to the systole peak position and amplitude on the blood pressure pulse at different places on the arterial tree, arteries are stiffer at the distending pressure,

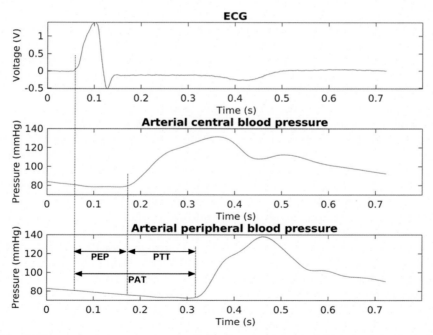

FIGURE 8.2 ECG, central aortic pressure, and peripheral pressure signals. Pulse arrival time, pulse transit time, and the preejection period are shown.

and the stiffness is different depending on the thickness and other properties of different arteries. As invasive methods for determining PTT can only be applied during surgeries, the noninvasive alternative methods typically rely on acquiring the arterial blood pressure signals at the two sites with handheld tonometers or Doppler ultrasound probes, followed by detecting the foot-to-foot time delay between the pulses of the collected signals (Zhang, 2012).

Pulse arrival time (PAT) is the time delay between the peak of the R-wave of an ECG waveform and the minimum of the pulse of the blood pressure waveform in the same cardiac cycle. PAT equals the sum of PTT and the *preejection period* (PEP). PEP is the time delay between the Q-wave of an ECG and the opening of the aortic valve. It represents the interval from the start of electrical stimulation of the ventricles until the mechanical opening of the aortic valve. PEP is shown in Fig. 8.2; its end corresponds to the minimum of the aortic blood pressure pulse, which again corresponds to the aortic valve opening—see Fig. 8.1. It has been shown that PTT is correlated with blood pressure, while the correlation between PAT and blood pressure is weaker because of the PEP. This is verified through an animal study (Zhang et al., 2011a) that shows a large variability of PEP between the subjects, as well as changes in PEP over time with different interventions. Therefore, it was concluded in Zhang et al. (2011a) that PAT might not be an adequate alternative for PTT as a marker of blood pressure. PAT and PEP are also shown in Fig. 8.2.

8.2.3 Surrogates of PTT and PAT

The previous section showed that PAT is measured using ECG and blood pressure signals. This section considers time delays between fiducial points of different signals, including ECG, PPG, PCG, and SCG signals, as shown in Fig. 8.3. Signals in Fig. 8.3 are obtained simultaneously from a healthy subject using sensors described in Appendix 1. In the experiment, ECG electrodes were placed in a lead I configuration, a PPG sensor was attached to a fingertip, and a microphone and an accelerometer were placed on the sternum.

The most common surrogate of PAT measured between ECG and blood pressure signals is PAT between ECG and PPG signals, as shown in Fig. 8.3 (Winokur, 2014). PAT_p is the time difference between the peak of the ECG R wave and the maximum of the PPG pulse and, PAT_f represents the time difference between the peak of the ECG R wave and the foot of the PPG pulse, and PAT_D is the time difference between the peak of the ECG R wave and the maximum slope of the PPG pulse. It is crucial to note that the PAT obtained using a PPG signal instead of a blood pressure signal conveys a different meaning since the PPG signal is related to blood flow in the peripheral vasculature, which is different from the blood pressure signal that is related to pressure pulse propagation, wave reflection, and other factors (Gao et al., 2016). In Gao et al. (2016), several PPG probes were placed on the body, and the delays between the instances of the selected fiducial points of the signals obtained from the probes were computed, resulting in a PPG-based surrogate for PTT that was referred to as PPG-PTT. It was shown in (Gao et al., 2016) that PPG-PTT could be used instead of PTT obtained from arterial waveforms for estimating blood pressure (Box 8.1).

Besides the combination of ECG and PPG, other sensors are also used to compute the surrogate of PTT and PAT. The accelerometer signal in z direction obtained from the subject's chest is shown in Fig. 8.3 as the bottom plot representing the SCG signal. The SCG signal has multiple peaks during one cardiac cycle. It has been shown in (Yang and Tavassolian,

---

BOX 8.1

C o n t r o v e r s i e s   i n   u s i n g   P A T   a n d   P T T   t o   e s t i m a t e   B P

Even though it has been shown that PAT is not an appropriate alternative for PTT (Zhang et al., 2011a) in estimating blood pressure, many researchers use PAT for this purpose. In addition, they make an additional assumption that the time difference between of the peak of the ECG R wave and the fiducial point on the PPG signal is similar to the time difference between the peak of the R wave and points on the blood pressure pulse, so that it is possible to use PPG signal as a surrogate to blood pressure signal in

determining the PTT. The research is going in multiple directions including:

- Computing PAT or PTT using different combinations of sensors.
- Compensating for the error in using PAT by using additional features from the pulse that are not temporal such as the ratio of amplitudes of different fiducial point of the PPG pulse.
- Developing machine learning models for estimating blood pressure based on temporal and other features.

2018) that the first maximum peak after the minimum in each cardiac cycle indicates the timing of the aortic valve opening event—please see the arrow pointing to the peak on the SCG waveform at the bottom of Fig. 8.3. Therefore, if that peak can be used instead of the ECG R peak, it would be possible to avoid the influence of PEP and compute PTT directly. Therefore, SCG technology shows great promise in obtaining a surrogate of the PTT when combined with some other sensor placed on a distal site, such as a PPG sensor.

PCG signal will be presented in more detail in Chapter 10. Here, it is important to notice several characteristic waves on the PCG signal at the bottom of Fig. 8.1. We will mention here only the first heart sound S1 wave that occurs at the start of the systole, and the second heart sound that occurs at the end of the systole. In A Foo et al. (2006), PCG and PPG signals were used to obtain the time difference between the S1 of the PCG and the upstroke of the corresponding pulse in the PPG. It has been shown, similar to the timing of the SCG peak, that the position of S1 corresponds to the time of the opening of the aortic valve. Therefore, S1 of the PCG signal has the potential to be used as a surrogate signal in computing PTT. PCG signal is the third plot in Fig. 8.3.

Other sensors, including radars (Ebrahim et al., 2019), capacitive ballistocardiography sensors (He et al., 2019), and bioimpedance sensors (Heydari et al., 2018), are also used to obtain surrogate PTT and PAT values.

8.2.4 Classifying temporal relationships between the signals

In this section, we will consider devices with multiple sensors that can collect different physiological signals simultaneously. Fig. 8.4 presents classifications based on the number of measurement points on the body and on the homogeneity of the sensors. It also shows the applications, as well as the advantages and disadvantages of different approaches.

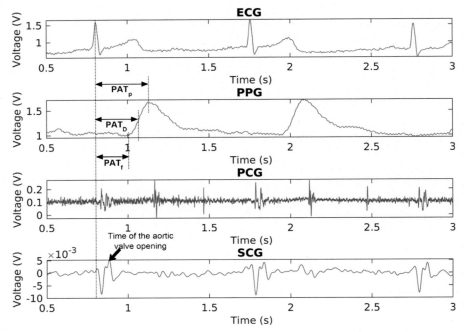

FIGURE 8.3   ECG, PPG, PCG, and SCG signal is z direction. Also, pulse arrival times measured between the R peak of the ECG signal and different fiducial points of the PPG signal are shown.

Devices can be designed to measure all the signals of interest at *a single point* on the body. This means that all sensors are placed in one device with a relatively small form factor and attached to a wrist, chest, or another place on the body. There are neither external wires nor wirelessly connected external sensors. Single-point devices include health and smartwatches measuring multiple signals, including PPG, temperature, and acceleration. Smartwatches are sometimes designed to measure ECG as well: one electrode is at the bottom of the watch touching the skin of the arm, and another electrode is on the watch's surface and needs to be touched with the other arm. These devices normally include multiple heterogeneous sensors. Therefore, they can be used to measure PAT between the ECG signal and, for example, the PPG signal, as shown in Fig. 8.3. However, the major disadvantage of these devices is that they cannot measure the time it takes for the signal to propagate between two points on the body—PTT. We will explain in Section 8.3 that PTT is inversely related to PWV and thus to blood pressure, and therefore, blood pressure estimated based on PTT may be more accurate than the one based on PAT.

Devices can be designed to measure the signals at *multiple points* on the body. These devices contain the same type of sensors placed at different points on the body or heterogeneous sensors. The same type of sensor can be a PPG, arterial tonometry pressure sensor, or oscillometric pressure sensor placed at multiple locations on the body. The main goal of these devices is to determine the PTT between the fiducial points of these signals. Then, the computed pulse transit time can be used to estimate the peripheral or central blood pressure or arterial stiffness.

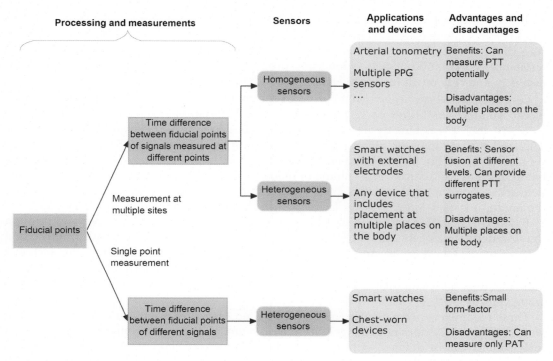

FIGURE 8.4 Classification of the devices based on the number of points on the body at which the device measures signals and the homogeneity of sensors.

It is common nowadays to have devices with heterogeneous sensors. These sensors include ECG, PPG, accelerometers, and pressure and temperature sensors. As there are several types of sensors placed at different points on the body, various PTT surrogates can be estimated. The main disadvantage of these devices is that they require placement at different locations on the body, which is much less convenient than having all the sensors at a single point.

The first part of Fig. 8.4 shows the processing and measurement steps. First, the fiducial points are extracted from the pulses. These fiducial points normally consist of the time instant and the amplitude of the minimum, maximum, the second minimum or maximum, and the maximum of the first derivative of the pulse. Then, the time difference between the fiducial points corresponding to the same heartbeat is computed to obtain PTT or PAT. After the PAT or PTT are computed, they can be used to estimate blood pressure, arterial stiffness, or other variables.

8.3 Modeling the relationship between PTT and blood pressure

8.3.1 Definitions

We will consider an artery of length l and the lumen area A. The volume of the artery is $V = Al$. *Resistance* of the blood vessel is proportional to the pressure difference between two points and inversely proportional to the flow through the vessel $R = \Delta p/Q$. This is equivalent to Ohm's law in electrical circuits, where the potential difference corresponds to the pressure difference, and the current corresponds to the flow. Based on Poiseuille's law, the resistance of the blood vessels is inversely proportional to the fourth power of the arterial radius r, and it is much higher in small vessels than in arteries.

C_a is the arterial *compliance* (in ml or m^3 over mmHg or Pa) which is the change in arterial blood volume (ΔV) due to a given change in arterial blood pressure (Δp), that is, $C_a = \Delta V/\Delta p$. It was already introduced when discussing the Windkessel model in Section 4.3.2 and the oscillometric blood pressure generation model in Section 5.5. The compliance represents the ability of arteries to expand during the systole period so they can accumulate blood. During the diastole, this volume of blood is discharged along the arteries. *Elastance* is defined as the reciprocal of compliance. When compliance is normalized with the volume of the organ, it is called *distensibility*. Therefore, distensibility is defined as $K = (\Delta V/\Delta p)/V$.

Blood is accelerated and deaccelerated within each beat, which can be observed through the first derivative of the flow $dQ(t)/dt$. The relationship between the pressure and the first derivative of the flow is $\Delta p(t) = LdQ(t)/dt$, where $L = \rho l/A$ represents the *inertance* [mmHg·s^2/ml] and ρ is the blood density. The inertance is important for large vessels and, together with compliance, represents an important factor in determining the pulse wave velocity.

These parameters normally depend on the volume. However, we usually consider that the length of the artery does not change with the pressure. Therefore, compliance, elastance and distensibility can be defined with respect to the lumen area. When we present these parameters with respect to the area, we will use the subscript "A." For example, compliance with respect to the area is $C_A = \Delta A/\Delta p$.

8.3.2 Model based on Moens–Korteweg equation

We will first introduce several equations that are usually applied in this field. The *Bramwell–Hill equation* relates pulse wave velocity with the area of the vessel as well as its area compliance C_A:

$$\text{PWV} = \sqrt{\frac{A}{\rho C_A}} = \sqrt{\frac{A}{\rho}\frac{dp}{dA}} \tag{8.1}$$

This equation is valid under the assumption that the vessel is an elastic thin cylindrical tube, the thickness and radius of the artery remain fixed as the blood pressure changes, the blood is incompressible and the blood flow is in the axial direction with a uniform velocity profile. *Thin-walled pressure vessels* have an outer diameter that is approximately equal to the inside diameter, and the ratio of the radius r to wall thickness h is: $r/h > 20$. It can also be shown that the area compliance for the tube can be obtained as (Waite and Fine, 2007):

$$C_A = \frac{2\pi r^3}{Yh} \tag{8.2}$$

where Y is the wall's Young's modulus, h is the thickness of the wall, r is the inner radius of the artery. By combining (8.1) and (8.2), we obtain the Moens–Korteweg equation that relates PWV with the dimensions of the vessel and the Young modulus of elasticity:

$$\text{PWV} = \sqrt{\frac{Yh}{2r\rho}} \qquad (8.3)$$

This equation gives the instantaneous pulse wave velocity for a particular arterial segment. However, it does not take into account the pulsatile flow of the blood, and it is derived for nonviscous fluid and does not consider reflections of the pressure wave. Even though the blood is a viscoelastic fluid, Eq. (8.3) approximates the PWV in arteries well.

Several corrections have been applied in the literature to the Moens–Korteweg equation. One assumption is that the thickness and radius of the artery remain fixed as the blood pressure changes. This assumption is addressed by including the Poisson ratio μ in Eq. (8.3). From Chapter 2, we know that the Poisson ratio is the ratio of change in the diameter over the ratio of change in length. Here, under increasing pressure, the artery will become narrower. If $\mu = 0.5$, no change in the volume of the material occurs for a very small strain, and therefore, a Poisson ratio around 0.5 is typically considered.

$$\text{PWV}_m = \sqrt{\frac{Yh}{2r\rho(1 - \mu^2)}} \qquad (8.4)$$

where PWV_m is the modified pulse wave velocity.

Example 8.1. This example is modified from Waite and Fine (2007).

The parameters of the carotid artery for two groups of subjects reported by Lin et al. (1999) are:

1. Normal subjects with the following average arterial parameters: $h = 1.22$ mm and $Y = 2.34 \cdot 10^5 \text{ N/m}^2$
2. Patients with coronary (and carotid) artery disease (CAD) have increased wall thickness and Young's modulus of elasticity: $h = 1.98$ mm and $Y = 4.18 \cdot 10^5 \text{ N/m}^2$, respectively.

Estimate the arterial compliance for both groups as well as PWV and PWV_m. Assume that $r = 4$ mm and that $\mu = 0.45$.

Solution. The solution includes applying Eqs. (8.2)–(8.4), and it is shown in Table 8.1. As seen in the table, the PWV for the patients with CAD is significantly higher than PWV estimated for the normal patient group. Also, modified PWV is higher than PWV, which does not include the Poisson ratio, by more than 10%.

As the blood pressure increases, the material in the artery walls becomes stiffer and resists strain. Therefore, Young's modulus of elasticity does not remain constant. In addition, smooth muscles in arteries can contract and consume energy in an effort to resist the strain. There have been a number of works to describe the relationship between elasticity and pressure. Hughes found the empirical relationship between pressure in the aorta and the modulus of elasticity:

TABLE 8.1   Compliance and pulse wave velocity for the carotid artery measured for normal subjects and CAD patients.

Population	Area compliance (m²/Pa)	PWV (m/s)	PWM$_m$ (m/s)
Normal	1.4×10^{-9}	5.9	6.5
CAD patients	5.1×10^{-10}	9.7	10.9

$$Y = Y_0 e^{\alpha P} \tag{8.5}$$

where Y_0 is the modulus of elasticity at zero blood pressure and α is a positive constant. Next, let us look at a segment of the artery of length l along which *the pulse wave velocity is constant.* Combining Eqs. (8.3) and (8.5) gives the relationship between the pressure and the PTT:

$$PTT = l/PWV = \sqrt{\frac{2r\rho}{hY_0}} \cdot le^{-\frac{\alpha p}{2}} \tag{8.6}$$

Even though Eq. (8.6) shows an inverse relationship between PTT and the pressure in an arterial segment, it is difficult to compute the pressure based on the measured PTT because the diameter of the artery, the thickness of the arterial wall, and parameters Y_0 and α are usually not known and are very difficult to estimate/measure without using equipment such as ultrasound. Therefore, as we will see in Section 8.4, there are various approximations of this formula.

PWV can be obtained in different ways starting from the Bramwell–Hill Eq. (8.1) if we have a way to model and estimate the compliance or the relationship between the area of the artery and the pressure.

8.3.3 Computational models

Computational models were introduced in Chapter 4. The choice of model depends on the required degree of detail and the desired focus. Therefore, we distinguish between several different models.

- The Windkessel model allows us to understand the effect of total arterial compliance on aortic pressure and cardiac output (Westerhof et al., 2010). Wave transmission cannot be described by the Windkessel model, and therefore, the model is less useful in the analysis of systems for continuous blood pressure measurements or arterial stiffness estimation that are based on PWV and PTT. However, the Windkessel model is also used as an integral part of more complex models, such as tube-load models.
- Distributed models and tube models allow for modeling local flows and pressures as well as the propagation of pressure waves in both time and space. Therefore, they can be useful for generating simulated waveforms of pressure and flow at different body sites or as interpretable models when estimating blood pressure.

Fig. 4.3 represents a 4-element *Windkessel model*. There are also 2-element Windkessel models that include only the peripheral resistance R and the arterial compliance C_a. A 3-element Windkessel model includes the characteristic impedance Z_0 as well, which introduces transmission concepts into the Windkessel model and provides for the correct behavior of the model at high frequencies (Westerhof et al., 2010). The total arterial inertance improves the behavior of the model at low frequencies (Westerhof et al., 2010) and is modeled as an inductor in a 4-element Windkessel model.

The Windkessel model only simulates the arterial system/subsystem with relatively high fidelity at its input. The pressure at the output of the Windkessel model $p_p(t)$ is the aggregate pressure of the rest of the arterial system and, therefore, does not represent well the pressure in the more distal part of this system.

The time constant RC_a determines the shape of the decrease of the aortic pressure. For example, in Fig. 4.4B, the pressure starts decreasing at around 200 ms after the flow $Q_a(t)$ is zero. Therefore, if it is possible to measure the aortic pressure, the decay time of the pressure signal could be used to obtain the time constant related to compliance and peripheral resistance. The increase in mean pressure mainly relates to an increase in peripheral resistance, while the increase in pulse pressure mainly results from decreased total arterial compliance.

Distributed models are normally 1-D models based on the Navier–Stokes equation and rely on dividing the arterial tree into small segments whose properties are known or their parameters need to be estimated. These properties include geometrical and mechanical parameters of the arteries. Distributed models can be used to simulate the propagation of pressure and flow, and therefore it is possible to obtain pressure or flow waveforms at different positions in the entire arterial tree. As these models include propagation delays, they can be used for studying arterial stiffness and estimating central or peripheral blood pressure based on the pulse transit time. Most of the work on these models has been focused on generating waveforms from artificial subjects at different positions in the arterial tree and then using these waveforms to develop algorithms for estimating arterial stiffness and blood pressure. Examples of these works and their applications are Alastruey et al. (2012) and van de Vosse and Stergiopulos (2011). Validation of 1D simulations against real measurement is shown in Alastruey et al. (2011). The virtual database approach was used to assess the accuracy of pulse wave analysis algorithms (Willemet et al., 2016) and blood pressure estimation (Huttunen et al., 2019).

Tube-load models are transmission line models built with multiple parallel tubes with loads representing arteries, and as such, they can model wave propagation and reflection with only a few parameters. The artery is presented as a tube connected to the Windkessel model of the peripheral arteries. As these models are much simpler than distributed models and still model propagation delays, they might be more suitable for estimating the properties of the arteries, which can then be used for estimating blood pressure.

8.4 Cuffless, continuous blood pressure monitoring

In this section, we describe models used for continuous blood pressure estimation as well as ways to estimate the parameters of these models. Then, we show some devices and solutions on the market used for measuring PTT or PAT. Instead of concentrating on machine-learning approaches, the focus will be on white-box models.

Some of the major differences between continuous monitoring blood pressure devices and other devices that we covered in Chapters 5–7 are as follows (Box 8.2):

- Continuous blood pressure devices are normally based on a mathematical model that relates features such as PTT to blood pressure.
- Often, parameter estimation of the models for continuous blood pressure devices needs to be done before or during their use (see Section 8.4.3), which is different from traditional devices that are calibrated before selling the device. In addition, some continuous blood pressure devices require frequent re-estimation.

8.4.1 Signals and noise characterization

There are various solutions for obtaining the time difference between fiducial points of signals. They are based on different sensors that can be placed on different points on the body to collect data to obtain PTT/PAT or other features—see Fig. 8.4. Therefore, it is important to specify what sensors are used in each device and what parameters are extracted from the signals acquired from sensors.

Major sources of error when computing the time difference between different signals are the following.

- The sampling rate of A/D converters where the resolution of PTT is higher when the sampling rate increases. For example, if the sampling rate is 100 Hz, then the time

---

BOX 8.2

Parameter estimation and calibration

In metrology, calibration is defined as a process of standardizing by determining the deviation of a measurement from an established standard of known accuracy. In continuous blood pressure measurement terminology, the device is initially calibrated by the manufacturer as explained in Section 8.4.2. However, very often, re-calibration or patient-specific adjustment of the parameters needs to be performed by the subject to determine the deviation of measurement based on a single or multiple measurements from an oscillometric device (Solà and Delgado-Gonzalo, 2019). The reference oscillometric measurements are obtained in uncontrolled settings, while the traditional calibration is performed in controlled settings. Therefore, in this chapter, we will not use the term calibration or re-calibration for setting the parameters of the model before using the device but for parameter estimation or adjustment.

difference between each sample is 10 ms. The sampling error can be represented using the uniform distribution. After computing the time difference, the error in the difference follows the triangular distribution.
- Noise that can cause misestimating the exact instance of fiducial points, such as the time instant of the trough of the pulse. Therefore, the time instant of the fiducial point can be detected incorrectly, resulting in an error in computing PTT.

To train models to be used in continuous blood pressure devices, one can use several databases with patient signals as well as signals generated using simulators. Two commonly used databases are:

1. The Medical Information Mart for Intensive Care (MIMIC) database includes arterial blood pressure, PPG, and ECG signals obtained from the intensive care unit (ICU) patients (Saeed et al., 2011). However, the issue with the database is reported in the literature—for some subjects, PPG and ECG are not properly synchronized. In addition, patients are from ICU, which means that medications and one or more co-occurring conditions are most likely present, but information about them is not included in the database.
2. PPG-BP Database contains PPG signals collected along with BP readings from 219 subjects admitted to the Guilin People's Hospital in Guilin, China (Liang et al., 2018). ECG signal is unavailable; therefore, blood pressure cannot be estimated from PTT or PAT but only from extracted PPG features. Three PPG pulses and reference blood pressure values are available per patient.

Virtual databases are described in Charlton et al. (2019) and Willemet et al. (2015). To form the database of virtual subjects in Willemet et al. (2015), the following parameters are varied between subjects: arterial pulse wave velocity (PWV), diameter of arteries, heart rate, stroke volume, and peripheral vascular resistance. It is applied for PWV simulation and estimating arterial stiffness. An advantage of using a virtual database is that we can look at the morphology of the signal at different places on the body. In contrast, the signals from the databases with patients are obtained with the sensors fixed at a single point on the body, such as, for example, a finger for PPG measurements.

8.4.2 Performance measures

There are several standards and guidelines for continuous blood pressure measurements. Researchers and companies often rely on standards for oscillometric blood pressure measurements described in Chapter 5. However, these standards assume that the subject is stationary and sitting and takes only a few measurements. In continuous blood pressure monitoring, blood pressure should be measured under different conditions, postures and activities while performing normal daily activities. Standards should include experiments under different conditions based on a reference device and the *device under test*. The reference device can be an oscillometric device or a continuous blood pressure monitoring device based on a volume clamp or oscillometric method that allows for beat-to-beat blood pressure estimation. Several types of assessments for calibrating continuous blood pressure monitoring devices are identified in Solà et al. (2020), including.

1. Baseline assessment in which the blood pressure is recorded by a reference device and the device under test during a static test which corresponds to the "sitting and relaxed" condition.
2. Blood pressure change intervention in which blood pressure is recorded by a reference device and the device under test while the subject is performing maneuvers described in Chapter 4, exercising, or having drug intervention to modify his or her blood pressure.
3. Body position intervention in which blood pressure is recorded for different postures, including sitting, standing, and supine postures.
4. Hydrostatic bias intervention, that is, relevant when the device under test is placed on the wrist or another part of the body that is mobile and can change the height relative to the heart level.
5. Test after a certain period from the initial calibration.

IEEE standard for wearable monitoring devices (IEEE Standards Association, 2014) partially addresses assessments 1, 2, and 5 for continuous monitoring. In all of these experiments, three measurements need to be performed with the reference device, as well as with the device under test. The standard suggests dividing the validation process into two phases, with 20 subjects in Phase 1 and 25 subjects in Phase 2. The subject population should include about the same number of male and female subjects and several groups of 5 subjects with different blood pressures. At least one subject should have darker skin. The mean absolute difference (MAD) between the measured and the reference values and mean absolute percentage difference (MAPD) should be used when reporting the accuracy. MAPD is defined as

$$\text{MAPD} = \sum_{i=1}^{n} \left| 100 \cdot \left(p_{dut,i} - p_{ref,i} \right) / p_{ref,i} \right| / n$$

where n is the number of measurements, $p_{dut,i}$ is the ith measurement of pressure using the device under test and $p_{ref,i}$ is the ith measurement of pressure using the reference device.

8.4.3 Parameter estimation

In Chapter 1, Fig. 1.8, we introduced the workflow for uncertainty quantification. In this workflow, the model is the most important and is defined in step A. Input parameters and their uncertainties are determined in step B. Parameter estimation is also performed in step B. Fig. 8.5 presents steps A and B for the continuous blood pressure measurement. Inputs to the model are PAT or PTT, as well as some other features that can be extracted from the input signal that we will mention later. The model estimates systolic and diastolic blood pressure. Parameter estimation and adjustments can be made in multiple ways depending on the type of the model and the training data, and it includes the following (Solà and Delgado-Gonzalo, 2019).

1. A person or patient-specific method is based on estimating model parameters for each subject separately. It involves the measurement of pairs of PTT or PAT and reference blood pressure values obtained using oscillometric or other methods before using the device. Several blood pressure reference points are needed, and the number of points

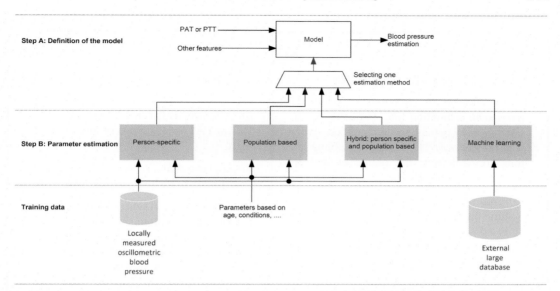

FIGURE 8.5 Parameter estimation based on the workflow for uncertainty quantification presented in Fig. 1.8. Four parameter estimation/adjustment types are shown based on the selected model and available training data.

should be equal to or greater than the number of model parameters. Very often, subjects have normal blood pressure and, therefore, they perform maneuvers (introduced in Chapter 4) that increase or decrease the blood pressure so that a larger range of values of PAT or PTT can be collected. In this case, each subject needs to perform several measurements using a reference device before or during the use of the continuous blood pressure measurement device (Box 8.3).

2. Estimation of model parameters based on population-based information relies on historical data, and it is based on assigning typical parameters to an entire class of subjects with a given pathology, age, and other characteristics. This information could be used to obtain all PAT-BP model parameters; however, these parameters are not personalized.

3. A hybrid method combines patient-specific and population-based approaches. Oscillometric blood pressure measurements are needed to determine some model parameters, while other parameters are obtained based on a person's basic information as well as hers/his medical history.

4. Machine learning models are trained using existing large databases, such as the MIMIC III database, so they do not require any patient-specific parameter adjustment or estimation.

The main disadvantage of the patient-specific and hybrid methods is the need for a reference device and the requirement to collect several reference blood pressure measurements before using the continuous blood pressure measurement device. In addition, the accuracy of the cuffless device generally deteriorates with time after the initial calibration, so frequent re-adjustment of the model parameters is required.

BOX 8.3

Controversies in patient-specific parameter estimation

The number of measurement points for parameter estimation and frequency of re-estimation in patient-specific methods is not well-understood. The number of points and time for collecting the points needed for estimation can be selected as:

- Fixed number of points at predefined time intervals
- Fixed number of points, but the patient is asked to perform some actions such as exercise, raising the arm that the sensors are attached to, changing posture or similar maneuver that would affect the blood pressure
- The time interval is adaptive and requires the patient to carry the

oscillometric device for some time so that the measurements can be done when PTT or PAT values are very low or very high.

Researchers analyzed several approaches for re-estimation, including:

- Periodic estimation when the model parameters are re-estimated against the reference oscillometric device at repeated time intervals, such as, for example, every day or every month.
- Adaptive estimation based on observing changes in some of the measured parameters.

The goal of future cuffless continuous blood pressure measurement devices is to calibrate the device using methods based on population or machine learning so that no reference measurements are needed before using the device.

8.4.4 Model-based blood pressure estimation

8.4.4.1 Blood pressure estimation based on the Moens–Korteweg equation

The model is presented at a central point in Fig. 8.5. The input to the model is PTT or PAT and potentially some other features, while the outputs are the estimates of systolic and diastolic blood pressure. In this section, we will discuss several simple models that are commonly used. We will start with the models based on PTT and then extend our analysis by adding other features.

In Eq. (8.6), we introduced the relationship between PTT and blood pressure. It is evident that this relationship depends on parameters that are difficult to estimate or measure for each patient, such as the thickness and the radius of the vessel. Next, we will show several ways to represent pressure as a function of PTT. From Eq. (8.6), we get

$$p = c_a + c_b \cdot ln \cdot (\text{PTT}) \tag{8.7}$$

where $c_a = \frac{2}{a}\ln \cdot \left(L\sqrt{\frac{2r \times \rho}{hY_0}}\right)$ and $c_b = -2/a$. Parameters c_a and c_b can be estimated from data using regression. The process of estimating parameters c_a and c_b will be explained later in Example 8.4.4.3.

The first-order Taylor series polynomial for Expression (8.5) is $Y = Y_0(1 + \alpha p)$. By plugging this into (8.3), we obtain

$$p = c_c + c_d/PTT^2 \tag{8.8}$$

where $c_d = \frac{2r\rho l^2}{ahY_0}$ and $c_c = -1/a$. Parameters c_c and c_d can be estimated as well from data using regression.

Both models (8.7) and (8.8) require at least two calibration points and accurate measurement and estimation of PTT.

There have been a number of works based on removing some assumptions from the model used to derive the Moens–Korteweg equation. The work in Ma et al. (2018) addressed the following assumptions: (1) the artery wall is thin such that it can be modeled as a thin shell, and (2) the thickness and radius of the artery remain fixed as the blood pressure changes. An experimental setup with external tubes was proposed and developed to perform the experiments. The result of the paper is a complex relationship between the pressure and the PWV (and therefore PTT) that was then simplified to the same model as in Expression (8.8).

8.4.4.2 Blood pressure estimation using surrogate signals

One of the studies that establishes that PPG-PTT can be used for noninvasive blood pressure is Gao et al. (2016). This is the animal study that includes measurement of PTT between the aorta and the femoral artery using ultrasound for the aorta (therefore obtaining blood flow) and an invasive catheter in the femoral artery. In addition, PPG is measured noninvasively at the proximal and distal sites. Invasive PTT based on blood pressure signals has a high correlation with diastolic, mean, and systolic blood pressure estimates (group average R^2 values between 0.86 and 0.91). PPG-PTT also correlated well with diastolic, mean, and systolic BP (group average R^2 values between 0.81 and 0.85). Previously, it was established that PAT does not correlate well with diastolic and mean blood pressure (Mukkamala et al., 2015). Consistent with previous findings, PAT correlated better with systolic BP (group average R^2 value of 0.70) than diastolic blood pressure (group average R^2 value of 0.49). This means that models (8.7) or (8.8) can be used with PTT calculated based on signals from PPG and other sensors obtained from two different points on the body and not necessarily based on PTT obtained from the blood pressure signals.

In addition, it is common to use PAT instead of PTT in the same models even though the error is introduced because of PEP time.

8.4.4.3 Example: Parameter estimation and uncertainty propagation in estimating continuous blood pressure using pulse arrival time

In this section, we will perform parameter estimation and uncertainty propagation through the model that relates systolic blood pressure and pulse arrival time (PAT) based on model (8.7). Parameter estimation is patient-specific, and it is done for one subject.

8.4.4.3.1 Data collection

Two signals are measured to calculate PAT: ECG and PPG signals. Fiducial points include the instant of the peak of the R wave of the ECG and the instant of the peak of the PPG pulse in the same cardiac cycle. Both signals are sampled using an A/D convertor with 2000 samples per second sampling rate. Systolic blood pressure (SBP) is calculated using the model (8.7) in which PTT is just replaced with PAT.

To determine coefficients c_a and c_b, a dataset that related systolic blood pressure with PAT is provided. The data is obtained from one healthy subject. The systolic blood pressure data is collected using the Finapres Medical System (https://www.finapres.com/), which provides beat-to-beat blood pressure estimates. For each pulse, PAT is calculated from ECG and PPG waveforms. About 8 min of data that includes systolic blood pressure estimates per beat and corresponding PAT values per beat are provided. The scattered plot of SBPs versus PAT values is shown in Fig. 8.6A.

8.4.4.3.2 Parameter estimation

Coefficients c_a and c_b are obtained by performing parameter estimation. For each pulse and its measured PAT, there is a reference value of SBP. Simple linear regression can be performed in this case. The coefficients values include $c_b = -95.7$ with 95% confidence interval bounds $\{-111.3, -80.2\}$ and $c_a = 17.7$ with 95% confidence interval bounds $\{2.2, 33.2\}$. Confidence intervals are very large, which is expected if we consider the spread of points around the regression line in Fig. 8.6A. Please note that all the units are in mmHg. From the confidence intervals, we can easily compute the standard deviation. Parameter estimation was performed using the function *fit* in Matlab, which provides the estimates of the linear coefficients and their confidence intervals.

8.4.4.3.3 Uncertainty propagation

The next step is to define uncertain parameters in the system and their distributions. The parameters include coefficients c_a, c_b, and the input PAT. The mean, the confidence intervals

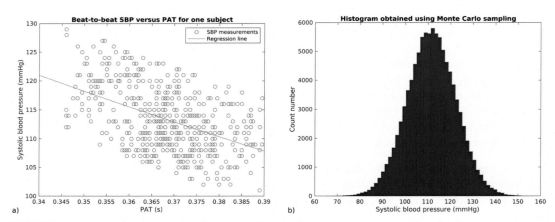

FIGURE 8.6   (A) systolic blood pressure versus pulse arrival time per pulse for one subject obtained during an 8-minute experiment. (B) A histogram obtained using Monte Carlo (MC) sampling of the coefficients c_a and c_b of Eq. (8.7) and PAT. The mean value of PAT is 0.375 s.

and the standard deviation of the coefficients c_a and c_b have already been estimated. As signals are sampled at 2000 samples per second (0.5 ms sampling period), the quantization error of fiducial points of both ECG and PPG signals can be represented as a uniform distribution over the intervals (-0.25 ms; 0.25 ms). The distribution of a random variable computed as a difference between two random variables that follow uniform distribution is a triangular distribution. However, in the simulation, we compute the uncertainties in PAT by sampling two uniform random numbers and subtracting them. For more details, please refer to the corresponding Matlab code shown on the book webpage. To compute the confidence levels for the systolic blood pressure, we set PAT to 0.375 s. $M = 100{,}000$ samples were simulated using Monte Carlo sampling of PAT and parameters c_a and c_b, and the histogram of computed SBP values obtained by propagating the samples through the model (8.7) was formed. The histogram is shown in Fig. 8.6B. The mean systolic blood pressure is estimated to be 111.6 mmHg with the standard deviation of 10.7 mmHg. Confidence interval bounds at 95% confidence are {90.5 mmHg, 133.0 mmHg}. The blood pressure estimates have an extremely large standard deviation, and as such, a system like this cannot be considered precise enough for home blood pressure monitoring.

8.4.4.3.4 Reducing the standard error

A common way to reduce the variability in the estimates is to average the estimates at the output. In this case, the standard error reduces as a square root of the number of samples being averaged if the errors are uncorrelated. For example, averaging 15 samples would reduce the standard error to 10.7 mmHg/$\sqrt{15}$ = 2.8 mmHg if the errors are uncorrelated. If the errors are perfectly correlated, then averaging does not help. As errors in estimating blood pressure are computed from the time series of PATs and the values of PATs in the time series are correlated, we can expect that the standard error will reduce by averaging but not as much as by the square root of the number of samples being averaged.

To summarize, the following steps have been performed

1. Uncertainty in PAT estimation was determined by taking into account the quantization error of A/D converters.
2. Confidence intervals in the values of the coefficients c_a and c_b given data are estimated. Coefficients are assumed to follow the normal distribution.
3. Uncertainty in estimating SBP is obtained by generating M samples from c_a, c_b, and PAT, propagating them through the model (8.7) and computing M values of systolic blood pressure, forming the empirical distribution based on these values and computing the mean and the confidence intervals from the empirical distribution.

Sensitivity analysis shows that coefficients c_a and c_b contribute exclusively to the large standard error. Even if the sampling rate is reduced, the situation is still the same due to very large systolic blood pressure variations versus pulse arrival time in the available dataset.

A similar analysis can be done for the relationship between PAT and diastolic pressure.

8.4.4.4 Models with additional parameters

Several proposed models take into account features such as heart rate or features extracted from PPG, ECG, or other signals in addition to PAT or PTT. These models are based on regression or relatively simple mathematical formulas. The parameters that are estimated usually have physical or physiological meanings. For example, if a feature is the heart rate and is a part of linear regression, then the estimated parameter related to the heart rate explains how much the heart rate affects the blood pressure.

The linear regression model is proposed by Baek et al. (2010) that relates blood pressure with PAT, averaged RR intervals obtained from ECG, and TDB, representing the time difference between the following two points on the PPG waveform *ms* and *dia* (see Fig. 6.17 of Chapter 6). It was shown experimentally that there is a high correlation between blood pressure and the selected parameters (heart rate and TDB). The model then has four unknown parameters $c_1, \ldots, c_4$:

$$p = c_1 + c_2\text{PAT}_\text{D} + c_3\text{RR} + c_4\text{TDB}$$

where PAT_D was introduced in Section 8.2.3, and it is the time from ECG R peak to the *ms* point on PPG pulse. It was shown in (Baek et al., 2010), that multiple regression provides more accurate blood pressure estimates than the linear regression based only on PAT in several cases, including experiments under dental anesthesia, in a vehicle, at the computer desk, and on the toilet. Interestingly, in the last three cases, ECG electrodes were embedded into the seats or the surrounding objects.

There have been several works that investigate different models for estimating systolic and diastolic pressures. An example is shown in Chapter 10 of Solà and Delgado-Gonzalo (2019):

$$\widehat{DBP} = c_{d1}\frac{l}{\text{PAT}} + c_{d2}$$

$$\widehat{SBP} = c_{s1}\frac{l}{\text{PAT}} + c_{s2} + c_{s3}\left(\frac{l}{\text{PAT}}\right)^2 y$$

Where $\widehat{DBP}$ represents the estimate of the diastolic pressure, $\widehat{SBP}$ represents the estimate of the systolic blood pressure, l is the distance between the heart and the measurement point, and y represents the amplitude of PPG or accelerometer (BCG) signals that need to be estimated from data. This model follows the Taylor approximation of the Moens–Korteweg equation. It includes the distance l between the heart and the measurement point, which is important because blood pressure measurements can be done at sites that are at different distances from the heart.

There have been several models that also include patients' age, sex, and other parameters into the models. Many of these models are covered in Solà and Delgado-Gonzalo (2019).

8.4.4.5 Models based on machine learning

Besides physiological models with interpretable parameters, there have been an increasing number of works based on machine learning models such as neural networks and random forests. The research has intensified with the appearance of freely accessible databases containing many waveforms and reference blood pressure values, such as Saeed et al. (2011)

and Liang et al. (2018). This is important because complex machine learning models can be trained by using these large datasets. Feature engineering is still commonly done, meaning that the features are first extracted from the waveforms and used as inputs to the machine learning models. However, there have been several recent works in which only the preprocessed waveforms are used as inputs to a deep learning model. Here, we will discuss the most common features extracted from ECG and PPG signals and then used as inputs to machine learning models. In addition, PAT or PTT can also be used as a feature.

An interesting approach is shown in Huttunen et al. (2019), where a machine-learning model based on the Gaussian process is developed for the database of thousands of virtual patients. The features are still mainly timing features, and PTT is calculated as the time between the aortic valve opening and the fiducial points of the pressure pulse on the carotid artery. The features include PTT_p, PTT_f, PTT_D, the time to the dicrotic notch as well as the heart rate. It has been shown that combining all these features gives better results than using models with only one feature.

The advantage of most machine learning methods is that they are not only based on PAT/PTT but include statistical, morphological, and timing parameters. In this section, we will show features extracted only from PPG and ECG signals, including:

- Time domain features include intervals between different fiducial points on the pulses and their ratios.
- Frequency domain features include frequencies and magnitudes on the fundamental and harmonics, as well as the area under the spectrum curve in the specified frequency ranges (Chowdhury et al., 2020).
- Statistical features include the mean, standard deviation, skewness, and kurtosis (Chowdhury et al., 2020).

A detailed list of physiological features extracted from ECG is given in Bird et al. (2020). We showed in Table 8.2 only those features that can be extracted from a single lead ECG signal. It was shown in Karaagac et al. (2014) that the TpTe interval and TpTe/QTc ratio are higher in cardiac patients than in the control group. Fiducial points that can be extracted from the PPG signal are shown in Fig. 6.17 in Chapter 6. Based on these fiducial points, a number of features can be extracted. These features are shown in Table 8.2.

8.4.5 Main processing stages

Fig. 8.7 shows the processing pipeline stages for obtaining PAT and PTT and extracting other features from the pulses. These steps include:

1. Filtering of signals acquired from sensors such as PPG and ECG. Bandpass filtering is commonly applied where low frequencies are removed to remove the baseline drift, and higher frequencies are removed to filter out high-frequency noise and AC interference.
2. Detecting pulses and estimating PAT or PTT and other features presented in Table 8.2.
3. Performing pulse/signal quality assessment, which can be done by using an additional accelerometer sensor to detect motion artifacts, followed by removing signal segments

TABLE 8.2   Fiducial point and indices or features of ECG and PPG signals commonly used in continuous blood pressure estimation models.

Signals	Fiducial points	Features
ECG	Time instances and amplitudes of P, Q, R, S, and T waves	RR intervals: the interval between the consecutive R waves PR intervals: the interval between the beginning of the P wave and the beginning of the QRS complex QTc interval or corrected QT interval: the ratio between QT interval and the square root of RR interval TpTe interval: the interval between the peak and the end of the T wave on the ECG TpTe/QTc ratio: the ratio of TpTe and QTc intervals
PPG	s, dia, dic, ms, a, b, c, d, e, CT, ΔT, areas (see Fig. 6.17 of Chapter 6)	Reflection index: the ratio of the magnitudes of the systolic and diastolic peaks: $RI = dia/s$ Augmentation index: $AI = s/dia$ Systolic amplitude: s Pulse width: the time interval measured between the rising and falling edge of the pulse at the half height of the systolic peak Peak–peak interval PPI: the time interval between s points of the two consecutive pulses Inflection point area ratio IPA: the ratio of the areas under pulse between the beginning of the pulse and the *dic* notch and the area from *dic* notch till the end of the pulse

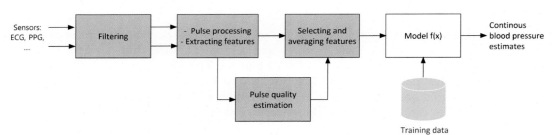

FIGURE 8.7   Processing stages of continuous blood pressure estimation.

during which the motion was detected. Assessment of the quality of PPG and ECG signals based on physiological feasibility, spectral characteristics and other features extracted from the signals were presented in Chapter 4. If the signals are of satisfactory quality, they can be used for extracting timing or other features.

4. Selecting important features
5. Fitting data to models presented in Section 8.4.4.

In addition to the earlier steps, adjusting the model for different real-life situations is described next.

8.4.6 Adjusting and improving models

8.4.6.1 Adjusting algorithms based on a subject's posture

Most traditional BP devices, such as oscillometric devices, assume and require that the sensor is positioned at the heart level. This is automatically achieved when the cuff is placed on the upper arm for the oscillometric blood pressure measurement. When the cuff is on the wrist, the subject is asked to move the arm to the position where the wrist is at the heart level. Another option would be to estimate and incorporate hydrostatic pressure into the blood pressure estimation algorithm to compensate for the difference in levels.

There have been several locations on the body where the devices for continuous blood pressure are placed, including the wrist as the most popular location due to the integration of continuous blood pressure measurement systems with smartwatches, chest, and neck. It was shown in Solà and Delgado-Gonzalo (2019) that changes in the vertical height of the sensor of 10 cm relative to the heart level could cause changes of 7 mmHg. Therefore, adjusting the blood pressure algorithm based on the posture is needed by incorporating the hydrostatic pressure $\rho g H$ into the blood pressure estimation algorithm, where ρ is the density of blood and H is the estimated vertical height relative to the heart level and g is the gravitational constant. To estimate hydrostatic pressure, the devices need to estimate H using, for example, gyroscopes and accelerometers (Luinge and Veltink, 2005).

In Butlin et al. (2018), it was shown that there is a significant decrease in blood pressure and cfPWV when changing the posture from sitting to supine. A change in cfPWV also means a short-term change in the arterial stiffness measured between the carotid and femoral arteries and, therefore, a change in blood pressure. However, no significant changes were observed in PWV between carotid to radial and carotid to finger with that posture change. This means it is necessary to research the relationship between blood pressure and PTT when the posture changes in case the device is placed on the body at a location that is not at the heart level.

8.4.6.2 Taking into account PEP

PAT consists of two components introduced in Section 8.2.1: the preejection period (PEP) and the PTT. While PTT is proportional to BP as described by the Moens–Korteweg and Bramwell–Hill models, PEP shows a different kind of relationship that is not well understood.

In Wong et al. (2011), PAT, PEP, and PTT and their relative change were measured for 22 healthy subjects before and after the exercise. It was shown that, on average, the change in the measured variables for all subjects after the exercise were: PAT: −13.1%, PEP: −34.6%, and PTT: −5.9%. All three variables have a negative correlation with increased blood pressure. Significant changes after exercise were observed in PEP and not in PTT. PEP was well correlated with SBP, but it was not correlated with DBP. However, it should be noted that the paper (Wong et al., 2011) analyzed the linear relationship between blood pressure and PTT, while the majority of the models rely on the nonlinear relationship between them. In Ebrahim et al. (2019), it was shown that removing PEP from PAT improves blood pressure estimation results by around 9%. The difference in the effects of PEP on blood pressure between Wong et al. (2011) and Ebrahim et al. (2019), as well as some other works, can be attributed to different sensors used in the measurements, as well as different ways of PEP estimation.

We showed in Section 8.2.3 that SCG and PCG signals have peaks close to the instant of the aortic valve opening and, therefore, can be used to obtain PEP or compute PTT surrogates. For example, this approach is used in A Foo et al. (2006) for the PCG signal and, more recently, in Ganti et al. (2021) for the SCG signal. A smartwatch design with ECG, PPG, and accelerometer sensors is presented in Ganti et al. (2021). Blood pressure measurements require touching the sternum with the smartwatch to obtain the SCG signal. Then, the PTT was calculated as the difference between the aortic valve opening point obtained from a specific peak of the SCG and the foot of the PPG signal obtained from the wrist. Highly accurate blood pressure estimation over 24 h was reported.

8.4.7 Circuits and devices

A generic system that includes an ECG, PPG, and accelerometer sensors is shown in Fig. 8.8. The sensors used for estimating PTT are connected to a simultaneous or multiplexed A/D converter. The output of the A/D converter is connected to a microcontroller which controls the operation of the whole device as well as stores data and results, and communicates with a smartphone or computer using a wireless network. Even though only one LED is needed to obtain a PPG signal, these solutions are often used for estimating SpO_2, and therefore, at least two LEDs are included. Available integrated solutions (described later) normally include the circuits for driving LEDs, as well as circuits for PWM modulation and for time-multiplexing (sequencing) of the LEDs. The accelerometer is needed to determine subjects' activities and, in cases when the device is placed on the chest to obtain the SCG signal. Other sensors could potentially be used, including temperature sensor, pressure sensor for tonometry applications, and bioimpedance circuitry.

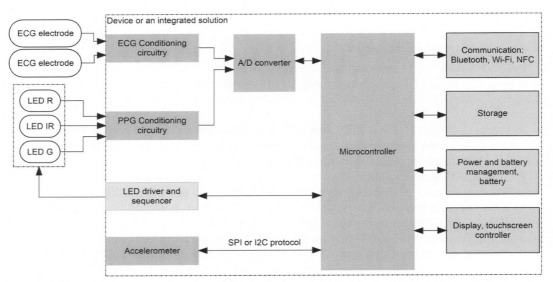

FIGURE 8.8 Generic architecture of a wearable device with ECG and PPG sensors, accelerometer, and processing circuitry that can be used for continuous blood pressure monitoring.

8.4.7.1 Academic solutions

Many multiparameters monitoring devices have been recently developed in academic settings. They come in different form factors, including smartwatches, skin patches, headphone-based devices, and chest and arm patches. The majority of these devices measure PPG and ECG and estimate blood pressure using PAT.

A smartwatch that can measure ECG, PPG, and SCG has been introduced in Section 8.4.6.2 (Ganti et al., 2021). In addition, the watch includes temperature, pressure, and humidity sensors so that it can potentially include environmental conditions in the model for blood pressure estimation. The estimation is based on PTT computed between fiducial points of SCG and wrist PPG sensors. The sampling rate is 125 samples per second. Even though it is not a low sampling rate, it would be better if it was higher because the blood pressure estimation relies on an accurate estimation of PTT. One of the disadvantages of the system presented in Ganti et al. (2021) is that each sensor has its A/D convertor. Here, it would be beneficial to use simultaneous A/D converters presented in Section 3.5.3.1 in Chapter 3.

Smart-skin solutions are very popular for continuous blood pressure measurements. The work by Li et al. (2020) describes a flexible PPG-based solution for measuring blood pressure. PPG LEDs and photodetectors are placed several centimeters apart, and the time delay between the fiducial points of the signals from these photodetectors is calculated. A new model for systolic and diastolic blood pressure estimation was developed that is similar to Eq. (8.8) but also includes DC and AC components of the absorbed light of the PPG signal. Very good agreement was shown between this model and the results obtained using invasive blood pressure measurements.

A smart-skin solution for neonatal intensive care was developed in Chung et al. (2019). It contains ECG and PPG sensors and the front-end circuits and estimates blood pressure based on estimated PAT. Interestingly, the system is powered up by near field communication (NFC) and does not use batteries, making it suitable for neonatal care.

8.4.7.2 Integrated industrial solutions

A number of companies provide solutions with integrated ECG and PPG electronics and microcontrollers. Regarding PPG electronics, LED drivers, photodiode, analog front-end, and the circuit that provide the sequence for turning the LEDs on and off are commonly included on the chip. In addition, very often, a temperature sensor and an accelerometer are integrated. These integrated solutions require additional electrodes for ECG and LEDs for PPG. Examples include AS7038RB from AMS AG, Austria (https://ams.com/as7038rb) and RT1025 from Richtek Technology Corporation, Taiwan (https://www.richtek.com/Design%20Support/Technical%20Document/AN057). These chips are suitable for different types of devices, including medical-grade ECG or PPG devices, and fitness and wearable devices.

8.4.7.3 Devices

Even though this chapter focuses on measurements based on time differences between the fiducial points of several signals, there have been several industrial solutions that rely only on extracting features from PPG signals and then using these features only to estimate blood pressure.

Several commercial systems have been developed and validated using studies with a relatively large number of subjects. BB-613WP device by Bio-Beat (https://www.bio-beat.com/) is worn on the chest or as a bracelet. The calibration/parameter estimation is patient-specific and requires using a reference oscillometric device at least once in 3 months. Its accuracy is validated on 1057 subjects during a static test. About half of the subjects performed exercises, and data points from the Bio-Beat device were collected with the oscillometric measurement after the exercise. Blood pressure estimates were compared against the reference oscillometric device, and high accuracy (less than 3.6 mmHg standard deviation) and a high interclass correlation coefficient (ICC > 0.9) were achieved (Nachman et al., 2020). Aktiia Bracelet is also based on PPG measurements only (Solà et al., 2020). It has been validated on 31 patients against the invasive radial blood pressure measurements, as well as on more than 85 patients against auscultation measurements by two nurses measuring blood pressure on the upper arm.

Companies producing smartwatches have started to incorporate blood pressure measurements. The devices mainly require periodic patient-specific model parameter adjustment with an oscillometric device.

8.4.8 New developments and directions

This section discusses sensor developments, miniaturization of sensors and devices, and improvements in algorithms, models, and methods. Several challenges, as well as technological and algorithmic/modeling developments, have been identified (see Fig. 8.9).

Long-term monitoring is a challenge for multiple reasons. First, person-specific continuous blood pressure monitors require repeated parameter estimation. Therefore, developing

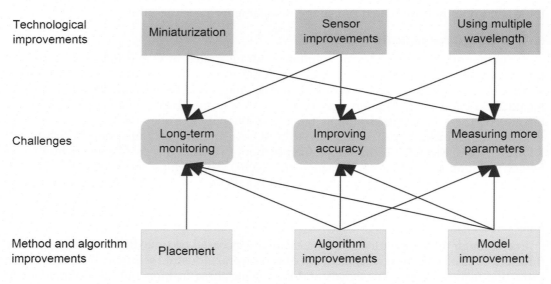

FIGURE 8.9 Challenges, technological and modeling developments related to continuous blood pressure measurements.

devices and systems that can perform measurements without reestimation is very important. Alternative ways of calibration and parameter estimation are summarized in Mukkamala et al. (2015). A potential approach is to develop calibration curves for people of different ages and conditions. This would require many measurements and correspond to the hybrid method presented in Section 8.4.3.

In addition, long-term monitoring requires an *ergonomic design*. Therefore, it is important to design devices that are not burdensome to wear. The devices should be visually appealing if they are exposed, and they should not interfere with normal life. *Battery life* is normally from 1 day to 1 week. Much work has been done on improving the battery life of these wearable devices.

The *accuracy of wearable devices* for continuous blood pressure measurements is a major issue at this point. As the measurement is done in an uncontrolled environment, it is necessary to account for motion artifacts, different postures, etc. The accuracy could be improved if more advanced *physiological models* were developed for continuous blood pressure estimation. Complex distributed models (Section 8.3.3) are used for data generation, but not much work has been done on fitting parameters and personalizing these complex models for continuous blood pressure estimation. Tube-load models represent a tradeoff between the complexity of distributed 1D models and simplifying assumptions in Hughes models and the Moens-Korteweg equation. A work on tube-load models for estimating blood pressure was described in Zhang et al. (2011b), showing that PTT estimation from tube-load models is better correlated with blood pressure than PTT obtained from foot-to-foot measured time-difference between two signals. The work on physiological models and black-box machine learning developed independently. We believe that deep learning models should take advantage of domain-level knowledge and models, and therefore we see as one of the major directions in this field the future work on merging physical and deep learning models. This work on *scientific machine learning or physics/mechanistic guided matching learning* has already started in other fields such as physics—please see Rackauckas et al. (2020).

Personalized models require entering personal information into the model, including age, sex, health conditions, and smoking habits. If the model can incorporate all these parameters, it should be able to provide more accurate estimates of blood pressure. Another personalized model is based on using historical information from the same person (Leitner et al., 2019). In continuous blood pressure devices, the blood pressure signal is a time series indicating a high correlation between the blood pressure values over time. Therefore, it makes sense to incorporate the estimated blood pressure values from the past when estimating future values.

Estimating blood pressure for a specific group of patients, such as patients with arrhythmias and obese patients, is very important. However, not much work has been not done on evaluating continuous blood pressure devices for these patients, as well as on developing models that are suitable for specific patient populations. In Liu et al. (2020), ECG and PPG signals were collected together with the invasive arterial blood pressure signals for 35 patients who had different types of arrhythmias. Traditional machine learning approaches such as random forest were used to estimate blood pressure. Much more work of this type is needed to address the needs of specific patient populations.

Recently, several research groups have looked into the effect of various biomarkers and cardiovascular parameters by using a *combination of different sensors* in a single device. Heart rate and blood pressure can be affected by variations in concentrations of biomarkers such as glucose or lactate. These variations are due to motion, stress, or intake of food, drinks, and drugs. We know that blood pressure changes over the time of the day; therefore, there is a potential to relate these changes to variations of biomarkers if they are all measured simultaneously and continuously. In addition to continuous blood pressure measurements, combined monitoring of biomarkers and cardiac signals has the potential for monitoring cardiac diseases, detection of septic shock by looking at a sudden drop in blood pressure accompanied by increased lactate levels in the blood, monitoring hypoglycemia- or hyperglycemia-induced hypotension or hypertension, and so on (Sempionatto et al., 2021). One important aspect when designing these devices is that patient compliance would increase if all the sensors were placed in a single device and all the measurements were done at a single point. In Sempionatto et al. (2021), a smart-skin sensor was proposed that can measure hemodynamic parameters such as blood pressure and heart rate, as well as the levels of the following biomarkers: glucose and lactate. In addition, it can measure levels of caffeine and alcohol. Ultrasonic transducers are used to monitor heart rate and blood pressure, while electrochemical sensors measure the levels of biomarkers. When the ultrasonic sensor is placed over the carotid artery, it is possible to obtain the lumen area and its changes. Measured lumen area changes and measured PWV (based on PTT measurements) can be used to estimate the pressure changes based on Expression (8.1).

8.5 Other applications of PTT/PAT

8.5.1 Arterial stiffness estimation

This section is based mainly on Bolic et al. (2021). Arterial stiffness has already been presented in Chapter 8, where we discussed how it could be determined from oscillometric signals with an emphasis on the analysis of pulses. Here, we will discuss arterial stiffness based on the time difference between two signals. The time difference is used to obtain pulse wave velocity, which is the established metric for estimating arterial stiffness and is considered more accurate than pulse analysis methods (Salvi et al., 2019).

8.5.1.1 PWV and arterial stiffness

PWV has already been introduced in Section 8.2.1. Often it is measured between two arterial sites simultaneously using two sensors or sequentially. Sequential measurements are based on using ECG as a reference. In this case, PAT between the ECG R peak and the fiducial point of the signal at the proximal site is computed (PAT_{pr}), and then PAT between the ECG R peak and the fiducial point of the signal at the distal site is computed (PAT_{di}). PTT is computed as $PTT = PAT_{di} - PAT_{pr}$. As we can also measure the length between the two measurement sites and adjust it using empirical formulas to correspond to the internal path length through the arteries l, we can compute PWV as $PWV = 1/PTT$. PWV is usually measured over several beats and then averaged. The standard for measurement of arterial stiffness is to measure PWV between specific sites on the carotid and femoral arteries (cfPWV) (DeLoach and Townsend, 2008).

The estimation of cfPWV can be done noninvasively using one of several commercial devices based on tonometry or PPG. Commercial devices for estimating cfPWV include Complior Analyse, PulsePen ET, PulsePen ETT, and SphygmoCor. It was shown in Salvi et al. (2019) that there is a high correlation between invasive aortic PWV and estimated cfPWV obtained from each of these devices.

Disadvantages of the assessment of arterial stiffness based on PWV include that it normally requires a trained operator to use one of the mentioned devices and that PWV is a surrogate measure of arterial stiffness and is not identical to the arterial stiffness itself.

8.5.1.2 Sensors

The advancements in the technologies are achieved through developments in sensor technology, miniaturization of sensors, integration of multiple sensors on a single chip, and integration of the electronics for amplification and filtering on the chip. Ultrasound devices measure aortic PWV based on the Doppler effect. Ultrasound wall tracking can be used to acquire the arterial pulse. New miniature ultrasound devices can achieve the goal of continuous monitoring of the arterial pulse from different body sites, including the carotid artery (Wang et al., 2018). In Wang et al. (2018), PTT is computed from two points using ECG as a reference, as explained earlier, and then used to compute PWV.

There has also been a surge in recent years of wearable devices with novel tonometric sensors for measuring and recording arterial pressure waveforms. A multiarray seven-channel 17×17 mm pressure sensor was developed to be placed over the radial artery (Roh et al., 2020). Capacitive and piezoresistive MEMS pressure sensors have been developed by major manufacturers, including Analog Devices, STMicroelectronics, Murata, and others. A MEMS pressure sensor element with three sensors for the radial artery was also recently developed (Kaisti et al., 2019). Both technologies described in Roh et al. (2020) and Kaisti et al. (2019) captured a high-fidelity arterial pulse wave.

8.5.1.3 Physical and machine learning models

In this section, we describe approaches based on a single-point measurement (such as PPG) or measurements using two sensors (such as PPG and ECG) and physical and machine learning models used to derive indices related to arterial stiffness from these signals.

In Wei (2013), the spring constant method was applied to the PPG signal to obtain a coefficient that was shown to be negatively correlated to cfPWV. The method is based on the fact that the radial vibration of the peripheral arterial wall can be modeled as an elastic spring. The spring constant, which represents the ratio between exerted force and displacement based on Hooke's law, was assumed to be proportional to the elasticity of the arteries. This means that lower values of the spring constant represent the deterioration of the elasticity of an arterial wall.

Several works have used machine learning regression models to predict PWV from tonometry or PPG signals measured at a single point. This would simplify the measurement of PWV because only a single measurement point is required. However, in this case, PWV cannot be measured, but it could potentially be learned with the help of machine learning models. In Alty et al. (2007), morphological and subspace features were extracted from the PPG signal and fed into a regression model based on a neural network. The standard deviation of the difference between the actual and predicted PWV was found to be 2.07 m/s. The tonometry signal obtained from the carotid artery was recently used to estimate PWV with a root mean square error of 1.12 m/s (Tavallali et al., 2018). The features used for machine learning were extracted from the carotid arterial pulse waveform, but additional clinical features were added, including blood pressure and age. In addition, in Alty et al. (2007), patients were divided into two classes based on whether PWV is lower or higher than 10 m/s, and then the extracted features were used to classify patients as being below or above the cut-off value for arterial stiffness.

8.5.2 ECG-assisted oscillometric blood pressure

In this section, we return to oscillometric blood pressure measurements described in Chapter 5. The idea behind oscillometric measurements is to extract the oscillometric waveform during cuff deflation and then detect pulses and use some feature from each pulse, such as the amplitude of the peak of each pulse, to produce the oscillometric envelope. The maximum of the envelope corresponds to MAP. Systolic and diastolic pressures are then estimated from the oscillometric envelope; however, the estimated MAP value is used as a starting point for their estimates in almost every algorithm. For example, in the maximum amplitude method described in Chapter 5, the MAP estimate is used to compute the SBP and DBP from the envelope of the oscillometric pulses using specified estimation ratios.

The system presented in Ahmad et al. (2010) allows for simultaneous ECG and oscillometric signal acquisition. The oscillometric waveform is obtained using a brachial cuff connected to a pressure sensor. A new mathematical model was proposed in Forouzanfar et al. (2015) that computes PAT between the ECG signal and the oscillometric signal during cuff deflation based on the system proposed in Ahmad et al. (2010). PAT values for each pulse are presented similarly to the oscillometric envelope. It was shown in Forouzanfar et al. (2015) that the envelope's maximum corresponds to the MAP. This finding is important because obtaining MAP from the regular oscillometric envelope is often inaccurate, and the envelope based on the maxima of the pressure pulses is quite flat around its peak value. In addition, when oscillometric measurements are performed on obese patients, excess fat tissue dampens the pulse amplitudes reaching the cuff and the pressure sensor. Furthermore, vessels lose elasticity and become rigid in atherosclerosis, weakening the pulse amplitudes. Thus, oscillometric blood pressure estimation algorithms have to deal with weak pulses in these patient populations. This can lead to inaccurate and unreliable MAP estimates, leading to inaccurate estimates of SBP and DBP. On the other hand, ECG is measured using electrodes and its peaks or pulses are not dampened or weakened under the above physiological conditions. Therefore, blood pressure obtained from a PAT-based oscillometric envelope might be beneficial in several scenarios.

The algorithm for computing MAP consists of the following steps (Ahmad et al., 2010):

1. Detect the R-peaks of the ECG Signal.
2. Detect the instances of the maximum slope (or other fiducial points) of each pulse of the oscillometric waveform.
3. Calculate PAT_D between the R peaks of the ECG and the maximum slope points of the oscillometric pulses.
4. Remove outliers in PAT_D values.

5. Interpolate PAT_D so that the number of points is equal to the sampling rate of ECG/ oscillometric pulse to obtain PAT_D envelope.
6. Filter the PAT_D envelope using a lowpass filter
7. Find the maximum point of the PAT_D envelope. The time instant of that maximum point corresponds to the time instant of the MAP value on the cuff deflation curve.

8.6 Summary

In this chapter, we discussed the devices that use multiple sensors and where the temporal relationship between the fiducial points of pulses obtained from different sensors during the same cardiac cycle carries physiological meaning. This temporal information is then used to estimate blood pressure or arterial stiffness noninvasively. We showed different models that relate blood pressure and PTT or PAT. Then, we described how these models are used in continuous blood pressure monitoring devices. Trends and research directions for continuous blood pressure and arterial stiffness monitors are also pointed out.

Even though this chapter deals with signals from multiple sensors, we focused mainly on fusion based on temporal information. However, there are several other ways to perform fusion between these different signals, which will be covered in Chapter 10.

8.7 Appendix 1—data collection system

Description of multibiomedical signal acquisition shown in Fig. 8.3:

1. ECG sensor
   SparkFun single lead heart rate monitor—8232 CE with 3 connector sensor cable Sensor Cable—Electrode Pads—CAB-12970—SparkFun Electronics.
   Datasheet: 8232 CE (sparkfun.com).
   The ECG electrodes are placed on the left arm, right arm, and right leg. The electrode on the right leg acts as an active ground.
2. PPG sensor
   SparkFun pulse sensor: Pulse Sensor—SEN-11574—SparkFun Electronics.
   PPG signal is collected by placing the optical PPG sensor on the fingertip.
3. PCG sensor
   Adafruit MAX4466 microphone module: MAX4465-MAX4469.pdf (adafruit.com).
   1063 Adafruit Industries LLC | Development Boards, Kits, Programmers | DigiKey
   For PCG signal acquisition, the microphone is placed on the sternum, close to the heart.
4. SCG sensor
   SparkFun ADXL377 3-axis accelerometer.
   ADXL377 (Rev. B) (analog.com).
   The accelerometer is placed on the sternum, close to the heart, for SCG signal acquisition.

All these sensors provide analog output and are digitized by an NI-USB6002 DAQ card USB-6002—NI connected to a computer. The signals are collected and visualized in MATLAB with NI-DAQ API. The units for these signals are in volts (V).

8.8 Problems

8.8.1 Simple questions

8.1. How does the blood pressure change with the hydrostatic pressure?

8.2. Is the value of PEP constant when a subject performs maneuvers or exercises? If not, why does it change?

8.3. What is the disadvantage of using the Windkessel model for continuous blood pressure measurements?

8.4. What is the advantage of using tube-load models for continuous blood pressure measurements?

8.5. Describe the relationship between the following parameters: PWV, PTT, PAT, arterial stiffness, and blood pressure.

8.8.2 Problems

8.6. How does the PAT/PTT-based blood pressure measurement depend on the A/D converter sampling rate? Is it necessary to sample PPG and ECG simultaneously (see Section 3.5.3.1)?

8.7. Design a circuit for continuous blood pressure monitoring that includes PPG and ECG sensors, conditioning electronics, driving circuitry as well as an A/D convertor. Show how one can compute the time difference between the PPG and ECG signals to obtain PAT in (a) hardware and (b) in software algorithmically.

8.8. (a) Draw the PPG waveform in Fig. 8.10 at the bottom relative to the ECG waveform shown in Fig. 8.10 at the top. Please make sure that your drawing is proportional and makes sense physiologically. Next, define and draw the pulse arrival time.

8.9. PPG-PTT is the time difference between the fiducial points of the PPG signals measured at two different places on the body: proximal and distal. In this problem, the goal is to design a schematic that will be used to measure PPG-PTT.

    (a) Start first with the block diagram of the PPG device. Then, explain the function of each block of the system.

    (b) Then, use the design in (a) as a black box. Use two PPG circuits and show the block diagram of the circuit that can measure PPG-PTT.

8.8.3 Simulations

8.10. This question is related to uncertainty propagation in continuous blood pressure measurement presented in Example 8.4.4.3.

    (a) Perform sensitivity analysis to determine the effect of coefficients c_a, c_b, as well as the input PAT on the blood pressure estimate.

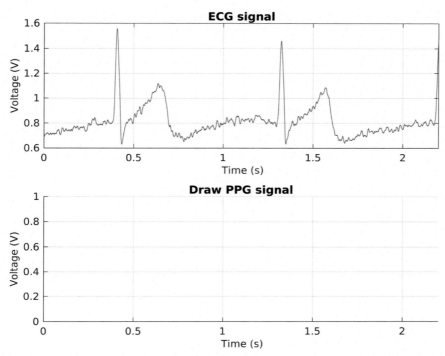

FIGURE 8.10 ECG signal is shown in the figure on the top. Draw the PPG signal below while ensuring that the timing between different fiducial points of the PPG signal and ECG waves makes sense physiologically.

(b) Reduce the standard deviation of the coefficients a and b five times. Analyze the effect of the parameters on the output in this case. How would reducing the sampling rate from 2000 samples per second to 200 samples per second affect the precision?

8.11. This question is related to parameter estimation in continuous blood pressure presented in Example 8.4.4.3.
  (a) Replace the model (8.7) with (8.8) and estimate the model parameters. What are the confidence intervals obtained for these parameters?
  (b) Add the heart rate to the model (8.8). Does it improve blood pressure estimation or not?

8.12. Model in Matlab, Python, or a language of your choice the dependence between blood pressure and pulse transit time by including hydrostatic pressure by following Eq. (8.3) from the paper (Poon et al., 2006). How does PTT change with increasing the distance between the sensor and the heart?

8.13. Obtain example waveforms from the virtual database http://haemod.uk/original for carotid and femoral pressure. Draw the pressure signals and observe the differences in morphologies.

8.9 Further reading

Please note that the following material is selected based on availability and the author's preferences and that many good resources might be omitted unintentionally. We consider references related to: (1) the theory needed to understand continuous blood pressure measurements, including biofluid mechanics and hemodynamics, (2) continuous blood pressure measurements, and (3) arterial stiffness.

The Moens—Korteweg equation and the theory around it are described in a number of books, including Waite and Fine (2007). Very detailed and accessible descriptions of pulse propagation and reflection, the Windkessel model, pulse analysis, mechanisms and devices for arterial stiffness, and central and peripheral blood pressure are given in Salvi (2017).

A comprehensive overview of continuous noninvasive blood pressure measurement methods is given in Solà and Delgado-Gonzalo (2019), including covering standards for continuous blood pressure measurements and calibration methods. Several excellent review papers describe machine learning models developed and applied for continuous blood pressure monitoring (Martinez-Ríos et al., 2021; Maqsood et al., 2022).

A detailed review of arterial stiffness and the relationship between arterial stiffness and hypertension from a medical point of view is shown in Boutouyrie et al. (2021). A description of devices used for assessing arterial stiffness is given in Salvi et al. (2019).

Acknowledgements

The author would like to thank Shan He, University of Ottawa, for performing experiments and collecting data shown in Figs. 8.3 and 8.6.

References
Modeling and blood pressure estimation

Ahmad, S., Bolic, M., Dajani, H., Groza, V., Batkin, I., 2010. Measurement of heart rate variability using an oscillometric blood pressure monitor. IEEE Transactions on Instrumentation and Measurements 59 (10), 2575—2590.

Alastruey, J., Khir, A.W., Matthys, K.S., Segers, P., Sherwin, S.J., Verdonck, P.R., et al., 2011. Pulse wave propagation in a model human arterial network: assessment of 1-D visco-elastic simulations against in vitro measurements. Journal of Biomechanisms 44, 2250—2258.

Alastruey, J., Parker, K.H., Sherwin, S.J., 2012. Arterial Pulse Wave Haemodynamics. 11th Int Conf Pressure Surges, Bedford, UK, pp. 401—442.

Baek, H.J., et al., 2010. Enhancing the estimation of blood pressure using pulse arrival time and two confounding factors. Physiological Measurements 31 (145).

Bird, K., et al., 2020. Assessment of hypertension using clinical electrocardiogram features: a first-ever review. Frontiers in Medicine, 7. https://doi.org/10.3389/fmed.2020.583331. (Accessed 3 April 2020).

Butlin, M., Shirbani, F., Barin, E., Tan, I., Spronck, B., Avolio, A.P., 2018. Cuffless estimation of blood pressure: importance of variability in blood pressure dependence of arterial stiffness across individuals and measurement sites. IEEE Transactions on Biomedical Engineering 65, 2377—2383.

Charlton, P.H., et al., 2019. Modeling arterial pulse waves in healthy ageing: a database for in silico evaluation of haemodynamics and pulse wave indices. American Journal of Physiology-Heart and Circulatory Physiology.

Chowdhury, M.H., et al., 2020. Estimating blood pressure from photoplethysmogram signal and demographic features using machine learning techniques. Sensors 20 (11), 3127. https://doi.org/10.3390/s20113127.

Chung, H.U., et al., Mar. 2019. Binodal, wireless epidermal electronic systems with in-sensor analytics for neonatal intensive care. Science 363 (947).

Ebrahim, M.P., Heydari, F., Wu, T., Walker, K., Joe, K., Redoute, J.-M., Yuce, M.R., 2019. Blood pressure estimation using on-body continuous wave radar and photoplethysmogram in various posture and exercise conditions. Scientific Reports, Nature 9 (16346).

A Foo, J.Y., Lim, C.S., Wang, P., 2006. Evaluation of blood pressure changes using vascular transit time. Physiological Measurements 27 (685).

Forouzanfar, M., Ahmad, S., Batkin, I., Dajani, H.R., Groza, V.Z., Bolic, M., 2015. Model-based mean arterial pressure estimation using simultaneous electrocardiogram and oscillometric blood pressure measurements. IEEE Transactions on Instrumentation and Measurement 64 (9).

Ganti, V.G., Carek, A.M., Nevius, B.N., Heller, J.A., Etemadi, M., Inan, O.T., June 2021. Wearable cuff-less blood pressure estimation at home via pulse transit time. IEEE Journal of Biomedical and Health Informatics 25 (6), 1926–1937.

Gao, M., Olivier, N.B., Mukkamala, R., 2016. Comparison of noninvasive pulse transit time estimates as markers of blood pressure using invasive pulse transit time measurements as a reference. Physiological Reports 4 (10).

Heydari, F., et al., 2018. Continuous cuffless blood pressure measurement using body sensors. In: 13th International Conference of Body Area Network (BodyNets) of EAI. Finland, Oulu.

He, S., Dajani, H.R., Meade, R.D., Kenny, G.P., Bolic, M., 2019. Continuous Tracking of Changes in Systolic Blood Pressure Using BCG and ECG. IEEE EMBC conference.

Huttunen, J.M.J., Kärkkäinen, L., Lindholm, H., 2019. Pulse transit time estimation of aortic pulse wave velocity and blood pressure using machine learning and simulated training data. PLoS Computational Biology 15 (8).

IEEE Standards Association, 2014. IEEE Standard for Wearable Cuffless Blood Pressure Measuring Devices. IEEE Standard, 1708-2014.

Karaagac, K., et al., 2014. Evaluation of Tp-Te interval and Tp-Te/QTc ratio in patients with coronary artery ectasia. International Journal of Clinical and Experimental Medicine 7 (9), 2865–2870.

Leitner, J., Chiang, P.-H., Dey, S., 2019. Personalized blood pressure estimation using photoplethysmography and wavelet decomposition. In: IEEE International Conference on E-Health Networking, Application and Services (HealthCom).

Li, H., et al., May 2020. Wearable skin-like optoelectronic systems with suppression of motion artifacts for cuff-less continuous blood pressure monitor. National Science Review 7 (5), 849–862.

Liang, Y., et al., 2018. A new, short-recorded photoplethysmogram dataset for blood pressure monitoring in China. Scientific Data 5, 180020.

Lin, W.W., Chen, Y.T., Hwang, D.S., Ting, C.T., Want, G.K., Lin, C.J., 1999. Evaluation of arterial compliance in patients with carotid arterial atherosclerosis. Zhonghua Yi Xue Za Zhi (Taipei) 62 (9), 598–604.

Liu, Z., Zhou, B., Li, Y., Tang, M., Miao, F., 2020. Continuous blood pressure estimation from electrocardiogram and photoplethysmogram during arrhythmias. Frontiers in Physiology 9 (11).

Luinge, H.J., Veltink, P.H., 2005. Measuring orientation of human body segments using miniature gyroscopes and accelerometers. Medical Biological and Engineering and Computing 43 (2), 273–282.

Ma, Y., et al., 2018. Relation between blood pressure and pulse wave velocity for human arteries. PNAS No 30 (115), 11144–11149 (44).

Maqsood, S., Xu, S., Tran, S., et al., 2022. A survey: from shallow to deep machine learning approaches for blood pressure estimation using biosensors. Expert Systems with Applications 197 (116788).

Martinez-Ríos, E., Montesinos, L., Alfaro-Ponce, M., Pecchia, L., 2021. A review of machine learning in hypertension detection and blood pressure estimation based on clinical and physiological data. Biomedical Signal Processing and Control 68.

Mukkamala, R., Hahn, J., Inan, O.T., Mestha, L.K., Kim, C., Hakan, T., 2015. Toward ubiquitous blood pressure monitoring via pulse transit time: theory and practice. IEEE Transactions on Biomedical Engineering 62, 1879–1901.

Nachman, D., et al., 2020. Comparing blood pressure measurements between a photoplethysmography-based and a standard cuff-based manometry device. Scientific Reports, Nature 10 (16116).

Poon, C.C.Y., Zhang, Y.-T., Liu, Y., 2006. Modeling of pulse transit time under the effects of hydrostatic pressure for cuffless blood pressure measurements. In: Proceedings of the 3rd IEEE-EMBS International Summer School and Symposium on Medical Devices and Biosensors, MIT, Boston, USA, Sept.4–6.

Rackauckas, C., et al., 2020. Universal Differential Equations for Scientific Machine Learning. arXiv preprint. https://arXiv:2001.04385.

Saeed, M., et al., 2011. Multiparameter intelligent monitoring in intensive care II (MIMICII): a public-access intensive care unit database. Critical Care Medicine 39 (952).

Sempionatto, J.R., et al., 2021. An epidermal patch for the simultaneous monitoring of haemodynamic and metabolic biomarkers. Nature Biomedical Engineering 5, 737–748.

Salvi, P., 2017. Pulse Waves: How Vascular Hemodynamics Affects Blood Pressure, second ed. Springer.

Solà, J., Delgado-Gonzalo, R., 2019. The Handbook of Cuffless Blood Pressure Monitoring: A Practical Guide for Clinicians, Researchers, and Engineers. Springer.

Solà, J., et al., 2020. Are cuffless devices challenged enough? Design of a validation protocol for ambulatory blood pressure monitors at the wrist: the case of the Aktiia Bracelet. In: 42nd Annual International Conference of the IEEE Engineering in Medicine and Biology Society (EMBC), pp. 4437–4440.

Townsend, R.R., et al., 2015. Recommendations for improving and standardizing vascular research on arterial stiffness: a scientific statement from the American Heart Association. Hypertension 66, 698–722.

van de Vosse, F.N., Stergiopulos, N., 2011. Pulse wave propagation in the arterial tree. Annual Review of Fluid Mechanisms 43, 467–499.

Waite, L., Fine, J., 2007. Applied Biofluid Mechanics. McGraw-Hill Companies.

Westerhof, N., Stergiopulos, N., Noble, M.I.M., 2010. Snapshots of Hemodynamics: An Aid for Clinical Research and Graduate Education, second ed. Springer.

Willemet, M., Chowienczyk, P., Alastruey, J., 2015. A database of virtual healthy subjects to assess the accuracy of foot-to-foot pulse wave velocities for estimation of aortic stiffness. American Journal of Physiology: Heart and Circulatory Physiology 309, H663–H6675.

Willemet, M., Vennin, S., Alastruey, J., 2016. Computational assessment of hemodynamics-based diagnostic tools using a database of virtual subjects: application to three case studies. Journal of Biomechanisms 49, 3908–3914.

Winokur, E.S., 2014. Single-Site, Noninvasive, Blood Pressure Measurements at the Ear Using Ballistocardiogram (BCG), and Photoplethysmogram (PPG), and a Low-Power, Reflectance-Mode PPG SoC. Ph.D. thesis. Massachusetts Institute of Technology.

Wong, M.Y., Pickwell-MacPherson, E., Zhang, Y.T., Cheng, J.C., 2011. The effects of pre-ejection period on postexercise systolic blood pressure estimation using the pulse arrival time technique. European Journal of Applied Physiology 111, 135–144.

Yang, C., Tavassolian, N., 2018. Pulse transit time measurement using seismocardiogram, photoplethysmogram, and acoustic recordings: evaluation and comparison. IEEE Journal of Biomedical Health Information 22, 733–740.

Zhang, G., et al., 2011a. Pulse arrival time is not an adequate surrogate for pulse transit time as a marker of blood pressure. J. Applied Physiology 111, 1681–1686.

Zhang, G., Hahn, J.-O., Mukkamala, R., 2011b. Tube-load model parameter estimation for monitoring arterial hemodynamics. Frontiers in Physiology 2 (72).

Zhang, G., 2012. Cardiac Output Monitoring Techniques by Physiological Signal Processing. Ph.D. thesis, Michigan State University.

Arterial stiffness

Alty, S.R., Angarita-Jaimes, N., Millasseau, S.C., Chowienczyk, P.J., Dec. 2007. Predicting arterial stiffness from the digital volume pulse waveform. IEEE Transactions on Biomedical Engineering 54 (12), 2268–2275.

Bolic, M., Dajani, H.R., Yoshida, M., Groza, V., 2021. Progress in the assessment of arterial stiffness. IEEE Instrumentation and Measurement Magazine 24 (2), 54–59.

Boutouyrie, P., Chowienczyk, P., Humphrey, J.D., Mitchell, G.F., 2021. Arterial stiffness and cardiovascular risk in hypertension. Circulation Research 128 (7), 864–886.

DeLoach, S.S., Townsend, R.R., 2008. Vascular stiffness: its measurement and significance for epidemiologic and outcome studies. Clinical Journal of American Sociology and Nephrology 3, 184–192.

Kaisti, M., Panula, T., Leppänen, J., et al., 2019. Clinical assessment of a noninvasive wearable MEMS pressure sensor array for monitoring of arterial pulse waveform, heart rate and detection of atrial fibrillation. Nature Digital Medicine 2 (39).

Roh, D., Han, S., Park, J., Shin, H., 2020. Development of a multi-array pressure sensor module for radial artery pulse wave measurement. MDPI Sensors 20 (33).

Salvi, P., et al., 2019. Noninvasive estimation of aortic stiffness through different approaches comparison with intra-aortic recordings. Hypertension 74, 117–129.

Tavallali, P., Razavi, M., Pahlevan, N.M., 2018. Artificial intelligence estimation of carotid-femoral pulse wave velocity using carotid waveform. Scientific Reports 8 (1014).

Wang, C., et al., 2018. Monitoring of the central blood pressure waveform via a conformal ultrasonic device. Nature, Biomedical Engineering 2 (9), 687–695. https://doi.org/10.1038/s41551-018-0287-x. Available: http://xugroup.eng.ucsd.edu/wp-content/uploads/2018/12/52_Nat_Biomed_Eng.pdf.

Wei, C.C., Jan. 2013. Developing an effective arterial stiffness monitoring system using the spring constant method and photoplethysmography. IEEE Transactions on Biomedical Engineering 60 (1), 151–154.

Acronyms and explanations

AASM	The American Academy of Sleep Medicine
AHI	apnea—hypopnea index
bpm	breaths per minute
CHF	congestive heart failure
COPD	chronic obstructive pulmonary disease
CSR	Cheyne Stokes respiration
FEV$_1$	forced expiratory volume that one exhales in 1 s of an FVC maneuver.
FVC	forced vital capacity
IMU	inertial measurement units
MEMS	micro-electromechanical system
Oronasal	related to both the mouth and the nose
OSA	obstructive sleep apnea
PLL	phase lock loop
PSG	polysomnograph
PVDF	polyvinylidene fluoride
RIP	respiratory inductance plethysmography

Variables used in this chapter include:

f_r	breathing rate
T_p	time instant of a detected peak on the breathing signal during one breathing cycle
T_t	time instant of a detected trough on the breathing signal during one breathing cycle
V	tidal volume
$\dot{V}$	flow rate
$\dot{V}_{AV}$	alveolar ventilation
V_I	in-phase component
V_Q	quadrature component
ε	strain
Z	impedance
θ	phase difference

9.1 Introduction

This chapter describes systems that can continuously monitor breathing outside the hospital. Continuous breathing monitoring devices are mainly used for monitoring breathing rate. Devices that perform pulmonary function tests include lung capacity measurements that require forced labored breathing and therefore are not used for continuous measurements.

Even though wearable devices for continuous monitoring of breathing cannot provide all the variables and indices available in clinical settings, they allow for monitoring quantities of interest over an extended period. We discuss what breathing variables can be measured and what indices can be estimated by wearable devices and point out some research directions.

The chapter starts with definitions of breathing rate, tidal volume, airflow, and some respiration variables. We also introduce common breathing patterns and describe several conditions, such as sleep apnea, where breathing is continuously monitored. Next, we look into the morphology of the signal and define fiducial points commonly extracted from the signal. A signal quality assessment is presented as well. To use data for developing new devices, the existing databases can be used or breathing waveforms can be simulated. Therefore, the databases of respiratory signals and simulators used to generate artificial breathing signals are described.

Table 9.1 shows the transducers we described in this chapter, either briefly or in detail. In addition, it shows the body location where the sensor is placed, as well as physical or physiological variables than can be measured (Massaroni et al., 2019b). For example, piezoresistive transducers are commonly attached to the chest and measure strain—expansion and contraction of the chest wall. Devices placed on the chest are used to measure changes in the chest volume during breathing. They include devices based on piezoresistive, inductive, or piezoelectric transducers or devices based on accelerometers or bioimpedance measurements. Devices based on temperature or pressure sensors are used to measure airflow next to the nose or mouth. All these devices show relative values of airflow or volume. We also show some ways to calibrate the devices to obtain absolute values of the volume or flow. In addition, we present several different ways to classify breathing patterns.

The code for recreating all the figures is in Matlab and can be downloaded from the book webpage.

TABLE 9.1 Devices and transducers covered or mentioned in this chapter along with references to research papers we discussed in this chapter.

Transducer, sensor or device	Monitoring location on the body	What is measured directly?	Portable, wearable	Covered in this chapter	Some applications	References
Piezoresistive	Chest	*Strain*	Yes	Briefly	Breathing rate	Chu et al. (2019)
Inductive (respiratory inductance plethysmography)	Chest	*Strain*	Yes	Briefly	Sleep studies, breathing rate	Cohen et al. (1994)
Piezoelectric	Chest	*Strain*	Yes	Briefly	Breathing rate	Lei et al. (2015)
Impedance (electrical impedance pneumography)	Chest	*Impedance*	Yes	Yes	Breathing rate	Van Steenkiste et al. (2020)

(Continued)

TABLE 9.1   Devices and transducers covered or mentioned in this chapter along with references to research papers we discussed in this chapter.—cont'd

Transducer, sensor or device	Monitoring location on the body	What is measured directly?	Portable, wearable	Covered in this chapter	Some applications	References
Accelerometer	Chest	Chest *movement*	Yes	Yes	Breathing rate	Rahmani et al. (2021)
Thermistor	Nose, mouth	Temperature changes due to *airflow*	Yes	Briefly	Sleep studies, breathing rate measurements	Liu et al. (2019)
Hot-film temperature sensor	In front of the nose	Temperature changes due to *airflow*	Yes	Briefly	Sleep studies, different in the flow between two nostrils	Jiang et al. (2020)
Air pressure sensor	Nose, mouth using cannula	Pressure changes due to *airflow*	Yes	Briefly	Sleep studies, breathing rate measurements	Massaroni et al. (2019)
Humidity sensor	Nose, mouth using mask	Changes in humidity due to *airflow*	Yes	No	Breathing rate	Zhong et al. (2021)
ECG electrodes or PPG sensor	Finger for PPG, normally Lead I configuration for ECG	*ECG or PPG*	Yes	Yes	Breathing rate	Liu et al. (2019)
Pneumotachograph	Mouth	*Flow*—measuring the pressure difference	Some[a]	No	Flow rate indices	—
Spirometer	Mouth	*Volume*	Some[a]	Briefly	Chronic obstructive pulmonary disease (COPD)	Exarchos et al. (2020)

[a]*There are portable home monitors, but the device is mainly used in clinical settings.*

9.2 What is measured?

The difference between ventilation or breathing and respiration is that *ventilation* represents the physical movement of air in and out of our lungs, while *respiration* represents gas exchange in the lungs and the capillaries. Therefore, in this chapter, we deal mainly with breathing. *The breathing* includes inhalation, which is the air movement from our mouth or nose to alveoli in the lungs, and exhalation, which is the air movement in the opposite direction. The inhalation is performed when the pressure within the alveoli is lower than the atmospheric pressure. Exhalation occurs when the pressure in the alveoli is higher than atmospheric pressure. Intrapulmonary pressure changes are the consequence of changes in lung volume. When the diaphragm is stimulated, the thorax expands, also causing the lungs to expand, decreasing the alveoli's pressure. When the alveoli pressure becomes smaller than the atmospheric pressure, inhalation occurs.

9.2.1 Breathing rate

Breathing or ventilation rate is one of four main vital signs, and it represents the rate at which breathing occurs, usually measured in breaths per minute. In practice, it is also called respiration rate. Human breathing rate is traditionally measured when a person is at rest and involves counting the number of breaths for 1 minute by observing how many times the chest rises. The typical respiratory rate for a healthy adult at rest is 12–20 breaths per minute. Breathing rate increases with stress, cognitive load, dyspnea, heat stimuli, physical effort, and different health conditions.

A significant increase in interest in ambulatory breathing rate monitoring resulted in many recent papers and reviews discussing new ways for continuous nonobtrusive measurements and the importance of measuring breathing rate (Nicolò et al., 2020; Scott and Kaur, 2020). In Nicolò et al. (2020), several main applications of breathing rate monitoring were identified, including:

- Presence of breathing—for example, for children below 1 year of age that are at risk of sudden infant death syndrome.
- Adverse cardiac events since elevated breathing rate is associated with the risk of cardiac arrest.
- Sleep apnea that is discussed in more detail in Section 9.2.4.
- Pneumonia—an infection of the lungs caused by bacteria or viruses. Breathing rate was identified as one of the variables for the diagnosis of the severity of pneumonia.
- Clinical deterioration—it has been shown that changes in breathing are a better predictor of clinical deterioration than changes in other vital signs (Subbe et al., 2003), as well as that a nocturnal breathing rate greater than 16 breaths/min is an independent predictor of long-term mortality for older adults living on their own.
- Dyspnea (shortness of breath)—one of the major symptoms of patients with chronic obstructive pulmonary disease (COPD) and other cardiorespiratory diseases. It has been noted that shortness of breath happens during the activities such as walking and is normally followed by an increase in breathing rate.

However, some of these applications, including monitoring sleep apnea and dyspnea, also require measurement of tidal volume, flow, or other variables besides monitoring breathing rate.

9.2.2 Volume and flow

Tidal breathing is the term used for restful unlabored breathing. *Loaded breathing* is breathing during efforts against occlusion. It is also used as a maneuver to support the development of healthcare respiratory monitoring devices. Labored breathing means that the subject needs to breathe as forcefully and rapidly as she/he can.

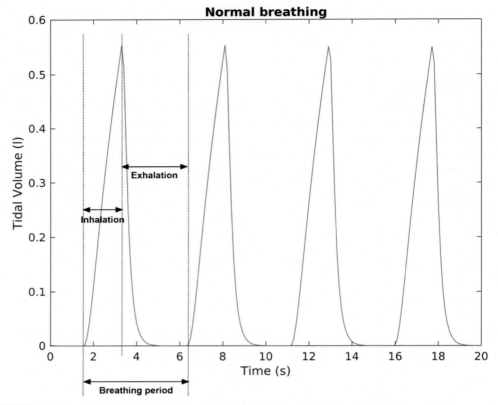

FIGURE 9.1   Tidal volume of a normal breathing signal simulated using PNEUMA 3.0 (Cheng et al., 2010).

9.2.2.1 Unlabored breathing

Let us first discuss variables that can be measured during normal unlabored breathing. *Tidal volume V* represents the volume of inspired or expired air during a normal respiratory cycle. It is normally given in liters. *Flow rate* $\dot{V} = dV/dt$ is proportional to the change of volume over time and is given in l/min.

Tidal volume over time is shown in Fig. 9.1. The figure is obtained from a simulator Pneuma with default parameters for simulating normal breathing (see Pneuma user manual (Cheng et al., 2013)). It shows a breathing period composed of inhalation, during which tidal volume increases and exhalation, during which the tidal volume decreases. Breathing pause during exhalation is normally included as a part of exhalation—for example, in Fig. 9.1, for the first breath, it starts from the time instant 4.5 s and lasts until the end of the exhalation period. A reciprocal of the breathing period is the *instantaneous breathing rate*.

Minute respiratory volume is the amount of air moved into the respiratory passage during 1 minute, and it is computed as the product of breathing rate and tidal volume. It is given in l/min and, therefore, actually represents the flow. The normal value is about 7.5 L/min.

Dead space refers to the part of the respiratory system that contains some air volume but where the gas exchange does not occur. This air is not used for gas exchange, and it is exhaled first. Gas exchange occurs only in the alveolar areas of the lungs. Anatomic dead space is the measurement of the volume of all the space of the respiratory system without considering the alveoli and closely related gas exchange areas. About 30% of the tidal ventilation could be wasted in dead space. *Alveolar ventilation* $\dot{V}_{AV}$ is equal to the product of breathing rate and the volume difference between tidal ventilation and anatomic dead space.

The volume of air that remains in the lungs at the end of a full expiration is called *residual volume*. It is about 1.2 L on average.

Example 9.1. Compute minute volume and alveolar ventilation if a subject breathes at 15 breaths/min and the tidal volume is about 500 mL. The volume of anatomic dead space is 100 mL.

Solution. Minute volume: $\dot{V} = 500\text{ ml} \cdot 15\text{ breaths/min} = 7.5\text{ l/min}$

Alveolar ventilation: $\dot{V}_{AV} = (500\text{ ml} - 100\text{ ml}) \cdot 15\text{ breaths/min} = 6\text{ l/min}$

Breathing is continuously monitored using devices called pneumographs. *Pneumographs* are devices that measure the velocity or force of the chest movement or impedance change during respiration, as well as estimate breathing rate. They will be discussed in detail in Section 9.4. The resulting waveform is called a pneumogram.

9.2.2.2 Labored breathing

In this section, we introduce indices obtained using spirometry during labored breathing. *Spirometry* is a pulmonary function test that measures the volume and airflow that can be inhaled and exhaled. Lung volume is measured using a *spirometer*. The recording of lung volume changes with time is known as a spirogram. Common measured/estimated indices are.

- Forced vital capacity (FVC): volume of air exhaled with force after inhaling as deeply as possible.
- Forced expiratory volume (FEV): the amount of air exhaled with force in one breath.
- FEV_1: the forced expiratory volume that 1 exhales in 1 second of an FVC maneuver—see Fig. 9.15 in Section 9.8.1.
- Forced expiratory ratio FEV_1/FVC: the ratio of FEV_1 to FVC and is normally about 70%—80%.
- Forced expiratory flow (FEF): air flow halfway through an exhale.
- Maximum voluntary ventilation (MVV): the maximal amount of air inspired or expired during 1 minute.
- Functional residual capacity (FRC): the volume of air present in the lungs.
- $\dot{V}_{maxFRC}$: the maximum flow at functional residual capacity.
- TPIF: tidal peak inspiratory flow.

The flow-volume loop is also examined because it shows the relationship between the flow and volume during the inspiration and expiration. Spirometry indices FVC, FEV_1, and FEV_1/FVC ratio are used to evaluate obstructive and restrictive respiratory impairments (Johnson and Theurer, 2014). In addition, the shape of the flow-volume loop changes in case of obstructive and restrictive respiratory impairments. Obstructive airway impairment (such as asthma) is characterized by airflow limitations and a decrease in the FEV_1/FVC ratio (Paraskeva et al., 2011). Spirometry alone cannot detect restrictive impairments (such as congestive heart failure) and require further tests. These impairments are characterized by a decreased FVC with normal or increased FEV_1/FVC ratio (Paraskeva et al., 2011).

For a complete spirometry procedure and analysis of results, please refer to Johnson and Theurer (2014).

Pneumotachograph is a device in which the flow rate is estimated indirectly by measuring the differential pressure across a small flow resistance. The resistance may consist of a wire screen or a series of capillary tubes the subject blows into. The flow rate is then proportional to the pressure difference measured across the flow resistance.

Spirometers and pneumotachographs are mainly used as clinical devices and therefore are not the focus of this book. For a detailed description of the operating principles of these devices, please refer to Mandal (2006).

9.2.3 Breathing patterns

Several breathing patterns are presented next. They include eupnea, apnea, Cheyne Stokes respiration, and Kussmaul breathing.

Eupnea represents normal breathing for an individual at rest with a frequency of 12–20 times per minute.

Apnea represents breathing cessation during which the volume of the lungs does not change. It can be caused by neurodegenerative illnesses, the use of medications like narcotics, and other reasons. It can happen during sleep (sleep apnea).

Cheyne Stokes respiration (CSR) is an abnormal breathing pattern that often appears in patients with stroke, heart failure, brain tumor, and traumatic brain injury. A Cheyne Stokes respiration cycle comprises a crescendo, apnea, and decrescendo phases, as shown in Fig. 9.2. The figure is obtained from a simulator Pneuma with default parameters for simulating central sleep apnea characterized with CSR with congestive heart failure (see Pneuma user manual section "External Intervention" (Cheng et al., 2013)). The total length of a Cheyne Stokes respiration cycle is around 45–90 s. In each CSR cycle, there is a phase of gradual increase in volume and frequency, followed by a phase of a gradual decrease in volume and frequency, and a phase of apnea lasting 10–30 s.

Kussmaul breathing is caused by deep and labored breathing. Normally, the breathing rate and tidal volume are increased so that the breathing rate is over 20 breaths per minute.

9.2.4 Applications

9.2.4.1 COPD

Chronic obstructive pulmonary disease (COPD) results in an airflow limitation due to inflammation in the lungs that makes normal breathing difficult. It includes symptoms like shortness of breath, wheezing, and coughing. Spirometry is used to measure airflow and airflow limitations. The prognosis of COPD is currently determined mainly based on FEV_1. Some composite markers have been defined for evaluating the severity of COPD. For

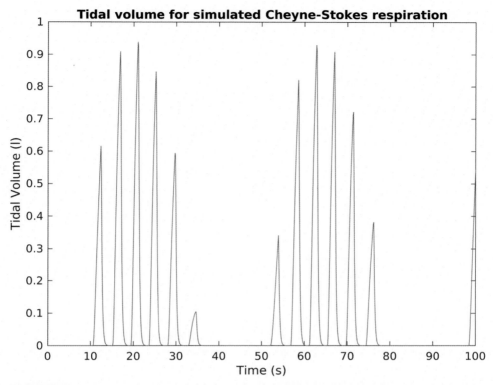

FIGURE 9.2 Cheyne Stokes respiration signal simulated using PNEUMA 3.0 (Cheng et al., 2010).

example, a marker based on a combination of the body-mass index, the degree of airflow obstruction measured using a spirometer and dyspnea level based on a predefined scale, and exercise capacity as assessed by the 6-min walk test was defined in Celli et al. (2004).

9.2.4.2 COVID-19

Monitoring applications for Covid-19 patients have been developed and applied in the following cases: (1) to assist in prescreening for detection of COVID-19 by following the changes in physiological variables, and (2) to track the progression of infection for confirmed COVID-19 patients during the recovery (Box 9.1). It was noted that out of more than 1000 patients hospitalized with symptomatic COVID-19, many had low blood oxygen saturation and tachypnea (breathing rate higher than 23 bpm); however, only 10% of patients reported shortness of breath (Chatterjee et al., 2021).

In Natarajan et al. (2021), the authors used the Fitbit device to monitor PPG and extract the breathing rate for more than 2000 people before and after they got infected by the virus. In a 7-day window starting 1 day before the symptoms were first present, it was shown that breathing rate increased during the night in 36.4% of symptomatic subjects by three breaths per minute or more at least for one night.

BOX 9.1

Integration of sensors into protective masks

With COVID-19 pandemics, it has become common for people to wear face masks. Research on integrating sensors into flexible face masks started much before COVID-19. For example, in Güder et al. (2016), a transducer based on graphite printed on paper was embedded into a textile mask and used to detect changes in moisture during inhalation and exhalation. However, the research has been intensified since 2020. Several conditions need to be satisfied for sensor systems to be a part of the mask, such as that the sensors need to be flexible, light, sensitive to breathing airflow, and capable of operating in environments with high levels of moisture. A complete implementation of a "smart" mask for COVID-19 that contains a self-powered pressure sensor (polymer electret material) and additional electronics for data acquisition and wireless transmission was presented in Zhong et al. (2021).

In Miller et al. (2020), a machine learning model is used to train the classifier to predict the infection by the SARS-CoV-2 virus based on subjects' breathing rates. As a result, 20% of positive subjects were identified 2 days before the symptoms were first present, and 80% of the subjects were identified 3 days after the appearance of the symptoms.

COVID-19 can cause discoordination of the sequence of inspirations and expirations (Machado and DeFina, 2020), and therefore, it is important to classify breathing patterns and not only estimate breathing rate when monitoring COVID-19 patients.

9.2.4.3 Sleep apnea

Sleep apnea events include both apnea, a cessation in breathing rhythm, and hypopnea, a reduction in the breathing amplitude. The American Academy of Sleep Medicine AASM defines apnea as the drop by more than 90% from the airflow baseline for 10 s or longer (Berry et al., 2015). Sleep apnea can be divided into:

- Obstructive sleep apnea (OSA), during which there is a continued or increased inspiratory effort throughout the apnea event,
- Central sleep apnea, in which there is no effort during the apnea event, and
- Mixed sleep apnea.

OSA is the most common.

The gold standard for monitoring and detecting sleep apnea in hospitals is a polysomnograph (PSG). PSG involves a number of sensors/systems for measuring respiratory events, cardiac events (ECG), sleep staging (electroencephalogram EEG), and movement events (electromyography EMG). Respiratory events include detections of apneas and hypopneas and respiratory effort (nasal pressure transducer, thermal airflow sensor, respiratory inductance plethysmography), and snoring (microphones). Oxygen saturation (SpO_2 monitors) is also measured.

The severity of sleep apnea is evaluated by measuring several derived indices. One of them is the apnea—hypopnea index (AHI). The AHI is calculated by dividing the total number of apnea and hypopnea events by the total hours of sleep. The severity of the OSA is then determined as (1) mild OSA: $5 \leq AHI \leq 15$, (2) moderate OSA: $15 < AHI \leq 30$, and (3) severe OSA: AHI > 30.

Home sleep apnea testing is used as either a prescreening tool for full PSG or to monitor a patient's progress during the treatment. Home sleep apnea testing appears in the AASM Manual for the Scoring of Sleep and Associated Events from 2015 (Berry et al., 2015). A home sleep apnea device includes methods for measuring respiratory flow and effort, oxygen saturation, and heart rate. Often, the device monitors snoring and provides body position and sleep/wake time estimates.

9.2.4.4 Monitoring breathing patterns for heart failure

Congestive heart failure (CHF) causes the heart to pump inefficiently. The worsening of the symptoms of heart failure is called decompensation. Recently, much effort has been put into developing methods for predicting the decompensation of CHF. Some existing methods of predicting decompensation rely on monitoring respiratory variables. However, continuous ambulatory monitoring of CHF using these methods has not been routinely done because current technology is relatively intrusive. The authors of Tobushi et al. (2019) proposed a "respiration stability index" as a metric to quantify irregular breathing that increases with increasing lung congestion, a common symptom of CHF. RSI was obtained from the reciprocal of the standard deviation of the range around the fundamental breathing component in the frequency spectrum. The lower respiration stability index indicates more broad spectral components and unstable respiration. Of course, continuous monitoring of breathing during sleep or during a part of the day is necessary to extract the breathing signal and detect changes in breathing that are indicative of CHF decompensation. Therefore, it is important to develop wearable devices that can be worn during the day without affecting one's daily activities.

9.2.4.5 Radiotherapy

Continuous breathing monitoring during radiotherapy uses wearable and contactless monitoring devices. Respiratory gating is a technique that relies on using a respiration monitoring device to predict the phase of the breathing cycle during radiotherapy. During radiotherapy of cancer patients, particularly the thorax and abdomen, the monitoring or mitigation of patient breathing motion is often an integral part of an accurate treatment delivery process (Keall et al., 2006). Using respiratory gating and breath-holding techniques during lung radiotherapy, for instance, can significantly reduce pulmonary and cardiac toxicities (Giraud and Houle, 2013). Other treatments might require patients to perform a deep-inspiration breath hold monitored using a breathing belt or other sensors. The solutions rely on patient-worn accessories (Wong et al., 1999) or relatively high-cost contactless monitoring devices based on cameras, radars, or other sensors (Fallatah et al., 2022).

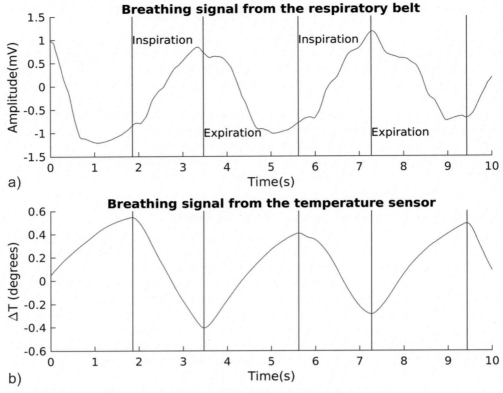

FIGURE 9.3 Waveforms extracted from the database by Shafiq and Veluvolu (2017) showing signals from (A) the respiratory belt and (B) the temperature sensor.

9.3 Features of the signal and the noise

9.3.1 Features extracted from the tidal volume or flow waveforms

It was pointed out in Noto et al. (2018) that in comparison with cardiac signals such as ECG, whose morphology is based on a predefined sequence of activities, breathing signals are more complex because people breathe at varying rates depending on their activities and psychological state, with the volume that can change almost per each breath. In addition, people may control their breathing patterns such as breathing faster, slower, or pausing for a short time. Additional complexity comes from the fact that the signals are acquired using different sensors measuring volume, flow, carbon dioxide, and so on. Therefore, the morphology of the signals might look very different depending on the sensor being used.

Fig. 9.3A shows the signal obtained from the breathing belt piezoresistive sensor. This signal is given in mV showing the expansion and reduction of the chest wall during breathing; therefore, it is related to the volume. Fig. 9.3B shows the signal extracted from the thermistor placed below one of the subject's nostrils. During the normal breathing cycle in an environment with room temperature, exhaled air is warmer than inhaled air. Therefore, the thermistor can measure the airflow in and out of the nostrils. Data collection of the signals presented in Fig. 9.3 was explained in Shafiq and Veluvolu (2017). It is shown in the figure that the inspiration period corresponds to chest expansion and temperature decrease of the temperature sensor and that the opposite events happened during the expiration.

In Noto et al. (2018), the nasal flow signal obtained using a nasal cannula attached to a piezoelectric pressure sensor was analyzed. Several fiducial points were identified, and indices or features were computed. Fiducial points include: inhale and exhale onsets, inhale peaks and exhale troughs, and duration of inhales, exhales as well as the duration of pauses between inhales and exhales. Based on these fiducial points, the following indices can be computed:

- *Interbreath interval* is the average time between inhale onsets
- *Inhale and exhale volumes* represent the integral of the airflow computed during the inhale and during the exhale
- *Duty cycle* is the ratio of the average inhale duration and average interbreath interval
- *Standard deviation* of instantaneous breathing rate and duty cycles,
- *Average peak flow rates and volumes.*

Software in Matlab that computes fiducial points and indices is made public by the authors of Noto et al. (2018).

The breathing signal is mainly affected by motion artifacts. The artifacts normally overlap in the frequency domain with the breathing signal. Therefore, simple filtering for removing motion artifacts normally does not work. The intervals with the artifacts can be identified, the breaths with the motion artifacts can be ignored, or more advanced signal processing methods can be applied to remove motion artifacts (Ansari et al., 2017). For some measurement modalities, such as bioimpedance, cardiac signals represent interference and can be filtered out.

9.3.2 Signal quality

The research on defining signal quality indices (SQI) for assessing the quality of respiratory signals is less developed than the research on SQI for ECG and PPG signals. One of the reasons might be that there are many different sensors for acquiring breathing signals and these acquired signals have different morphology. In addition, some of them are based on measuring flow and some on measuring volume. Sometimes, the breathing signal is extracted from the cardiac signals. Also, breathing signals are used for different applications with their own requirement for signal quality.

In Løberg et al. (2018), the authors were interested in evaluating portable respiratory effort belts for sleep apnea monitoring. Therefore, the main criteria included: sensitivity and specificity of detecting each breath and hypopnea. Detection of hypopnea requires the ability to distinguish each breath even when the breathing signal is weak. The study showed that all portable devices could detect each breath with high sensitivity.

A comprehensive study on SQI for the impedance pneumography described in Section 9.4.3.3 was presented in Charlton et al. (2021). The breathing signal is divided into 32-s segments, and a binary quality index is estimated for each segment, together with the breathing rate for good-quality segments. The quality assessment algorithm was based on a combination of SQIs covered in Chapter 4. They involve template matching mixed with the analysis of physiological feasibility. The algorithm consists of the following steps:

1. Identify each breath by detecting fiducial points.
2. Analyze the physiological feasibility of each breath by considering the duration of the breath relative to the other extracted breaths. Multiple criteria can be used here: (1) the normalized standard deviation of breath intervals is smaller than 0.25 s, which permits only moderate variation in the durations of detected breaths; (2) at least 60% of the interval is occupied by valid breaths.
3. Template matching is done by computing a template breath based on the detected breaths and then computing the correlation between each breath and the template. If the average correlation coefficient is larger than 0.75, the segment is accepted as a high-quality segment.

This approach has several issues when compared with similar approaches used for cardiac signals. The variation of the breath durations is much larger than the variation of the beat durations in cardiac signals. Therefore, a better approach might be to adjust for the duration between the signals by applying dynamic time warping and then form the template and correlate the template against the time-adjusted breaths.

In Moeyersons (2021), a deep learning method is applied to detect the signal quality of the signal obtained using impedance pneumography. The features are extracted from the autocorrelation function (ACF) and the spectral power ratios. The ACF is used to evaluate the periodic components within a time series. If the amplitudes of the first and the second peaks of the autocorrelation function are close to one, this means that the breathing is regular (close to periodic).

Similar to the method in Chapter 4, the signal-to-noise ratio based on the spectral power ratio can be computed. The power of the signal is computed in the range f_{min} to f_{max}, where these frequencies correspond to the frequencies where the power spectrum drops by ± 3 dB relative to the power at the fundamental breathing frequency. This power is divided by the power in the frequency range between 0.05 and 0.7 Hz, which corresponds to 3−42 breaths per minute. The value of the spectral ratio closer to one means that the signal is periodic and that there is not much noise.

Two signals from the database introduced by Shafiq and Veluvolu (2017) are shown on the left of Fig. 9.4. Fig. 9.4A shows the waveform obtained using a temperature sensor after the exercise. The amplitudes of the breaths are higher at the beginning than at the end. The SQI indicator based on the power spectrum ratio is close to 1. The ACF function is shown in Fig. 9.4B. It shows regularity, but the signal is not periodic. Fig. 9.4C shows the waveform obtained using a temperature sensor while the subject performed a number of maneuvers, including coughing, deep breathing, and fast breathing. The ACF shows that the signal is not periodic.

9.3.3 Databases

Data have been collected for different purposes; therefore, available datasets are very diverse regarding patient population, applied sensors, and types of breathing. The dataset

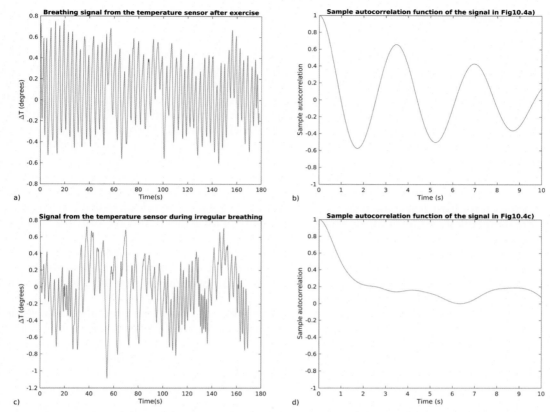

FIGURE 9.4 Figures on the left show waveforms extracted from the database by Shafiq and Veluvolu (2017), and on the right, corresponding ACF. (A) Waveform for the nasal temperature sensor after exercise and (B) its ACF. (C) Waveform for the nasal temperature sensor during irregular breathing maneuvers, and (D) its ACF.

presented in Shafiq and Veluvolu (2017) includes several healthy participants with four types of breathing: normal breathing, postexercise breathing, irregular breathing, and apnea maneuver. The irregular breathing consisted of coughing as well as paced, deep, and erratic breathing. An interesting aspect of this dataset is the large number of sensors used for data collection, including a respiratory belt for chest motion, a thermistor for airflow, and a VICON IR camera that detects the motion of 16 markers placed on the subject's torso. Waveforms from this dataset are shown in Figs. 9.3 and 9.4.

The CapnoBase dataset includes up to 2 min waveforms from 59 children and 35 adults obtained during surgeries. Waveforms include CO_2 gas concentration, flow, and nasal pressure signals for both ventilated and spontaneously breathing subjects (Karlen et al., 2010).

A comprehensive dataset of sleep-disordered breathing that includes thousands of patients performing PSG is presented (Chen et al., 2015). The data set is called The Multi-Ethnic Study of Atherosclerosis (MESA). PSG recordings include thoracic and abdominal respiratory inductance plethysmography, airflow measured by thermocouple and nasal pressure cannula, finger pulse oximetry together with several other physiological signals. A comprehensive overview of the databases used for sleep apnea research is presented in Mostafa et al. (2019).

9.3.4 Simulators

In this section, we first present mechanical breathing simulators that physically emulate breathing. IngMar Medical company released ASL 5000 Breathing Simulator that can simulate the breathing airflow and be applied in the product development and testing for ventilators, continuous positive airway pressure (CPAP), and other respiratory therapy devices (ASL 5000 Breathing Simulator). They also developed QuickLung Breather products that can simulate the airflow of several breathing patterns, including eupnea, Cheyne Stokes, Kussmaul, Biot's, and others (QuickLung Breather). Modus QA company developed a Quasar respiratory motion phantom (QUASARTM) that can simulate vertical chest wall movement. All mentioned devices were developed for breathing and tumor research.

Other types of simulators generate data. In Alinovi et al. (2017), researchers developed a breathing pattern simulation model based on a continuous-time Markov chain. This model is designed for their research on the video-based monitoring of breathing, which can simulate normal breathing and respiratory pauses. The quality of the generated waveform is evaluated by comparing the similarity of the feature distribution between the generated and real waveforms. They proposed a set of features for evaluation and used the Kullback—Leibler divergence to evaluate the feature similarity. To the best of our knowledge, other works on breathing simulators, except (Alinovi et al., 2017), did not provide a method to assess the signal quality except by visual comparison.

A model in Simulink that simulates airflow at the mouth or alveolar pressure and volume is shown in Marconi and De Lazzari (2020). In addition, the model allows for varying quantities such as resistance and pressure drop across the elements of the respiratory system and compliance of the chest wall. As such, this model is suitable, in our opinion, for biomedical engineering students to explore the effects of the parameters of the respiratory system on the design of the device.

There are also several breathing simulators available covering different aspects of breathing. One of the most comprehensive models is Pneuma (Cheng et al., 2010). It models respiratory and cardiovascular systems, central neural control, and interactions between them. It simulates physiological responses under various conditions, such as the sleep—wake cycle, Cheyne—Stokes respiration in chronic heart failure, and obstructive sleep apnea. It can also be used to research the effect of the Valsalva and Mueller maneuvers and the application of CPAP and other interventions on physiological parameters. The GUI in Matlab is very intuitive, and it allows for setting a large number of parameters. The code is provided, but its complexity makes it difficult to master and modify it.

9.4 Breathing rate estimation

Techniques for continuous monitoring of breathing were classified in Liu et al. (2019) based on:

1. the respiratory modulation on other physiological signals,
2. volume changes and body movement, and
3. airflow changes.

We use the same classification in this section. To measure breathing rate, the exact values of the amplitude of the breathing signal are not needed. We are interested only in the variation of the signal over time. Therefore, techniques two and three rely only on measuring changes in volume and airflow. Some efforts to calibrate them to obtain the absolute values of volume and flow are described in Section 9.5.3. However, there are several challenges in doing the calibration. For example, for respiratory belts, the signal amplitude depends on the location of the belt on the body, the posture, and very often on how tight the belt is. Therefore, the calibration can be easily invalidated. Consequently, the output of these sensors in this section is not referred to as tidal flow or volume but as a breath or *breathing signal*.

Techniques 2 and 3 require sensors for detecting the volume or flow changes. Technique 1 does not use special sensors for monitoring breathing but relies on sensors for acquiring cardiac or other physiological signals. The breathing signal is then extracted from these physiological signals using signal processing methods.

9.4.1 Breathing rate estimation algorithms

Each of the techniques that will be presented in the next sections requires extraction of the breathing signal. The signal can be extracted from different sensors based on different methods, but it is processed similarly to estimate the breathing rate f_r. After the signal is extracted, the first step is to filter the signal using a bandpass filter to remove the baseline drift at low frequencies and the cardiac signal and higher frequency noise at high frequencies. For example, a bandpass filter with a pass band of 0.08–0.7 Hz (4.8–42 bpm) could be used. After filtering, the signal is either processed in the time or frequency domain. Here, we will show only one algorithm in each domain.

Zero-crossing detection method directly estimates the frequency of a pseudo-periodic signal by measuring the number of negative-to-positive and/or positive-to-negative transitions of a time series in a given time window. A very similar method used here is to detect the number of local maxima and/or minima of a time series in a given window. For a signal in Fig. 9.4A, the breathing rate is estimated every 5 s using overlapped 15 s windows, and the breathing rates estimated using time and frequency-based methods are shown in Fig. 9.5A. In each time window, peaks and troughs are detected. In Fig. 9.5B, a 15 s segment between 115 and 130 s was shown with the detected peaks (red stars) and troughs (yellow stars). The mean time difference between the peaks and between the trough is computed, and the average of these two is used as a breathing rate. If the time instants of detected N peaks are T_{pi} where $i = 1, ..., N$ and the time instants of detected M troughs in each window are T_{tj} for $j = 1, ..., M$ then the breathing rate f_r is computed as

$$f_r = 0.5 \left[\frac{1}{\frac{1}{N-1} \sum_{i=2}^{N} (T_{pi} - T_{p(i-1)})} + \frac{1}{\frac{1}{M-1} \sum_{i=2}^{M} (T_{ti} - T_{t(i-1)})} \right] \cdot 60$$

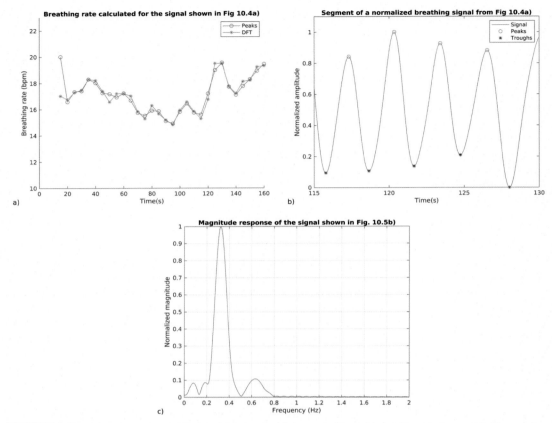

a)

b)

c)

FIGURE 9.5   (A) Breathing rate calculated for the signal shown in Fig. 9.4A for 15 s windows with 10 s overlap. (B) The method for peak detection that shows the segment of the signal from 115 to 130 s with identified peaks and troughs. (C) The magnitude response of that signal.

The first term in the brackets corresponds to the average breathing frequency in Hz obtained based on the time differences between detected peaks, while the second term corresponds to the average breathing frequency in Hz obtained based on the time differences between detected troughs. Of course, breathing rate can be computed only based on detected peaks. However, we also found it useful to detect troughs because the difference in the number of detected peaks and troughs can be used as an additional signal quality metric. If the number of peaks or troughs differs by more than one, then this might mean that some peaks or troughs were not detected properly due to noise or motion artifacts. Also, if the window is short (15 s), then only a small number of breaths are detected, and therefore the error in breathing rate estimation could be reduced if both peaks and troughs are used. In the example in Fig. 9.5B, the breathing rate is estimated to be 19.5 breaths per minute, while it would be smaller if only time differences between peaks were used to compute the breathing rate. This method based on detecting peaks/troughs is very flexible because it works well even if the breathing rate changes over time. However, it is sensitive to noise and peaks caused by motion artifacts.

In the frequency domain, the signal is first windowed using Hamming window, and then the discrete Fourier transform is applied (Oppenheim and Schafer, 1998). Next, the frequency that corresponds to the maximum peak of the spectrum is selected. Fig. 9.5C shows the signal spectrum from Fig. 9.5B. The maximum magnitude is at a frequency of 0.33 Hz, corresponding to 19.8 breaths per minute.

The breathing rate computed over the overall 160 s interval is shown in Fig. 9.5A.

9.4.2 Extracting breathing rate from cardiovascular signals

This technique relies on the fact that breathing modulates other physiological signals, including ECG, PPG, oscillometric blood pressure, BCG, and SCG (Charlton et al., 2016). Modulations include baseline changes due to breathing, amplitude modulation (AM), and frequency modulation (FM). Fig. 9.6A shows an example of baseline modulation of the ECG signal where the peak-to-peak amplitudes of consecutive R waves are constant, but the whole waveform is shifted up and down synchronously with inhalation and exhalation,

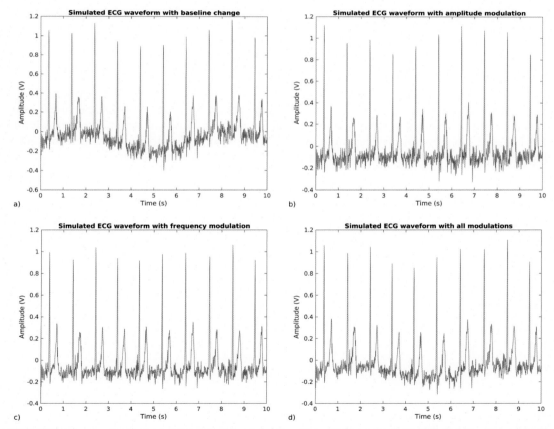

FIGURE 9.6 Simulated noisy ECG signal modulated with breathing with (A) baseline drift, (B) amplitude modulation, (C) frequency modulation, and (D) all three modulations.

respectively. The simulated heart rate is 60 beats per minute while the simulated breathing rate is nine breaths per minute for all cases in Fig. 9.6. We added white Gaussian noise with a standard deviation of 0.05 V. Fig. 9.6B shows AM modulated ECG signal. Here, the amplitudes of ECG waves are modulated with the breathing signal. Frequency modulation is shown in Fig. 9.6C. There, the timing of heartbeats is modulated by breathing. The ECG signals frequency modulation or respiratory sinus arrhythmia was introduced and modeled in Chapter 7. Changes in the ECG R—R interval over time are synchronous with respiration so that the R—R interval shortens during inspiration and prolongs during expiration. Here, the respiration signal can be derived from RSA. ECG signal modulated with all three modulations is shown in Fig. 9.6D.

As mentioned earlier, the estimation of breathing rate from the ECG signal is based on signal processing. The breathing signal can be extracted from the physiological signals mainly based on two methods: fiducial points detection and filtering. Breathing rate estimation based on fiducial points includes the following steps:

1. Detecting fiducial points on the physiological signal. An example of fiducial points could be the peaks of R waves on the ECG signal. The time differences between these points are often irregular, and the average time difference is much larger than the sampling period.
2. Extracting the breathing signal from the fiducial points detected in step 1 and resampling it to the sampling frequency of, for example, 10 Hz or more.
3. Estimating the breathing rate based on the techniques that were already introduced.

Fig. 9.7 shows the extracted breathing signals from the simulated ECG waveform shown in Fig. 9.6D. The signal based on ECG R peak amplitudes is shown in Fig. 9.7A. Since the heart rate is 60 beats per minute, we obtain one point per second. This time series is interpolated to 100 Hz, and the waveform is shown in Fig. 9.7A. Please note that there is no need for such a high number of interpolation points and that about five samples per second would be

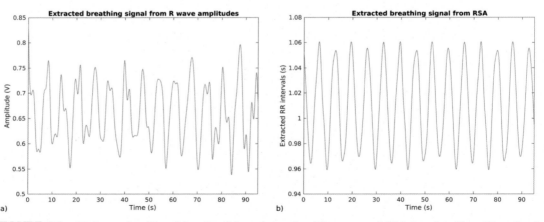

FIGURE 9.7   (A) Interpolated breathing signal from the peaks of R waves and (B) interpolated breathing signal from the RSA. Both of these signals are extracted from the signal whose segment is shown in Fig. 9.6D.

enough. If we consider single breaths in the breathing signal, we can see that they do not have a regular shape due to baseline drift and FM modulations. In Fig. 9.7B, the waveform based on time differences between the consecutive R peaks is interpolated to 100 Hz. Please note that signals in Fig. 9.6 are simulated; therefore, the breathing waveform in Fig. 9.7B is smoother than the waveforms obtained from real ECG signals. In both cases, the estimated breathing rate using the peak detection algorithm is 9 breaths per minute.

9.4.3 Chest wall movement

Chest wall movement can be measured by several methods, including:

- Detecting movements through the use of accelerometers and gyroscopes,
- Performing strain measurements that include resistive, capacitive, and inductive transducers,
- Detecting changes in the impedance.

9.4.3.1 Accelerometers

In this section, we will briefly introduce accelerometers, explain their use in smartphones, and show how they are used to obtain the breathing signal when the accelerometer device is placed on the chest. *Inertial measurement units* (IMUs) comprise three-dimensional accelerometers, gyroscopes, and potentially other sensors such as magnetometers. In fields that do not deal with human subjects, the IMU is commonly attached to an object to provide an estimation of the object's orientation in space. The piezoresistive and capacitive transducers are mainly used in designing inertial sensors based on MEMS.

In this section, we will discuss only accelerometers and their use in detecting chest wall movement. *Accelerometers* are sensors that measure proper acceleration or "g-force." Accelerometers placed in smartphones can be used for extracting the breathing signal and measuring breathing rate. An example of a setup is shown in Fig. 9.8, together with the directions of the accelerometer planes. It can be seen that the x axis is in the direction from the head to the feet, the y-direction is to the side of the subject, and the z axis is pointing upwards, away from the subject's chest. The accelerometer shows zero acceleration if the object does not move or moves with constant velocity. This is, however, not the case for the acceleration in the direction of z coordinate. If the device is placed on the table and does not move, it will show 0 m/s^2 in x and y direction but 9.81 m/s^2 in z direction. The tri-axis accelerometer comprises parts that bend or move when they accelerate and when gravity pulls on them. Since gravity is pulling these parts even when the chip is on a flat surface and does not move, the accelerator shows 9.81 m/s^2 in the z direction. Also, positive values of acceleration indicate an increase in velocity, while negative values indicate a decrease in velocity.

Accelerometers have been used for monitoring breathing for several decades. A recent review (De Fazio et al., 2021) describes the use of inertial sensors for estimating breathing rate. The major issue when monitoring breathing using IMU sensors is movement artifacts. Several proposed solutions include two IMU sensors placed on the thorax and the abdomen. The signals from the thorax and abdomen are correlated, and this can be taken advantage of when removing the uncorrelated noise. Besides being placed on the thorax and abdomen, several researchers extracted breathing signals by placing the accelerometer on the lower neck.

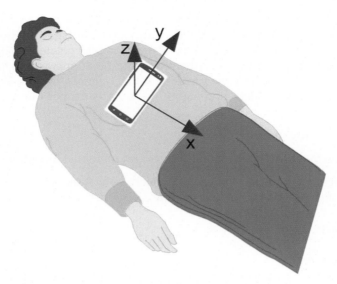

FIGURE 9.8   A smartphone containing an accelerometer placed on the subject's chest and abdomen. The co-ordinates of the accelerometer are shown as well. The pictures of the person and the phone are extracted from the SmartDraw software package.

Besides being used for direct monitoring of breathing, accelerometers were used in other devices, such as breathing belts, to detect movement artifacts.

In research, commercial off-the-shelf single-chip accelerometers or IMUs have been used mainly as components of off-the-shelf boards, such as Shimmer or researchers' own printed circuit boards. Smartphones with accelerometers were also used in research. In smartphones, accelerometers and gyroscopes are normally embedded in a single chip. For example, Samsung Galaxy S9 used an STMicroelectronics LSM6DSL 3D accelerometer and 3D gyroscope. The accelerometer in LSM6DSL has a full scale that can be set to ±2 g, ±4 g, ±8 g or ±16 g. In Rahmani et al. (2021), the components recently used in research related to designing devices for monitoring breathing were reviewed. They include IMU chips from STMicroelectronics, Analog devices, TDK-InvenSense and NXP Semiconductors.

Fig. 9.9 shows the accelerometer waveforms obtained from an accelerometer in the Samsung Galaxy Android smartphone. A healthy subject placed the phone on his chest for about 45 s in the same setup as shown in Fig. 9.9. As can be seen, the largest acceleration is in x direction because of the movement of the abdomen during breathing. As expected, the accelerator shows a value of around 9.81 m/s^2 in the direction of z axis. We subtracted 9.81 m/s^2 in order to show the acceleration in z axis on the same figure as the acceleration in x and y axes.

9.4.3.2 Measuring strain

Measuring changes in the volume of the chest or abdomen during breathing is called plethysmography. Most devices for monitoring respiration are based on a respiratory belt that a subject would place around the thorax or the abdomen. High-level block diagrams of several devices that measure strain are shown in Fig. 9.10. The figure also shows the sensor and the

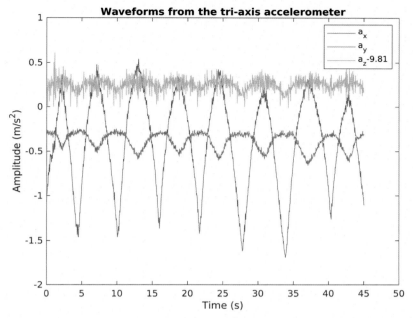

FIGURE 9.9 Waveforms obtained from a smartphone accelerometer showing an acceleration in *x*, *y*, and *z* directions.

body location where the sensor is commonly placed. All the devices indirectly measure chest cavity expansion or contraction by measuring the strain ε.

Resistive breathing belts or patches use piezoresistive transducers, such as strain gauges, connected to a Wheatstone bridge (Fig. 9.10 on the top). Therefore, the strain is converted into the resistance change, which is then converted to the voltage at the bridge output. This voltage is further amplified by amplifiers in the front-end of the circuit. A detailed front-end circuit for acquiring the breathing signal from the resistive transducer is shown in Figs. 4.16 and 4.18 in Chapter 4. A waveform obtained from a commercial resistive breathing belt is shown in Fig. 9.3A.

In novel works on piezoresistive transducers for breathing rate estimation, the transducers are not placed in the breathing belt (De Fazio et al., 2021). In Chu et al. (2019), two piezoresistive metal thin films were attached to the chest and the abdomen, and they were used to measure the breathing rate and to estimate the tidal volume (described in Section 9.5.3.1). The thin film was set in a silicone elastomer substrate and taped on the skin. The length of the sensors is small—around 2.1 cm and the breathing signal obtained this way was of high quality. In Nguyen and Ichiki (2019), the piezoresistive transducer was mounted to the eyeglasses' nose pad. The transducers detected a change in pressure due to the skin vibration during breathing. In smart textiles, piezoresistive transducers that are not based on traditional semiconductors or metal are used. New piezoresistive nanoscale materials, including graphene, carbon nanotubes, and metal nanowires nanoparticles, have been developed—please see De Fazio et al. (2021) for more information. These materials are more stretchable and/or more sensitive than traditional strain gauges.

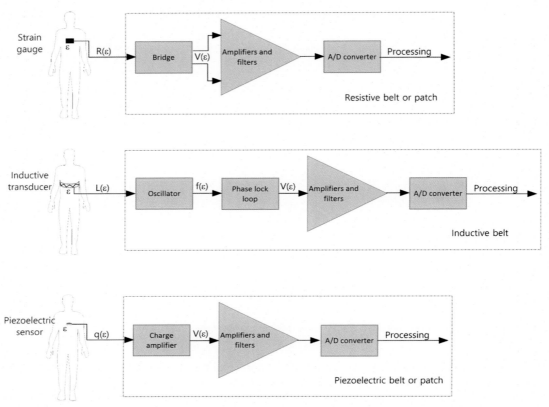

FIGURE 9.10   Block diagrams of devices for acquiring the breathing signal based on expansion/contraction of the chest cavity. All sensors measure the strain ε. The devices are based on changing the resistance of the transducer with breathing (top block diagram), changing the inductance (middle block diagram), and changing the charge of the piezoelectric transducer (bottom block diagram). *The figure is modified based on Massaroni et al. (2019b).*

Respiratory inductance plethysmography (RIP) is basically a standard for measuring breathing rate in sleep apnea studies. American Academy of Sleep Medicine recommended RIP technology for measuring respiratory effort. A number of commercial RIP belts exist. A high-level block diagram of a RIP-based system is shown in Fig. 9.10 in the middle. The sensor is an expendable belt with an isolated wire sewn into it. When the band is placed around the subject's body, the wire forms a coil with an inductance that depends on the enclosed cross-sectional area (Cohen et al., 1994). Variations in the chest's volume result in elongating and shortening the wire and, therefore, in variations of the inductance of the belts. The ends of the wire are connected to an oscillator circuit that generates a sine wave whose frequency depends on several parameters, including the inductance of the wire. Hence, the change in the inductance results in changes in the oscillators' sine wave frequency. This is why it is shown in the figure that the frequency f of oscillations depends on strain $f(\varepsilon)$. This oscillator is connected to a component called a phase lock loop (PLL). The PLL provides the DC signal whose amplitude is proportional to the oscillator frequency at the input. The voltage signal is then filtered, amplified, and converted into a digital domain by an A/D converter. A detailed description of the circuit is given in Cohen et al. (1994).

Piezoelectric transducers can be used as a part of a breathing belt, or they can be taped on the skin. The transducer is connected to a charge amplifier which gives voltage at the output that can then be further amplified and digitized—see the bottom plot of Fig. 9.10. Nowadays, a thin sheet of polyvinylidene fluoride (PVDF) is often used. PVDF materials are thin, flexible, light-weight, and inexpensive, and they can generate electrical signals when they are mechanically stimulated. Patch-type solution based on PVDF was proposed in Lei et al. (2015) for monitoring breathing rate.

One of the reasons why RIP belts are preferred to piezoelectric belts during sleep studies is that the piezoelectric transducer is located only in a small section of the belt. Therefore, if the subject is lying on top of the transducer, this might affect the signal obtained from the sensor. With the RIP belt, the sensor is the wire that encompasses the subject's body; therefore, the signal quality is much less affected by the subject's posture. The problem with all breathing belts is that the belt should not be too tight or loose to obtain a signal of acceptable quality.

9.4.3.3 *Measurement of impedance changes*

9.4.3.3.1 Introduction

Transthoracic electrical impedance changes with breathing (Ivorra, 2003). Transthoracic means that the measurement is done on the chest wall. Measurements of lung volume due to changes in transthoracic electrical impedance are called *electrical impedance pneumography*. Electrical impedance, when measured on the body, is commonly called *bioimpedance*. Bioimpedance is measured using electrodes where a small current at a high frequency is induced into the body, and the voltage is measured between the electrodes. The same electrodes used for measuring ECG can also be used for measuring bioimpedance, which is a significant advantage for systems requiring cardiorespiratory measurements.

With inhalation, the chest expands and there is an influx of air in the lungs. The expansion increases the conduction path for the current, which is one reason the impedance increases. The impedance increase also occurs due to larger air volume in the lungs. The opposite happens during exhalation. Therefore, measuring transthoracic bioimpedance can allow for noninvasive monitoring of respiratory rate, changes in tidal volume and functional residual capacity, and changes in lung fluid due to congestion and edema. In Grenvik et al. (1972), changes in transthoracic bioimpedance are measured simultaneously with the changes in lung volume using a spirometer. Around 96% correlation was achieved on average for all the subjects. The mean chest impedance was 517 Ω at the functional residual capacity.

In addition to measuring respiration, bioimpedance changes are also due to blood volume changes; therefore, it is possible to use it for cardiac monitoring. *Impedance plethysmography* measures the impedance changes in arms and legs, while transthoracic impedance cardiography measures the impedance changes on the chest due to stroke volume.

Bioimpedance can be measured using a single frequency or over a span of frequencies. The technique used for measuring bioimpedance over a range of frequencies is called *bioimpedance spectroscopy* (Caytak et al., 2017). Bioimpedance measurements done with many electrodes, commonly at a single frequency, are called *electrical impedance tomography*. This section focuses only on measurements at a single frequency between only two points.

Bioimpedance is measured by introducing current into the body at a high frequency and measuring voltage. Impedance changes modulate the measured voltage so that it is possible to extract the impedance changes by demodulating the signal. A demodulation circuit was mentioned in Chapter 3, Section 3.3.5, when we discussed a circuit used to extract changes in capacitance when using a capacitive transducer.

Many studies were done regarding the optimal electrode placement for electrical impedance pneumography (Dolores et al., 2019). Electrode placement is important since thoracic bioimpedance measurement is a combination of impedances of different body tissues, such as the lungs, the heart, blood, and air. Simulations showed that placement of the electrodes in the middle of the thorax provides a relatively large contribution of the lung impedance to the total measured impedance (Dolores et al., 2019). The number of electrodes used is two or four. The two-electrode configuration is called *bipolar,* while the four-electrode configuration is called *tetrapolar.* The tetrapolar configuration uses two leads for injecting the excitation current *i* and the other two to measure the generated voltage *V*. The current electrodes are also called drive electrodes, while the voltage electrodes are called measurement electrodes. The bipolar configuration uses the same electrode as both drive and measurement electrodes. The tetrapolar configuration with two measurement electrodes and two drive electrodes placed on the subject's chest is shown in Fig. 9.11. The block diagram of the circuit will be explained later.

9.4.3.3.2 Computing the impedance

The electrical impedance is $Z = V/i$ where i is the input current and V is the measured voltage. As a complex number, the impedance can also be presented as $Z = |Z|\cos\theta + j|Z|\sin\theta$ where $|Z|$ is the magnitude of the impedance, and θ is the phase shift between V and i.

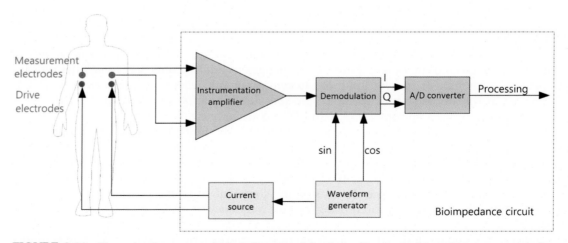

FIGURE 9.11 Electrode placement and block diagram of the device for electrical impedance pneumography. Light orange blocks correspond to the driving circuitry, while light blue blocks represent the analog front-end circuit and the A/D converter.

The other way to compute the impedance is through demodulation using mixers or multipliers. Multiplication of the measured signal is done with sine wave (in-phase components) and cosine signals (quadrature component) of the same frequency as the frequency of the current. Let us assume that the input current is $i(t) = I_0 \sin(2\pi ft)$ and the measured voltage is $V(t) = V_0 \sin(2\pi ft + \theta)$. The in-phase component is computed as

$$V_I(t) = V(t) \cdot sin(2\pi ft) = V_0 \, sin(2\pi ft + \theta)\sin(2\pi ft) = \frac{V_0}{2}[\cos(\theta) - cos(2\pi \cdot 2ft + \theta)]$$

After lowpass filtering with cutoff frequency f, the harmonic at the frequency $2f$ is removed and $V_I(t)$ becomes

$$V_I = \frac{V_0}{2}\cos(\theta)$$

The quadrature component is obtained in a similar way: $V_Q(t) = V(t) \cdot \cos(2\pi ft)$. After lowpass filtering, $V_Q = V_0 \sin(\theta)/2$. Next, the magnitude of voltage is computed as $|V| = \sqrt{V_I^2 + V_Q^2}$.

Finally, $|Z| = |V|/|i|$.

The phase difference can be obtained by dividing V_Q and V_I: $V_Q/V_I = \tan(\theta)$

$$\theta = \tan^{-1}\frac{V_Q}{V_I}$$

Example 9.2. Find the impedance if the input current is $i(t) = 10 \sin(2\pi ft)$ µA and the measured voltage is $V(t) = 3.5\sqrt{2} \sin(2\pi ft + \pi/4)$ mV. Let us assume that $f = 10$ kHz.

Solution 1. Computing the impedance by definition $Z = V/i$

We use a small current of 10 µA and a frequency of 10 kHz. These are the values that are commonly used in bioimpedance measurements. By dividing V over i we obtain $|Z| = 350\sqrt{2}\ \Omega$, which is about 500 Ω. The phase shift between the two signals is $\theta = \pi/4$. Therefore the impedance is $Z = |Z|\cos\theta + j|Z|\sin\theta = (350 + j \cdot 350)\ \Omega$.

Solution 2. Demodulation

The other way to compute the impedance is through demodulation using mixers or multipliers. Multiplication of the measured signal is done with sine (in-phase components) and cosine signals (quadrature component) of the same frequency as the frequency of the current.

$$V_I(t) = V(t)\sin(2\pi ft) = 3.5\sqrt{2}\sin(2\pi ft + \pi/4)\sin(2\pi ft)\ \text{mV}$$

$$= \frac{3.5\sqrt{2}}{2}[\cos(\pi/4) - \cos(2\pi \cdot 2ft + \pi/4)]\text{mV}$$

After lowpass filtering with cutoff frequency f, $V_I(T)$ becomes

$$V_I = \frac{3.5\sqrt{2}}{2}\cos(\pi/4)\text{mV} = 3.5\,\text{mV}$$

Quadrature components are obtained similarly: $V_Q(t) = v(t)\cos(2\pi ft)$. After low-pass filtering, $V_Q = \frac{3.5\sqrt{2}}{2}\sin(\pi/4)\text{mV} = 3.5\,\text{mV}$
Finally:

$$\left|Z\right| = \sqrt{\left(V_I^2 + V_Q^2\right)} / |i| = 3.5\sqrt{2}\,\text{mV} / 10\,\mu A = 350\sqrt{2}\,\Omega$$

The phase difference

$$\theta = \tan^{-1}\frac{V_Q}{V_I} = \tan^{-1}(1) = \pi/4$$

So, we got the same values using the demodulation technique as with the computation of the impedance by definition.

9.4.3.3.3 Circuits

The demodulation approach is the most common when implementing bioimpedance circuits in practice. The circuit includes a signal generator (we covered direct digital synthesis in Chapter 3) that generates sine and a cosine wave and a voltage-to-current converter to provide current to the electrodes—see Fig. 9.11. The voltage at the input of the analog front-end is mixed with generated signals to obtain in-phase and quadrature components. These components are then lowpass filtered and then sampled by an A/D converter. The final computation of the impedance is performed in software on the processor. Some industrial solutions, such as ADS129xR from Texas Instruments, provide multiple options for generating signals, including sine and quadrature signals. Other solutions allow demodulation in the digital domain after A/D conversion—for example, ADAS1000, the ECG, and bioimpedance chip from Analog Devices. In this case, the same electrodes for ECG measurements can be reused for measuring breathing rate. The signal is digitized first using a relatively high-speed A/D converter, and then the demodulation is done in the digital domain.

There are several issues to be considered when designing the circuit. The change of the impedance with breathing is quite small compared to the static impedance of the thorax, electrodes, and cables. For example, the Analog Devices chip ADAS1000 was designed to detect changes of 200 mΩ, assuming that the total static impedance will be smaller than 5 kΩ. In Gupta (2011), it was mentioned that changes in the impedance are up to 1 Ω. Range of impedance changes during respiration between 1 and 2 Ω was reported in Van Steenkiste et al. (2020). Other issues are related to:

- The quality of the current drivers is determined by quantization errors, current source matching, output impedance, and frequency range.
- Issues with the electrodes and their connection include electrode polarization, wiring crosstalk, leakage currents, and stray capacitance.

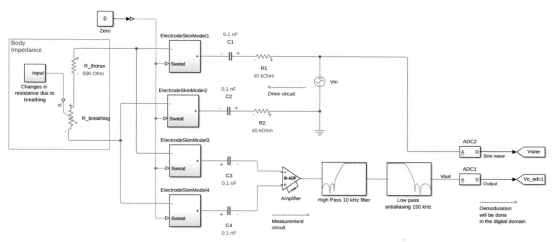

FIGURE 9.12 Model of a bioimpedance system with four electrodes. The demodulation is done in software in the same way as in Example 9.2, and it is not shown in the figure.

The performance of the front-end circuit depends on other factors, including CMRR of the amplifier, filtering, analog-to-digital sampling dynamic range and quantization errors, time source jitter, and phase accuracy.

The block diagram of the bioimpedance demodulation circuit is shown in Fig. 9.12. The fixed impedance of the thorax is modeled to be resistor R_thorax = 500 Ω. The impedance due to breathing R_breathing changes as a sine wave at the amplitude of 5 Ω. We selected 5 Ω (and not 1 Ω as explained before) to visualize changes due to breathing on the modulated signal. These two impedances are connected in series and represent the impedance of the thorax. The breathing is simulated as a sine wave at the rate of 12 breaths per minute. Four electrodes are connected to the chest and are modeled as wet electrodes with the same parameters as the wet electrodes introduced in Chapter 2. Two drive electrodes and their electrode-skin interface are labeled as ElectrodeSkinModel1 and ElectrodeSkinModel2. The drive comes from the voltage source set at a frequency of 32 kHz and an amplitude of 1 V peak-to-peak. It connects to the resistors R1 and R2 (40 kΩ) and capacitors C1 and C2 (0.1 nF). The purpose of the resistor is to limit the current through the body. The measured peak-to-peak current through the resistor R_thorax is 14.5 μA, the current considered safe at 32 kHz. Actually, the maximum allowable peak-to-peak current below 1 kHz is 10 μApp, where *pp* means peak-to-peak. At frequencies higher than 1 kHz and lower than 100 kHz, the peak-to-peak current is determined as *frequency* · 10 μApp/1000 Hz—please check the ANSI/AAMI ES1-1993 for more details about the safe current limits (Association for the Advancement of Medical Instrumentation, 1993). Capacitors C1 and C2 block any direct current from flowing into the chest from the drive side. Capacitors C3 and C4 serve the same purpose on the measurement side. The signal is then amplified and passed through the highpass filter with a 10 kHz cutoff frequency and the lowpass filter to prevent aliasing. A/D conversion is performed at 1 MHz using a 16-bit A/D converter. This circuit is described in the datasheet of the Analog Devices chip ADAS1000 (Analog Devices, 2018).

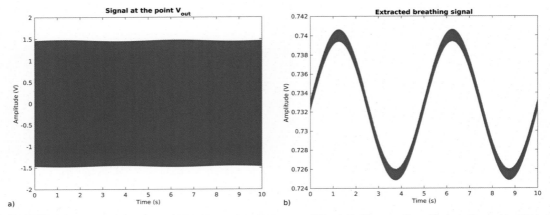

FIGURE 9.13 (A) Amplified and filtered signal at the output of Fig. 9.12. Very small oscillations due to breathing are visible by observing the amplitude of the signal. (B) Extracted respiratory signal after demodulation in the digital domain.

The demodulation, in this case, is done in the digital domain. In the digital domain, demodulation is performed in the same way described in Example 9.2.

The extracted signal Vout at the output of the lowpass antialiasing filter is shown in Fig. 9.13A. This signal is then digitized. In the digital domain, it is multiplied with the sine and cosine waveforms to obtain the I and Q components, and then both components are passed through a lowpass filter with a cutoff frequency of 7 kHz. The magnitude of the signal is shown in Fig. 9.13B.

If the input of the amplifier is connected to the drive electrodes instead of the measurement electrode, we have a bipolar or two-electrode system. In this case, the extracted signal is much smaller, and it is more difficult to extract the breathing signal from it.

9.4.4 Temperature and pressure measurement

Oronasal (or nasal) thermistors and air pressure transducers are commonly used as the surrogate of airflow measurements during sleep studies. In sleep studies, the signal obtained from a temperature transducer is used to detect the presence of airflow and obtain the breathing rate. The nasal pressure transducer is more sensitive to airflow restriction and is used for detecting the occurrence of hypopneas. Detecting hypopneas is done by observing the morphology of the signal—resistance to airflow leads to signals that look less like sine-wave signals.

Signals obtained from temperature and pressure transducers are normally not calibrated to provide the absolute value of the airflow. However, they are very reliable in estimating the breathing rate.

The most common *thermal transducers* include thermistors, thermocouples, and pyro-electric sensors. During breathing, the temperature of the exhaled air is normally higher than that of the inhaled air. This temperature difference between the exhaled and inhaled air can be used to obtain the breathing signal and, from it, the breathing rate. The signal

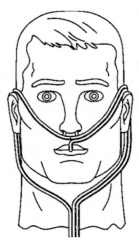

FIGURE 9.14   Oral and nasal cannulas. *The figure is modified from Science Kids—the figure is free to use and modify.*

obtained from a thermistor placed close to the subject's nose is shown in Fig. 9.3B and Fig. 9.4. In these figures, the temperature difference during normal breathing is about 0.5°C.

To obtain a breathing signal using *air pressure transducers*, a nasal/oral cannula is used. The openings of the nasal cannula are placed just inside the nostrils, and the oral cannula needs just to touch the lips—see Fig. 9.14. As the subject breathes, the air enters the cannula and causes pressure changes that are measured using a pressure transducer. A mobile device with all the electronics for sensing, data acquisition, and communication integrated into commercial headphones was developed and described in Massaroni et al. (2019a). The cannula is connected to the differential pressure sensor on one side and the nostril on the other side. The sensor was a differential pressure sensor SDP610, Sensirion AG Switzerland, with a pressure range up to ±125 Pa and an accuracy of 0.04% of the full-scale.

The front-end circuits for thermistors and piezoresistive pressure transducers can be implemented in a very similar way shown in Chapter 4, Fig. 4.18. The transducer is placed in the bridge since its resistance changes with pressure or temperature changes. Next, the signal from the bridge is amplified, filtered, and then digitized.

Problems with the thermal sensors are that they are less sensitive in detecting hypopneas than nasal pressure transducers. Also, thermal sensors have problems in very warm rooms or when a fan is directly oriented toward the subject. The tube can be squeezed or pressed by the limbs for nasal cannulas.

Tube-free solutions for obtaining the breathing signal based on sensing changes in temperature during breathing have been developed. They normally require sticking the sensor on the upper lip or placing it in the nostril. *Hot-film sensors* are based on a heated wire so that the temperature of the wire is normally higher than the body temperature; therefore, these sensors are less affected by environmental conditions. However, these systems have higher power consumption because of the energy required to heat the wire. In Jiang et al. (2020), the sensor placed on the upper lip contains three transducers. One is directly heated and used for measuring the airflow from both nostrils together. The other two transducers are placed directly below each nostril and are not directly heated. However, their temperature is increased due to the proximity of the hot wire sensor above the body temperature. These sensors are used to obtain the breathing signal from each nostril separately. That system can be used to detect apnea and hypopnea events by using a hot-wire sensor. In addition, it can be used to detect asymmetric breathing conditions for the right and left nostrils. The power consumption is high (about 60 mW) because of the need to heat the wire.

9.5 Tidal volume and flow estimation

This section describes devices and methods used to obtain tidal volume or flow rate outside hospitals. We start with portable spirometers, which are devices that have already been commercialized and that show good agreement in comparison with clinical spirometers. Next, we describe research directions in which flow or volume and the indices commonly estimated by spirometers are extracted using smartphones or wearable devices. Also, we presented approaches where impedance or pressure measurements on the chest are converted into air flow or volume.

9.5.1 Portable spirometers

Portable spirometers are used for monitoring several conditions, including asthma and COPD. They commonly provide FVC, FEV1, PEF, and some show flow/volume loop and volume over time. Some devices also include other sensors like pulse oximetry (e.g., Spirobank Smart by MIR). Devices commonly comply with the standards, including ISO 26,782 for spirometry and ISO 23,747 for peak expiratory flow meters. Some portable spirometers provide feedback to the patient regarding the quality of the test performance in the form of recommendations to exhale faster, for example. This is important because, in hospitals, the patient is normally coached on how to properly breathe when using the spirometer. The accuracy of the devices is reported in different ways—see Carpenter et al. (2018) for the review. Devices should report volume and flow accuracy, as well as the accuracy of estimated indices. For example, the error in FEV1 can be presented in liters or percentages (e.g., ± 0.1 L or $\pm 5\%$). The accuracy of some of the devices was validated by independent studies. For example, it was found that for FEV_1, FVC, and FEV_1/FVC, the Pearson correlation and interclass correlation coefficients between the Air Next spirometer by NuvoAir, Sweden and the reference clinical spirometer were greater than 0.94, exhibiting excellent agreement between the two spirometers (Exarchos et al., 2020).

9.5.2 Spirometers that do not require a mouthpiece

Due to the invasive nature of measuring volume and flow using mouthpieces of face masks, there have been several works in which pervasive devices such as smartphones were used to estimate the indices obtained using traditional spirometers. These indices include FVC, FEV1, and FEV1/FVC ratio. One of the first smartphone-based spirometers was developed in 2012 (Larson et al., 2012). The phone's microphone was used to record the signal, and the phone was kept at arm's length. A set of processing and machine learning steps was developed to convert the pressure signal from the microphone into the flow and then estimate FVC, FEV1, and FEV1/FVC ratio. Two transfer functions were proposed to convert the microphone pressure signal into the flow signal. The first transfer function compensates for the pressure losses over the distance from the mouth to the phone. Another transfer function converts the pressure drop at the lips into the flow. After that, the features were extracted from the flow waveform, and two regression models were developed: (1) to estimate quantities of interest and (2) to estimate the shape of the flow/volume curve. Very good agreement against the reference spirometer was obtained, especially for normal subjects. Another example of estimating FVC, FEV1, and FEV1/FVC indices using smartphone microphones was described in Zubaydi et al. (2020). The subject was supposed to keep the smartphone 5 cm away from the mouth and forcefully exhale. The signal is converted into the flow signal with less processing than in Larson et al. (2012). Then the flow signal is multiplied with the constant corresponding to the mouth's cross-sectional area to obtain the volumetric flow rate, which was then used to obtain the abovementioned indices. Very high agreement was obtained against the state-of-the-art spirometer.

There are several major limitations of this technology. The obtained signal has a relative amplitude and needs to be converted into volume or flow by developing a mathematical model or a transfer function. The transfer function that performs that conversion needs to be developed or learned if machine learning methods are applied. The parameter estimation of that model needs to be performed before the device is delivered to the patients. An error can result if the measurement is done differently than the ones originally used for the parameter estimation. Another issue is that the expiration becomes less audible over time. It was reported in Larson et al. (2012) that often, after 3 s, the signal is often not audible. Therefore, FVC is less accurate than other measures. This problem was not reported in Zubaydi et al. (2020)—one reason might be that the phone was kept only at a 5 cm distance from the mouth.

9.5.3 Volume and flow extracted from pressure or impedance signals

Tidal volume and flow estimation are commonly done using devices, such as a spirometer, that require the subject to wear a face mask or a mouthpiece and to blow the air directly into it. However, it is possible to estimate the tidal volume or the flow from the wearable devices with two piezoelectric belts or by using electrical impedance pneumography. Measured changes of impedance or pressure provide only a relative measure of volume change; therefore, a mathematical model or a transfer function needs to be developed to map the impedance or pressure into the tidal volume, and the parameter estimation needs to be performed to estimate the parameters of that model. The relationship between measured impedance or pressure and volume depends on the type of breathing and the contribution of abdominal or chest wall movement to the volume.

We are unaware of the works that first provided automated breathing pattern classification (covered in the next section) before attempting to estimate the tidal volume or airflow. However, several papers examined the agreement between the flow indices obtained using traditional and reference techniques and found that the agreement is lower if the subjects are not breathing normally (Malmberg et al., 2017).

9.5.3.1 Relationship between displacement and volume

One of the first works that related changes in the ribs and abdomen due to breathing with the volume is Konno and Mead (1967). Estimating the volume requires acquiring displacement signals from both the chest wall and the abdomen. The chest wall and abdomen move synchronously. However, it is possible to inhale mainly by moving the rib cage or by moving the abdomen. So, if we are able to measure the displacement of the abdomen and the rib cage, it should be possible to estimate the volume after calibration.

In Chu et al. (2019), a method for computing respiratory rate and estimating tidal volume was developed. Two strain sensors were connected to the thorax and abdomen and measured the local strain changes as the subject was breathing. Based on these measurements, the transfer function was designed as a regression function that related changes in the strain during breathing in the abdomen and chest to the changes in the volume. The volume data are the reference signal obtained by processing the airflow signal from a commercial spirometer. Parameter estimation of the model/transfer function was required for each subject where the subject first needs to breathe at different intensities while the reference data are collected using a spirometer.

A regression function was developed that maps the signals obtained from the breathing belts placed on the chest and abdomen into the respiratory airflow waveforms. The method was tested on five subjects; three had no respiratory-related symptoms, one had sleep apnea, and the last one had COPD. The method also requires parameter estimation for each patient separately (Seppänen et al., 2016).

9.5.3.2 Relationship between impedance and volume

Several studies related the changes in the impedance of the lungs with the volume changes. The relation between volume and bioimpedance changes is linear during normal breathing (Grenvik et al., 1972). In the study of airflow using electrical impedance pneumography (Malmberg et al., 2017), it was determined that the agreement between the electrical impedance pneumography and the airflow was affected by lower and upper airway obstruction, which was not the case for the subjects with normal breathing. Interestingly, no calibration or parameter estimation was necessary because the extracted flow indices did not depend on the amplitude (absolute measurements of the flow). The paper concludes that the agreement of tidal flow profiles obtained using electric impedance pneumography and pneumotachograph is satisfactory in those subjects (infants) who were stable and clinically without signs of bronchial obstruction at the time of the study. It was hypothesized in Dolores et al. (2019) that thoracic bioimpedance changes seem to be a combination of volume and other thoracic changes, especially when the subject does not breathe normally. The volume (using a spirometer) and chest motion (using an accelerometer) were measured together with bioimpedance, and inspiratory muscle force was estimated during loaded breathing and divided into 12 levels (Dolores et al., 2019). It was shown that the agreement between the bioimpedance and combined volume and acceleration is higher than the agreement between the bioimpedance and the volume or acceleration separately over the whole range of estimated muscle force levels. It was also shown that the agreement is better when a nonlinear model is used than in cases of using a linear model, especially for higher levels of inspiratory muscle force. The study results show that in the case of restrictive breathing common for respiratory patients, it is possible to monitor tidal volume continuously if both impedance and accelerometer signals are collected and used to estimate the volume.

9.6 Breathing pattern classification

Several types of classifications were published in the literature. For example, *heart failure decompensation* based on breathing patterns requires the classification of apneas, Cheyne—Stokes breathing, normal breathing, and so on. Fekr et al. (2016) proposed a respiration pattern classification method based on the signal collected from two accelerometers mounted on the subject's chest and abdomen. A support vector machine (SVM) algorithm with seven statistical features was used to classify eight respiratory patterns (normal, bradypnea, tachypnea, Kussmaul, apnea, Biot's, sighing, and Cheyne—Stokes breathing).

Classification in *respiratory gating* for radiation treatment normally includes breath hold, normal breathing, and irregular breathing (Fallatah et al., 2022).

In *sleep apnea*, we can classify respiratory events, the type of sleep apnea, and its severity. Respiratory events can be classified as normal breathing, apneas, hypopneas, and flow limitation events. Hypopneas require that the airflow amplitude is less than 50% of the surrounding baseline and that the event lasts for more than 10 s. Flow limitation is the restricted flow rate during the inhale or exhale cycle. Sleep apnea types are central and obstructive. Another possible classification includes the following classes: normal, apnea, coughing, Miller, sighing, and yawning on 30-s segments based on signals collected from an accelerometer and a gyroscope (McClure et al., 2020).

Traditional or deep learning algorithms were used for classification—we will not describe them here. Deep learning methods for detecting sleep apnea are described in Mostafa et al. (2019). Instead, we mention some commonly extracted features before applying traditional *machine learning models for classification*. The features can be divided into time-domain and short-term energy-domain features (Han, 2021). Examples of time domain features are amplitude variance, amplitude maximum, amplitude minimum, amplitude range, peak variance, peak average, peak maximum, and the number of peaks. Here, the peak represents the amplitude of the maximum value of the normalized breathing signal during one breathing cycle. Amplitude is the difference between the peak and the trough obtained during one breathing cycle. Short-term energy features include energy variance, maximum, minimum, and range, where the energy is computed on a segment that can be 15—30 s. Other features of interest include time-frequency domain features and statistical features (Han, 2021).

9.7 Conclusion

In this chapter, we described systems mainly used for monitoring breathing rate. As this book focuses mostly on wearable devices, we only mention clinical devices for tidal volume and airflow measurements, such as spirometers and pneumotachograph. We focused mainly on obtaining the breathing signal and estimating the breathing rate from cardiac signals, strain piezoresistive transducers (Chapter 4), and chest impedance measurements. We also mention new directions, such as obtaining airflow and tidal volume using wearable devices and ways to classify breathing patterns based on acquired breathing signals.

In Chapter 5, we identified several drivers and challenges for improvements in SpO_2 technology. Similar drivers and challenges can be defined here. They include miniaturizing the system, improving current sensors and developing new sensors, designing devices for non-obtrusive long-term monitoring, improving algorithms for breathing rate estimation to deal with motion artifacts and monitoring breathing during activities, measuring more parameters or estimating/extracting more information from the same signal, determining proper placement of the device and then designing the device fitted for that location, and improving mathematical models. Some of these challenges and research directions will be further elaborated on in the next Chapter.

Even though we introduced some commercial and research simulators, we did not describe mathematical models of breathing. This is very important in developing devices because it would allow us to model different breathing conditions. However, several simulators, such as PNEUMA, already provide that; therefore, we decided not to go into mathematical breathing models in this chapter.

We did not cover many important topics, including the measurement of carbon dioxide CO_2 and breathing rate using microphones. Capnography refers to the evaluation of CO_2 in expired gases. This method allows the continuous recording of CO_2 values in patients and is mainly used in hospitals even though some home-monitoring solutions exist. Capnography requires using a nasal cannula or, more frequently, the face mask or the mouthpiece. Transcutaneous carbon dioxide is a noninvasive and continuous method that measures the partial pressure of CO_2 by placing the sensor on the skin. This is possible because carbon dioxide diffuses through the body tissues and skin. However, to increase the blood supply to the capillaries, the temperature under the sensor needs to be increased. Using microphones for the measurement of breathing rate, snoring, and so on for sleep apnea is very important.

9.8 Problems

9.8.1 Simple questions

9.1. In the spirometry waveform shown in Fig. 9.15, do the following:
   (a) Estimate FEV_1/FVC
   (b) Define indices that have "?" next to them.
9.2. Compute breathing rate from the waveform shown in Fig. 9.3.
9.3. Define tidal volume, flow rate, minute respiratory volume, dead space, alveolar ventilation, and residual volume.
9.4. Define the following terms: dyspnea, eupnea, apnea, hypopnea, and apnea-hypopnea index.

9.8.2 Problems

9.5. Describe the difference in computed impedance in case two and in case four electrodes are used.

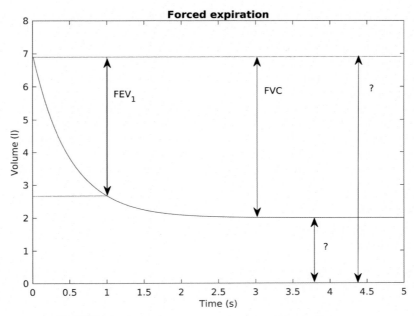

FIGURE 9.15 Volume versus time during forced exhalation.

9.6. Your goal is to design a breathing rate monitor using bioimpedance measurements where the four electrodes are placed on the chest. You will use a current source that injects current to the body ($I_0 = 1\ \mu A; f = 1$ kHz). The mixer (multiplier) and the filter are used to extract the $\sin(\theta)$ and $\cos(\theta)$ for the bioimpedance measurements.
   (a) Describe the operation of your bioimpedance breathing rate circuit, starting from the current source to processing the signal to extract the breathing rate.
   (b) Draw the block diagram of the circuit. Show the processing steps of the algorithm for breathing rate estimation.
   (c) Design a second-order lowpass filter that will be placed after the multiplier to extract the sine and cosine components. What should be the cutoff frequency of that filter?
9.7. By looking at Fig. 9.7A, it is obvious that the extracted waveform has multiple peaks for a single breath. Analyze the peak detection algorithm in Matlab to understand how one can correctly estimate breathing rate in this case.
9.8. Install an app on your phone that can collect signals from the phone's IMU. An example of the app is Physics Toolbox Suite Pro or Physics toolbox sensor suite from www.vieyrasoftware.net. Place the phone on your chest while in a supine posture so that the back side of the phone is on your chest. Collect data using different sensors (accelerometer, gyroscope, and magnetometer). Observe the signals you get and see if a signal resembles breathing or a cardiac signal. Collect data multiple times while measuring your heart rate by placing three fingers on your radial artery and controlling your breathing rate (e.g., at 12 breaths per minute).

(a) Plot the obtained breathing signal and compute the error between the estimated breathing rate and the reference breathing rate.

(b) Analyze the signals from the different axis of the accelerometer and comment on the quality of the signal.

(c) Analyze the signals from the gyroscope and magnetometer and comment on the quality of the signal.

(d) Describe the sensor that gave the best respiratory signal. For example, if the sensor that provides the clearest signal is the accelerometer, describe the features of the accelerometer chip features used in the phone (if you cannot find the exact one, then any sensor that you find for Android or iOs phones).

(e) Repeat steps (a)–(d) to extract the cardiac signal and estimate the heart rate.

9.8.3 Simulations

All problems are related to modifications of the Matlab code/Simulink models provided on the book web page.

9.9. Breathing rate estimation algorithms:

(a) Analyze the effects of window size and overlap on the accuracy of the breathing rate estimation algorithms based on peak detection and based on FFT shown in Fig. 9.5.

(b) Modify the code for computing breathing rate based on peaks to use only peaks (but no troughs). How does that affect the breathing rate estimation?

9.10. Extracting breathing signal from ECG:

(a) Replace the ECG signal with the PPG signal and estimate the breathing rate from it.

9.11. IMU:

(a) Instead of using accelerometer data, process data from the gyroscope and try to estimate breathing rate from it.

9.12. Breathing rate estimation using bioimpedance measurements:

(a) Modify the simulation by turning the indicator for sweating to 1. This will allow for much better contact between the electrodes and the skin. Observe the amplitude of the signal at the output.

(b) Reduce the amplitude of the impedance change from 5 to 1 Ω and then to 0.2 Ω. Observe the signal at the output.

(c) Remove electrodes 3 and 4 (blocks ElectrodeSkinModel3 and ElectrodeSkinModel4) from the simulator and connect capacitors C3 and C4 directly to the first two electrodes (C3 to the + input of ElectrodeSkinModel1 and C4 to the + input of the ElectrodeSkinModel2). Compare the output of the two-electrode system with the output of the four-electrode system. Why is there such a big difference in the results?

(d) Add noise source to the signal delta_R by uncommenting the noise term in the line when delta_R is formed. How is the standard deviation of the noise affect the results?

(e) Modify the circuit to implement demodulation in hardware instead of processing it in software. You would need to remove the highpass filter and replace an anti-aliasing filter with the lowpass filter with the cutoff frequency in the order of 100 Hz. Use mixer (multiplier) blocks from Simulink in this case. You would need to convert the signal from Simscape to Simulink to multiply it. A lowpass filter after the mixer can be implemented in Simulink or in Simscape.

9.9 Further reading

Additional resources related to different topics covered in this chapter are listed below. Please note that the following material is selected based on availability and the author's preferences and that many good resources might be omitted unintentionally.

For the physiology of respiration, an excellent book is Lumb (2017).

Several reviews on techniques and the importance of measuring respiratory rate have been recently published. They include Liu et al. (2019), Nicolò et al. (2020), and Scott and Kaur (2020). Excellent high-level descriptions of different sensors and circuits for monitoring breathing rates are presented in Massaroni et al. (2019a, b).

Several open-source software packages for processing breathing signals exist. Rspiro provides the implementation of spirometry equations in R language (https://cran.r-project.org/web/packages/rspiro/index.html). BreathMetrics is a Matlab toolbox for analyzing respiratory recordings (Noto et al., 2018) introduced in Section 9.3.1. NeuroKit2 (Makowski et al., 2020) is an open-source Python toolbox developed for neurophysiological signal processing. It provides models for several physiological signals, including the breathing signal. According to the toolbox descriptions, the breathing signal is collected by a respiratory belt, and the program can detect breaths and distinguish between inspiration/expiration phases.

An excellent overview of bioimpedance measurement applications is given in Naranjo-Hernández et al. (2019).

A very simple explanation of accelerometers and their operation is given in Accelerometer Technical Note at https://www.sciencebuddies.org/science-fair-projects/references/accelerometer.

References

Alinovi, D., Ferrari, G., Pisani, F., Raheli, R., 2017. Markov chain modeling and simulation of breathing patterns. Biomedical Signal Processing and Control 33, 245–254.

Analog Devices, 2018. ADAS1000/ADAS1000-1/ADAS1000-2: Low Power, Five Electrode Electrocardiogram (ECG) Analog Front End, Data Sheet, Analog Devices. Rev C.

Ansari, S., Ward, K.R., Najarian, K., 2017. Motion artifact suppression in impedance pneumography signal for portable monitoring of respiration: an adaptive approach. IEEE Journal of Biomedical and Health Informatics 21, 387–398.

ASL 5000 Breathing Simulator," IngMar Medical. https://www.ingmarmed.com/product/asl-5000-breathing-simulator/(accessed Jan. 22, 2021).

Association for the Advancement of Medical Instrumentation, 1993. Safe Current Limits for Electromedical Apparatus. ANSI/AAMI ES1-1993.

Berry, R.B., et al., 2015. The AASM manual for the scoring of sleep and associated events: rules, terminology and technical specifications. American Academy of Sleep Medicine 2 ver. 2.

Carpenter, D.M., Jurdi, R., Roberts, C.A., Hernandez, M., Horne, R., Chan, A., 2018. A review of portable electronic spirometers: implications for asthma self-management. Current Allergy and Asthma Reports 18 (10), 53.

Caytak, H., Boyle, A., Adler, A., Bolic, M., 2017. Bioimpedance spectroscopy processing and applications. In: Book Chapter, Encyclopedia of Biomedical Engineering Edited by Roger Narayan. Elsevier.

Celli, B.R., et al., 2004. The body-mass index, "airflow obstruction, dyspnea, and exercise capacity index in chronic obstructive pulmonary disease. The New England Journal of Medicine 350, 1005−1012.

Charlton, P.H., Bonnici, T., Tarassenko, L., Clifton, D.A., Beale, R., Peter, J.W., 2016. An assessment of algorithms to estimate respiratory rate from the electrocardiogram and photoplethysmogram. Physiological Measurement 37 (4), 610−626.

Charlton, P., Bonnici, T., Tarassenko, L., Clifton, D.A., Beale, R., Watkinson, P.J., Alastruey, J., Mar. 2021. An impedance pneumography signal quality index: design, assessment and application to respiratory rate monitoring. Biomedical Signal Processing and Control 65 (102339).

Chatterjee, N.A., Jensen, P.N., Harris, A.W., et al., 2021. Admission respiratory status predicts mortality in COVID-19. Influenza Other Respiratory Viruses 15 (5), 569−572.

Chen, X., Wang, R., Zee, P., Lutsey, P.L., Javaheri, S., Alcántara, C., Jackson, C.L., Williams, M.A., Redline, S., 2015. Racial/ethnic differences in sleep disturbances: the multi-ethnic study of Atherosclerosis (MESA). Sleep 38 (6), 877−888.

Cheng, L., Ivanova, O., Fan, H.H., Khoo, M.C., 2010. An integrative model of respiratory and cardiovascular control in sleep-disordered breathing. Respiratory Physiology Neurobiology 174 (1−2), 4−28.

Cheng, L., Ivanova, O., Fan, H.H., Khoo, M.C., 2013. Simulation of State-CardioRespiratory Interactions: PNEUMA" User's Guide. University of Southern California, Release 3.0. https://bmsr.usc.edu/files/2013/02/PNEUMA-User-Guide-Release-3.0.pdf. (Accessed 3 November 2021).

Chu, M., Nguyen, T., Pandey, V., et al., 2019. Respiration rate and volume measurements using wearable strain sensors. Npj Digital Medicine 2 (8).

Cohen, K.P., Panescu, D., Booske, J.H., Webster, J.G., Tompkins, W.J., 1994. Design of an inductive plethysmograph for ventilation measurement. Physiological Measurements 15, 217−229.

De Fazio, R., Stabile, M., De Vittorio, M., Velázquez, R., Visconti, P., 2021. An overview of wearable piezoresistive and inertial sensors for respiration rate monitoring. Electronics, MDPI 10 (2178).

Dolores, B.-A., Groenendaal, W., Catthoor, F., Jané, R., 2019. Chest movement and respiratory volume both contribute to thoracic bioimpedance during loaded breathing. Scientific Reports 9 (20232).

Exarchos, K.P., Gogali, A., Sioutkou, A., et al., 2020. Validation of the portable Bluetooth® Air Next spirometer in patients with different respiratory diseases. Respiratory Research 21 (79).

Fallatah, A., Bolic, M., MacPherson, M., La Russa, D.J., 2022. Monitoring respiratory motion during VMAT treatment delivery using ultra-wide band radar. Sensors, MDPI.

Fekr, A.R., Janidarmian, M., Radecka, K., Zilic, Z., May 2016. Respiration disorders classification with informative features for m-health applications. IEEE Journal of Biomedical and Health Informatics 20 (3), 733−747.

Giraud, P., Houle, A., 2013. Respiratory gating for radiotherapy: main technical aspects and clinical benefits. ISRN Pulmonology 2013. Article no. 519602.

Grenvik, A., Ballou, S., McGinly, E., Cooley, W.L., Safar, P., 1972. Impedance pneumography: comparison between chest impedance changes and respiratory volumes in 11 healthy volunteers. Chest 62, 439−443.

Güder, F., Ainla, A., Redston, J., Mosadegh, B., Glavan, A., Martin, T.J., Whitesides, G.M., 2016. Paper-based electrical respiration sensor. Angewandte Chemie International 55 (19), 5727−5732.

Gupta, A.K., February 2011. Respiration Rate Measurement Based on Impedance Pneumography," Application Report. Texas Instruments, no. SBAA181.

Han, Z., 2021. Respiratory Patterns Classification Using UWB Radar. M.Sc. thesis. University of Ottawa.

Ivorra, A., 2003. Bioimpedance monitoring for physicians: an overview," report, Centre Nacional de Microelectrònica Biomedical. Applications Group 11 (7).

Jiang, T., Deng, L., Qiu, W., et al., 2020. Wearable breath monitoring via a hot-film/calorimetric airflow sensing system. Biosensors and Bioelectronics 163. Elsevier.

Johnson, J.D., Theurer, W.M., 2014. A stepwise approach to the interpretation of pulmonary function tests. American Family Physician 89 (5), 359−366.

Karlen, W., Turner, M., Cooke, E., Dumont, G.A., Ansermino, J.M., 2010. CapnoBase: signal database and tools to collect, share and annotate respiratory signals. Annual Meeting of the Society for Technology in Anesthesia (STA), West Palm Beach 25.

Keall, P.J., Mageras, G.S., Balter, J.M., et al., 2006. The management of respiratory motion in radiation oncology report of AAPM Task Group 76. Medical Physics 33 (10), 3874—3900.

Konno, K., Mead, J., 1967. Measurement of the separate volume changes of rib cage and abdomen during breathing. Journal of Applied Physiology 22, 407—422.

Larson, E.C., Goel, M., Boriello, G., Heltshe, S., Rosenfeld, M., Patel, S., 2012. SpiroSmart: using a microphone to measure lung function on a mobile phone. ACM Conference of Ubiquitous Computer 280—289.

Lei, K.F., Hsieh, Y.Z., Chiu, Y.Y., Wu, M.H., 2015. The structure design of piezoelectric poly(vinylidene fluoride) (PVDF) polymer-based sensor patch for the respiration monitoring under dynamic walking conditions. Sensors, MDPI 15 (8), 18801—18812.

Liu, H., Allen, J., Zheng, D., Chen, F., 2019. Recent development of respiratory rate measurement technologies. Physiological Measurement 40. IOP Publishing.

Løberg, F., Goebel, V., Plagemann, T., 2018. Quantifying the signal quality of low-cost respiratory effort sensors for sleep apnea monitoring. In: Proceedings of the 3rd International Workshop on Multimedia for Personal Health and Health Care, pp. 3—11.

Lumb, A.B., 2017. Nunn's Applied Respiratory Physiology, eighth ed. Elsevier.

Machado, C., DeFina, P.A., May 2020. Abnormal respiratory patterns in Covid-19. European Journal of Neurology.

Makowski, D., et al., Jul. 06, 2020. NeuroKit2: a Python toolbox for neurophysiological signal processing. PsyArXiv. https://doi.org/10.31234/osf.io/eyd62.

Malmberg, L.P., Seppä, V.-P., Kotaniemi-Syrjänen, A., Malmström, K., Kajosaari, M., Pelkonen, A.S., Viik, J., Mäkelä, M.J., 2017. Measurement of tidal breathing flows in infants using impedance pneumography. European Respiratory Journal 49 (2).

Mandal, N.G., 2006. Respirometers including spirometer, pneumotachograph and peak flow meter. Anaesthesia and Intensive Care Medicine 7 (1), 1—5.

Marconi, S., De Lazzari, C., 2020. In silico study of airway/lung mechanics in normal human breathing. Mathematics and Computers in Simulation 177, 603—624.

Massaroni, C., Nicolò, A., Girardi, M., et al., 2019a. Validation of a wearable device and an algorithm for respiratory monitoring during exercise. IEEE Sensory Journal 19 (12), 4652—4659.

Massaroni, C., Nicolò, A., Presti, D.L., et al., 2019b. Contact-based methods for measuring respiratory rate. Sensors, MDPI.

McClure, K., Erdreich, B., Bates, J.H.T., McGinnis, R.S., Masquelin, A., Wshah, S., 2020. Classification and detection of breathing patterns with wearable sensors and deep learning. Sensors, MDPI 20 (22), 6481.

Miller, D.J., Capodilupo, J.V., Lastella, M., et al., 2020. Analyzing changes in respiratory rate to predict the risk of COVID-19 infection. PLoS ONE 15 (12), e0243693.

Moeyersons, J., March 2021. Advanced Tools for Ambulatory ECG and Respiratory Analysis. Ph.D. thesis. Technische Universiteit Eindhoven.

Mostafa, S.S., Mendonça, F., Ravelo-García, A.G., Morgado-Dias, F., 2019. A systematic review of detecting sleep apnea using deep learning. Sensors 19 (22), 4934.

Naranjo-Hernández, D., Reina-Tosina, J., Min, M., 2019. Fundamentals, recent advances, and future challenges in bioimpedance devices for healthcare applications. Journal of Sensors, Hindawi 2019. Article no. 9210258.

Natarajan, A., Su, H.W., Heneghan, C., 2021. Measurement of respiratory rate using wearable devices and applications to COVID-19 detection. Npj Digital Medicine 4 (136).

Nguyen, T.V., Ichiki, M., 2019. MEMS-based sensor for simultaneous measurement of pulse wave and respiration rate. Sensors, MDPI 19 (4942).

Nicolò, A., Massaroni, C., Schena, E., Sacchetti, M., 2020. The importance of respiratory rate monitoring: from healthcare to sport and exercise. Sensors, MDPI 20 (21), 6396.

Noto, T., Zhou, G., Schuele, S., Templer, J., Zelano, C., 2018. Automated analysis of breathing waveforms using BreathMetrics: a respiratory signal processing toolbox. Chemical Senses 43, 583—597.

Oppenheim, A.V., Schafer, R.W., 1998. Discrete-time Signal Processing, second ed. Prentice Hall.

Paraskeva, M.A., Borg, B.M., Naughton, M.T., April 2011. Spirometry. Australian Family Physician 40 (4).

QUASARTM Respiratory Motion Phantom (pRESP) - Breathing Phantom," Modus Medical Devices. https://modusqa.com/products/quasar-respiratory-motion-phantom-presp/, last accessed on January. 22, 2021.

QuickLung Breather," IngMar Medical. https://www.ingmarmed.com/product/quicklung-breather/(accessed January. 22, 2021).

Rahmani, M.H., Berkvens, R., Weyn, M., 2021. Chest-worn inertial sensors: a survey of applications and methods. Sensors, MDPI 21 (2875).

Scott, J., Kaur, R., 2020. Monitoring breathing frequency, pattern, and effort. Respiratory Care 65, 793–806.

Seppänen, T., Alho, O.-P., Vakkala, M., Alahuhta, S., Seppänen, T., 2016. Respiratory effort belts in postoperative respiratory monitoring: pilot study with different patients. In: Proceedings of the 9th International Joint Conference on Biomedical Engineering Systems and Technologies (BIOSTEC 2016), 5, pp. 76–82.

Shafiq, G., Veluvolu, K.C., 2017. Multimodal chest surface motion data for respiratory and cardiovascular monitoring applications. Scientific Data, Nature. Dataset. https://figshare.com/collections/Multimodal_chest_surface_motion_data_for_respiratory_and_cardiovascular_monitoring_applications/3258022. (Accessed 8 May 2021).

Subbe, C.P., Davies, R., Williams, E., Rutherford, P., Gemmell, L., 2003. Effect of introducing the modified early warning score on clinical outcomes, cardio-pulmonary arrests and intensive care utilization in acute medical admissions. Anaesthesia 58, 797–802.

Tobushi, T., et al., 2019. Blood oxygen, sleep disordered breathing, and respiratory instability in patients with chronic heart failure: prost subanalysis. Circulation Reports 1, 414–421.

Van Steenkiste, T., Groenendaal, W., Dreesen, P., et al., 2020. Portable detection of apnea and hypopnea events using bio-impedance of the chest and deep learning. IEEE Journal of Biomedical and Health Information 24 (9), 2589–2598.

Wong, J.W., Sharpe, M.B., Jaffray, D.A., et al., 1999. The use of active breathing control (ABC) to reduce margin for breathing motion. International Journal of Radiation and Oncology Biology and Physics 44 (4), 911–919.

Zhong, J., Li, Z., Takakuwa, M., et al., 2021. Smart Face Mask Based on an Ultrathin Pressure Sensor for Wireless Monitoring of Breath Conditions," Advanced Materials. Wiley.

Zubaydi, F., Sagahyroon, A., Aloul, F., Mir, H., Mahboub, B., 2020. Using mobiles to monitor respiratory diseases. Informatics, MDPI 7 (56).

Acronyms and explanations

AFE	analog front-end
BLE	Bluetooth Low Energy
DC/DC converter	converter that converts one value of DC voltage into another
FFT	fast Fourier transform
ISM	industrial, scientific, and medical
LDO	low dropout regulator
MCU	microcontroller unit
PCG	phonocardiogram
SCG	seismocardiogram
SOC	the state of charge of a battery
sps	samples per second

Variables used in this chapter include:

f_{clk}	clock frequency
I	current
I_{AFE}	current through the analog front-end
I_{LED}	LED current
I_{proc}	processing current
I_Q	quiescent current
P_{acq}	the power consumption of the sampling system and sensors
P_{comm}	power consumption required for communication
P_{prc}	power consumption of data processing
P_{sys}	power consumption of system-level processing
Q	capacity of the cell
V_{dd}	supply voltage
$z(t)$	state of charge of the battery

10.1 Introduction

The FDA defines remote and *wearable patient monitoring devices* as: "1. noninvasive remote monitoring devices that measure or detect common physiological parameters and, 2. noninvasive monitoring devices that wirelessly transmit patient information to their health care provider or other monitoring entity." The devices are remote because the measurement does not need to be done in a hospital or a clinic. In this chapter, we will use the term wearable devices for brevity instead of the term wearable patient monitoring devices. *Wearable devices* require direct contact with the patient to measure physiological variables. Please note that wearable patient monitoring devices are medical devices. This is different from *wearables* used as *wellness* or self-enhancement devices because these devices cannot be used for medical purposes and normally have no strict data integrity and accuracy requirements (Ravizza et al., 2019.

Fig. 10.1 shows a block diagram of a wearable device. It includes sensors/transducers, a microcontroller unit (MCU), a power and battery management unit, user interface blocks including speakers and touchscreen display, memory, a real-time clock module, and a communication unit. Some of the blocks will be explained in Section 10.4.2 when we discuss

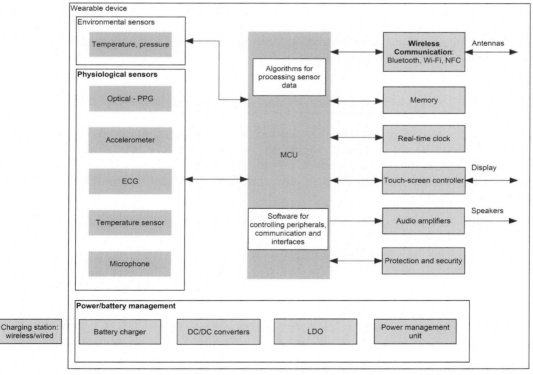

FIGURE 10.1 Block diagram of a modern wearable device. Blocks with labels in bold letters are discussed in more detail in this chapter.

power consumption. We will discuss blocks shown in Fig. 10.1 in bold letters: sensors, power management, and communication. Regarding the processing of signals, we can consider three cases.

- Processing can be done in real time on the device itself by the MCU,
- Only some processing can be done on the MCU (e.g., to determine if there is a motion to turn on the wearable device) while the raw or preprocessed sensor data is transferred to a computer or a smartphone, where it is further processed, or
- The overall processing is done outside of the wearable device.

Sensors and sensing modalities will be discussed in Section 10.2. In Section 10.2, we will also describe sensing modalities that were not covered in the rest of the book, including seismocardiography, phonocardiography, and tonometry. Several classifications of wearable devices will be presented in this chapter. A device can utilize multiple sensors in different ways, including combining the data using sensor fusion and using some sensors just for noise removal or assessing signal quality. Next, classifications are based on how wearables are charged and how they communicate. We also classify wearables based on their form factor. One of the most important features of wearable devices is battery life. Therefore, we will discuss several techniques for extending the battery life in Section 10.4.

Nowadays, wearable devices monitor vital signs and other physiological variables, the activity of subjects, and detect falls. However, the devices have the potential to do much more in collaboration with machine learning performed outside of the device, including recognizing the subject's daily activities and performing context-aware monitoring that is adjusted based on the recognized activity.

10.2 Sensors and sensing modality

This section will start with the classification of sensors/transducers used in wearable devices for cardiorespiratory monitoring. Next, we will introduce seismocardiogram (SCG), phonocardiogram (PCG), and arterial tonometry signals and related devices. These are some of the traditional methods that physicians have used for a long time by doing the following.

- Listening to the heartbeat on the chest wall (related to PCG),
- Detecting the pulse on the radial artery (related to arterial tonometry),
- Detecting the heartbeat vibration on the chest wall (related to SCG).

The major application of these signals that we will show is cuffless blood pressure monitoring. Please note that there are a number of other applications—but they will not be covered here.

10.2.1 Classification

Fig. 10.2 shows a potential classification of sensors. The physiological sources that generate the signal being measured are the heart and vascular system on one side and the lungs and the other organs of the respiratory system on the other side. They are causing electrical,

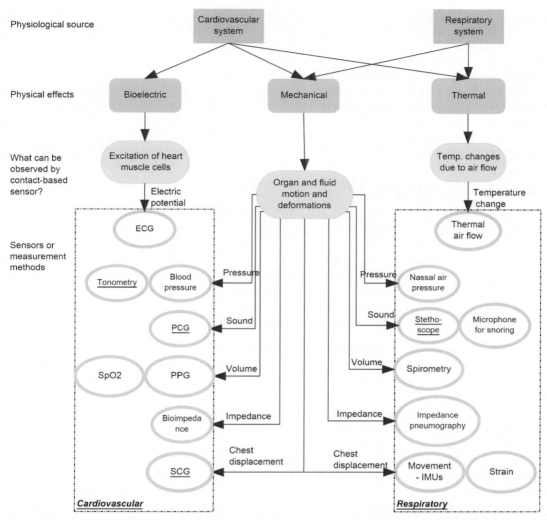

FIGURE 10.2   Physiological source and physical effects that can be measured using contact-based wearable sensors. *The figure is modified and extended from (Leonhardt et al., 2018).*

mechanical, thermal, magnetic, and chemical effects to be observed. Only the former three were considered here. We show sensing modalities related to cardiovascular monitoring on the left side of Fig. 10.2 and modalities related to monitoring breathing on the right side of the figure.

Bioelectric effects due to the excitation of heart muscle cells during the cardiac cycle can be observed through an ECG signal (Chapter 7). Mechanical effects include organ motions and fluid displacement. They can be observed by monitoring the displacement of the body surface and by measuring pressure, volume, sound, or impedance. Oscillometric methods are commonly used for measuring blood pressure. Measuring the arterial pulse waveform is done using pressure transducers. SpO_2 estimation and PPG signal are obtained by observing blood volume changes during superficial blood perfusion. They were described in Chapter 5. The cardiac signals obtained from the chest using microphones (PCG) or measuring chest displacement (SCG) are described later in this chapter.

Bioimpedance measurements were discussed in Chapter 9 when we introduced techniques for measuring breathing rates. Bioimpedance measurement can also be applied to monitor impedance changes due to blood flow. In addition, changes in the temperature in front of the nostrils are used for monitoring respiration rates. Other methods for acquiring breathing signals based on changes in chest volume and air temperature were described in Chapter 9.

10.2.2 Seismocardiogram

Accelerometers and their use for measuring the breathing signal were introduced in the previous chapter. It was explained there that the accelerometer detects the movement of the chest wall during breathing. We also showed the breathing signal obtained from the smartphone accelerometer. The SCG signal was shown in Chapter 8 when it was presented relative to ECG, PPG, and PCG signals. Here we will talk more about the SCG signal. Potential applications of SCG include cuffless blood pressure monitoring, heart rate estimation, heart rate variability estimation, monitoring left-ventricle function, coronary blood ventricular filling, cardiac valve closure, and others (Taebi et al., 2019).

10.2.2.1 Introduction to seismocardiography

Seismocardiograph is a device that measures cardiac-induced vibrations on the surface of the chest. The vibrations are normally measured using an accelerometer. These vibrations are caused by several factors, including cardiac muscle contraction, cardiac valve movement, and blood flow turbulence. Several studies analyze the temporal relationship between SCG and other commonly acquired physiological signals, such as ECG. Simultaneous recording of SCG and ECG indicated that SCG waves correspond to known physiological events, including mitral valve opening (MO), aortic valve opening (AO), the peak of rapid systolic ejection (RE), aortic valve closure (AC), and isovolumic movement (IM)—see Fig. 10.3. Please refer to the Wiggers diagram for more details about the events and timing of their occurrence during the cardiac cycle (Fig. 8.1). SCG waveform, together with several labeled waves, is shown in Fig. 10.3. The signal was filtered using a bandpass filter with the cutoff frequencies of 0.5 and 16 Hz. However, to obtain a better resolution, a much wider bandwidth (e.g., up to 90 Hz) should be used (Munck et al., 2020a). In addition, we also showed the ECG signal in Fig. 10.3, so one can observe the temporal relationship between SCG and ECG waves.

Changes in SCG morphology occur when changing posture, during and after exercising. The location of the SCG sensor on the chest is very important and affects the signal quality.

Besides measuring SCG at a single point, multichannel SCG is also possible. Data from multiple SCG sensors can be collected simultaneously and used to form a map of chest surface vibrations. In Munck et al. (2020a), chest surface vibration was analyzed using 16 three-axis accelerometers placed in a 4×4 grid on the chest. One of the study's goals was to analyze the instant of the aortic valve opening and closing over the chest area.

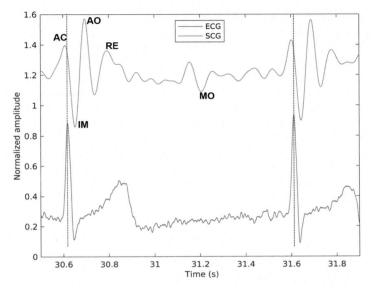

FIGURE 10.3   SCG and ECG waveforms. The data collection was described in the appendix of Chapter 8.

10.2.2.2 Databases and waveform segmentation

Data from Munck et al. (2020a) and the Matlab code for initial processing (Munck et al., 2020b) were made publicly available. Data was collected from 13 healthy male subjects for 10 min using 16 three-axis accelerometers, a breathing belt, and ECG at the sampling rate o0f 500 samples per second. Another database also contains respiratory, ECG, and accelerometer data. The data were collected from 20 healthy subjects for 1 h (García-González et al., 2013). The goals were to evaluate if the timing information obtained by processing the SCG signal could be used to replace RR intervals obtained from the ECG signal.

Common SCG signal processing techniques include bandpass filtering, motion artifact removal, and breathing signal removal. Segmentation of the SCG waveform is somewhat more complex than the segmentation of the ECG waveform due to the larger complexity of the SCG signal. The dataset of 66 subjects that includes ECG data together with a one-dimensional accelerometer SCG was explained in Wahlström et al. (2017). Dataset and the Matlab code for processing the SCG signals were made freely available. In Wahlström et al. (2017), the processing is based on modeling the SCG signal using a hidden Markov model and then estimating the heart rate and heart rate variability as well as the position of the waves on the SCG signal.

10.2.2.3 SCG measurements

As the seismocardiograph measures the vibrations caused by heartbeats, it is commonly measured by accelerometers and gyroscopes. Depending on the sensors used, SCG signals might consist of one or more axial and rotational components. For example, an accelerometer with a single axis can measure the SCG component in z-direction. On the other side, a triaxial accelerometer and triaxial gyroscope can provide information about motion due to heartbeats in three axial and rotational directions. The SCG sensors are most commonly placed on the sternum or left lower chest.

An example of an accelerometer used for SCG measurements is the ADXL335 MEMS accelerometer from Analog Devices. The ADXL335 is a triaxial accelerometer with a measurement range of ± 3.6 g and a sensitivity of 300 mV/g.

Smartphone accelerometers are commonly used for collecting SCG signals. SCG data collection using smartwatches is less popular. In Ganti et al. (2021), a custom watch was designed to collect PPG and SCG signals by placing the watch on the subject's chest. Then, the pulse transit time was computed between the AO wave on the SCG waveform and the foot of the PPG pulses. As a result, satisfactory accuracy of blood pressure estimation based on the pulse transit time was obtained for diastolic and mean blood pressure but not for systolic.

For reviews on sensors and data collection methods for SCG, please refer to Taebi et al. (2019) and Rahmani et al. (2021).

10.2.3 Arterial tonometry and pulse pressure waveform

The physical examination of the human arterial pulse has been performed for centuries. Tonometry means measuring pressure. Pressure is measured by placing a pressure transducer on the top of superficial arteries. To measure the pressure, the artery might need to be pressed and flattened.

10.2.3.1 Introduction to arterial tonometry

Arterial tonometry is a technique for obtaining an arterial pulse waveform. It is based on applying a pressure transducer on the skin on the top of the artery and then indirectly measuring the pressure inside the artery. The measurement gives relative values of the arterial pressure at the measurement site. Therefore, in arterial tonometry, we often work on analyzing pulse morphology as the absolute pressure values in the artery are unavailable. The waveform obtained in this way is called *pulse pressure* or *arterial waveform*. The most common measurement sites include the skin on the top of radial, femoral, and carotid arteries. *Arterial applanation tonometry* is based on collecting the pulse pressure waveform in a situation where the pressure is applied to the artery so that the artery is flattened. Applanation means flattening the artery.

The pulse wave velocity can be measured by performing the arterial applanation tonometry at two sites (femoral and carotid, for example) and then calculating the pulse transit time. The pulse wave velocity is then used for estimating arterial stiffness. Applanation tonometry normally requires a technician/nurse to gently press the artery just enough to obtain the arterial pulse. This procedure is not normally done at home and, therefore, will not be considered in this chapter. Arterial tonometry devices for arterial stiffness evaluation are not considered wearable devices.

Estimating blood pressure noninvasively can be done by using the pulse pressure waveform. Other applications of arterial tonometry include arterial waveform analysis, detection of arrhythmias, and heart rate monitoring. Arterial waveform analysis can be used to estimate relevant indices for assessing arterial stiffness. In addition, by using an appropriate model or a transfer function, it is possible to estimate the central blood pressure from the arterial waveform.

10.2.3.2 Transducers and measurements

Fig. 10.4A shows an example of a system in which pressure is applied to the radial artery. Then, pressure variations in the artery are measured using a transducer and front-end electronics. Commonly, the transducer and the electronics are embedded in a hand-held enclosure. The transducer is placed at the top of the enclosure where the enclosure is touching the skin. Then, the whole device is used to press the skin. Newer designs are in the form of wristbands. Fig. 10.4B shows the flattening of the artery after applying pressure on the skin.

Piezoelectric transducers are commonly used. In Wang and Lin (2020), a piezoelectric transducer was attached to an elastic wrist strap. The sensitivity of the transducer was 2 mV/mmHg. No additional external pressure (except the pressure from a relatively tight elastic band) was applied to collect the signal. It was recommended in Ciaccio and Drzewiecki (2008) to use a transducer with a flat surface at the interface with the skin. In addition, the material in which the transducer is enclosed should be much stiffer than the skin to be able to press the skin and the artery below the skin.

A major issue in arterial tonometry includes not placing the transducer directly over the artery. In addition, the strong dependence on the position of the sensor over the top of the artery makes arterial tonometry devices very sensitive to motion artifacts. With traditional arterial tonometry devices, the subjects were supposed to stay still during the measurements. Due to the dependence on the position of the transducer over the artery, devices often contain several transducers. Another reason for having multiple transducers is noise cancellation. The design with two piezoelectric transducers was proposed in Ciaccio and Drzewiecki (2008). One transducer is placed on the top of the artery to collect the arterial pulse, while the other is placed further away from the artery. Therefore, the signal collected by the second transducer is not affected by arterial pulsations but by all other sources of noise and interference. Therefore, it is used for noise suppression in the signal collected by the first transducer.

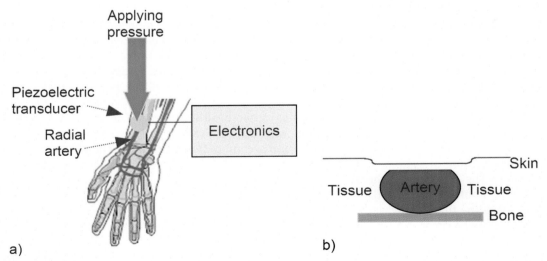

a)

b)

FIGURE 10.4 (A) Arterial tonometry is based on applying pressure on the skin over the artery and then measuring the pressure using a transducer. (B) Flattening the artery after applying pressure on the skin.

10.2.3.3 Cuffless blood pressure estimation

Estimating blood pressure based on arteria tonometry was described in Wang and Lin (2020). Initial calibration using an oscillometric device was required. A piezoelectric sensor was used because it is sensitive and provides a voltage at the output proportional to the applied pressure. At the time of the calibration, the trough of the pulse represents the diastolic pressure and the peak represents the systolic pressure. For the next pulses, the difference in the peak values of the pulses and the peak value of the initial pulse was used to update the systolic pressure obtained for the initial pulse. The interval of only 30 min was used to perform measurements and evaluate the accuracy of blood pressure estimation. This method is problematic because the error accumulates over time; therefore, frequent recalibrations or repeated parameter estimations are required. However, the device can be considered wearable because it includes a wristband with a piezoelectric transducer and electronics. Another solution for cuffless blood pressure estimation was based on simultaneous measurement of PPG at a finger and pulse pressure waveform over the top of the radial artery. Three strain sensors were used and placed in a way to form a triangle. The elastic strap was placed over the wrist so that the enclosure with transducers was pressed by the strap (Wang et al., 2018).

10.2.4 Phonocardiogram

Auscultation is a key screening exam that represents listening to cardiac and respiratory sounds using a stethoscope. These sounds can be recorded and stored as the phonocardiogram waveform. It was pointed out in Fang and T O'Gara (2019) that "accurate auscultation provides important insight into many valvular and congenital heart lesions." We will briefly discuss what PCG is, methods of PCG measurements, databases for PCG, PCG segmentation, and cuffless blood pressure measurement using PCG.

10.2.4.1 Introduction to phonocardiogram

PCG is the record of audible vibrations of the heart and blood circulation. The PCG signal is obtained by a transducer that converts the sound signal into an electrical signal, commonly done by a microphone placed on the chest surface. The frequency range of the PCG signal is between 5 and 600 Hz. Moreover, 120 Hz and above is considered a high-frequency sound, 80–120 Hz is considered a middle-frequency sound, and below 80 Hz is considered a low-frequency sound.

Fig. 10.5 on the top shows the PCG signals of two healthy subjects. The signal on the right was obtained with a much higher sampling rate. In each cardiac cycle, two main heart sounds are expected to occur, namely the first (S1) and second (S2) heart sounds. The S1 corresponds to the beginning of systole, while S2 corresponds to the end of systole and the beginning of diastole. The duration of the systole (S1–S2) is shorter than the duration of the diastole (S2–S1). Looking at Fig. 10.5 on the top left, eight heartbeats are recorded, and in each beat, S1 is the first sound, and S2 follows it. In addition, the third (S3) and fourth (S4) heart sounds may be present. S3 is present 120–180 ms after S2, as shown in the bottom left plot of Fig. 10.5. S4 is a small peak presented just before S1, as shown in the bottom right plot of Fig. 10.5. In addition to these sounds, the signal might contain murmur generated by flow turbulence often associated with pathological conditions. PCG signal collected simultaneously with PPG, ECG, and SGC signal is shown in Fig. 8.1 and Fig. 8.3.

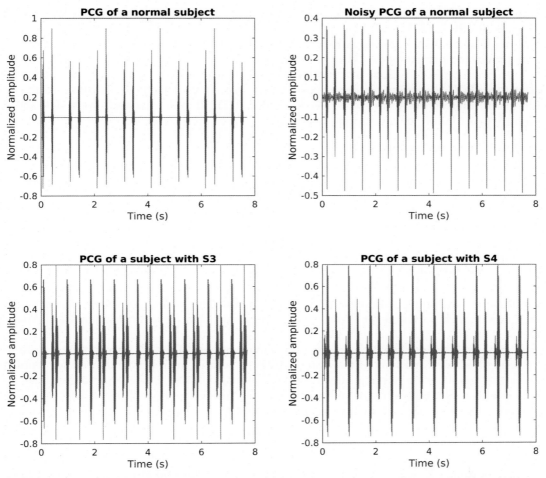

FIGURE 10.5 PCG signals from four different subjects. At the top, waveforms from two healthy subjects are shown. At the bottom, subjects with the S3 sound (left) and S4 sound (right) are shown. The waveform on the top right is from Bentley et al. (2011), while the other waveforms are from (Pinsky and Wipf).

10.2.4.2 Databases and waveform segmentation

Several public *datasets* are available, including PASCAL (Bentley et al., 2011) and the PhysioNet database for the classification of heart sound recordings (Liu et al., 2016). These datasets have been used for working on heart sound segmentation and classification challenges. PASCAL database allows for classifying the signals as normal, murmur, extra heart sound, artifacts, and others. The largest pediatric dataset on cardiac murmur classification is presented in Oliveira (2021). More than 5000 recordings have been collected from the four auscultation locations of 1568 patients. A comparison with other public databases of PCG signals was provided in Oliveira (2021).

Segmentation means detecting the events of interest in the signal, including S1 and S2. Segmentation steps are very similar to the segmentation steps of the ECG signal. They include preprocessing, time-frequency transformation and squaring of the signal, envelope detection,

detection of the peaks and then merging close peaks, as well as rejecting peaks that are not physiologically feasible. In Liu et al. (2016), Matlab code for segmentation and classification was made available. In addition, the segmentation method based on hidden Markov models was shown in Springer et al. (2016), and the Matlab code is also available on PhysioNet.

Significant work was done in developing machine learning and deep learning models for classifying PCG waveforms either as normal or abnormal or detecting particular health conditions of interest. For example, in Tseng et al. (2012), S3 and S4 sounds were detected to identify heart failure or myocardial ischemia.

10.2.4.3 PCG measurements

Since PCG is the record of audible vibrations of the heart and blood circulation, a sensor that converts audio into an electrical signal needs to be used. PCG measurements can be done by placing an electronic stethoscope or a microphone on the chest.

10.2.4.3.1 Digital stethoscopes

Electronic or *digital stethoscopes* convert the sound waves obtained through the chest piece into an electrical signal. After amplification, filtering, and noise cancellation, the signal is played on the headphones (Pariaszewskaa et al., 2013). The digitized signal can also be recoded and transferred to a computer for further analysis. Several companies develop digital stethoscopes, including 3M Littman, Welch-Allyn, and Think-labs. Also, several chip manufacturers provide solutions for digital stethoscopes, including Texas Instruments and Analog Devices.

Digital stethoscopes can include one or two microphones. The first microphone is used for detecting audible variations of the heart, while the second microphone is used for ambient noise that can later be removed from the PCG signal by noise cancellation algorithms. An example of a microphone used commonly is an electret microphone. This widely used inexpensive capacitive microphone can keep a constant charge because of the electret material. When the plate used in the microphones moves, its capacitance changes. These microphones can provide a high-frequency range (for example, from 10 Hz to 15 kHz) and have high-enough sensitivity for collecting the PCG signal—for example, 6 mV/Pa. Microphone sensitivity is normally given in mV per sound pressure where the sound pressure is given in Pa. MEMS microphones based on a diaphragm and capacitive sensing also include electronics in the chip, and as such, are becoming more popular in smartphones and smartwatches. The output of the MEMS microphone can be analog or digital, depending on whether the A/D converter is a part of the MEMS chip.

Digital stethoscopes contain an audio amplifier or precision operational amplifier to amplify the signal at the output of the microphone. These amplifiers have low noise and distortion, a high slew rate, and relatively high open loop gain. An example of an audio op-amp is the OPA161x1 audio operational amplifier by Texas Instruments. The amplified signal is filtered using an antialiasing lowpass filter (cutoff frequency of 800 Hz) and converted into a digital domain. The A/D converter's sampling frequencies are from 8000 samples per second to 44,100 samples per second, while the resolution of the A/D converter varies, starting with 16 bits. After processing and filtering in the digital domain, the processed digital signal is converted back to the analog domain by an audio D/A converter. A single speaker for both earpieces or two speakers, one for each earpiece, could be used.

Audio CODEC is a device or a chip that encodes and decodes the audio signal. It integrates several filters, amplifiers, and other components. Audio CODEC commonly provides stereo D/A and A/D converters, standard audio processing such as notch filtering and deemphasis, high-power amplifiers at the output for headphones, and so on. As such, the audio CODEC chip is a very attractive solution for digital stethoscopes. The signal can further be processed by an external processor/microcontroller. An example of an audio codec device is TLV320AIC3106 stereo audio CODEC by Texas Instruments.

In addition to collecting and processing the PCG signal, digital stethoscopes can be integrated with other sensors. An example is the New DUO by Eko (https://www.ekohealth.com/), which provides ECG and PCG signals.

10.2.4.3.2 PCG measurement using audio devices

Devices that collect audio signals from the chest can be specialized devices or already existing systems, such as smartphones. In Thoms et al. (2019), a low-cost device for capturing heart sounds was developed that consists of a bell, a tube, and a small microphone connected to a smartphone for processing. In Mondal et al. (2018), modified Bluetooth headphones that contain a microphone were used. A small custom rubber bell was attached to the microphone to improve the signal quality. The signal is transferred by using Bluetooth communication to the smartphone. In Fig. 8.3, we showed the PCG waveform collected using an inexpensive off-the-shelf microphone placed on the chest. The microphone was directly connected to the National Instruments data acquisition board.

10.2.4.4 Cuffless blood pressure estimation using PCG

There have been many attempts to use the PCG signal for cuffless blood pressure measurements. They differ based on features extracted from the PCG signal, as well as based on other signals used together with the PCG signal. We will mention several works, including the following:

- Only a PCG signal was used for estimating blood pressure (Omari and Bereksi-Reguig, 2019). It was found that S2—S1 duration correlates well with the pulse arrival time (PAT). Therefore, the S2—S1 interval was used as an input to a neural network together with several other features such as height, weight, heart rate, sex, and S1—S2 interval. Less accurate blood pressure estimation was obtained than in the case PAT was used to estimate blood pressure. However, the approach is promising because blood pressure was estimated using only a single sensor.
- PCG and PPG signals were collected, and PTT was computed between the S1 peak of PCG and different fiducial points on the PPG pulse during the same cardiac cycle (Esmaili et al., 2017). The results were compared against the blood pressure estimated based on PAT computed between ECG R peak and PPG fiducial points. It was found on a very small number of subjects that PAT is more accurate for systolic blood pressure estimation, while PTT was more accurate for diastolic blood pressure estimation. The fitting was done for each subject separately.

10.3 Classifications of wearable devices

Wearable devices can be classified in a number of ways. Here, we discuss classifications based on the number of sensors used in the device, frequency of measurements, storing and communication, form factor, and power consumption.

10.3.1 Multimodal monitoring devices

With a significant reduction in the price of hardware and with many low-power components available, it is possible to design devices that include multiple transducers/sensors. These devices are referred to as *multisensor* or *multimodal monitoring devices*. King et al. (2017) presented a high-level overview of sensor fusion for wearables. Multimodal monitoring devices can be classified in multiple ways—our classification is shown in Fig. 10.6.

Multiple sensors in wearable devices allow for monitoring several variables to:

- Using multiple sensors in applications where multiple metrics or indices are required to make clinical decisions. Examples include sleep apnea monitoring (breathing, SpO$_2$, and others) and syncope detection.
- Providing redundancy in the measurement of a single parameter to improve sensitivity, reduce motion artifacts, or provide signal quality assessment
- Allowing for understanding the context in which the measurement is taken. For example, an accelerometer can detect the subject's movement, which can be used to infer the actions and posture of the subject. Likewise, humidity and temperature sensors can detect air humidity and temperature and provide information about environmental conditions. Algorithms for processing physiological variables can therefore be adjusted based on this contextual information.

Sensors can be *homogeneous* and *heterogeneous*. For example, beat-to-beat continuous blood pressure estimation based on the time difference between fiducial points of two signals can be done with the signals of the same type (e.g., PPG) at two *different measurement points* or with

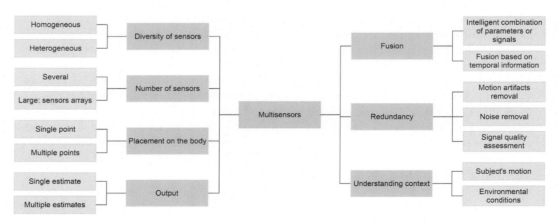

FIGURE 10.6   Classifications of multimodal monitoring wearable devices.

signals of different types at the *same measurement point* (e.g., SCG and PPG in a smartwatch). Normally, it is simpler for a patient to have a device that requires contact only at a single point. However, often this is not possible, and devices require multiple contacts with the body. An example of a single-point device is a wrist-worn wearable device.

The number of sensors in multimodal monitoring devices can be from several to several hundred. A large number of sensors can be integrated into a sensor array. An example of the sensor array is a multichannel SCG system we have already mentioned (Munck et al., 2020a). However, much larger sensor arrays exist, either attached directly to the body or as part of smart clothes.

Sensor fusion means the derived estimate or quantity of interest is obtained by processing signals from multiple homogeneous or heterogeneous sensors measuring the same physiological event. For example, in the paper on SCG signal segmentation using HMM (Wahlström et al., 2017) mentioned in Section 10.2.2.2, signals from a three-axis accelerometer and a three-axis gyroscope are combined to segment the SCG signal and estimate the heart rate. Multiple sensors are normally used for the continuous measurement of blood pressure. Sensor fusion is done for one or more of the following reasons:

- Reducing the noise by combining multiple signals through, for example, weighted sum.
- Utilizing temporal information among signals. This approach was presented in Chapter 8, where we showed how PTT could be used for estimating blood pressure.
- Combining features extracted from different signals. For example, cuffless blood pressure estimation algorithms include some indices extracted from PPG and ECG pulses in addition to pulse arrival time.
- Separating a signal from a mixture of signals.

The outcome of the fusion algorithm can be:

- A single derived index obtained by combining results from several sensors, such as blood pressure estimate based on PAT obtained using ECG and PPG sensors
- Multiple derived (estimated) indices. For example, the number of sensors used to monitor sleep apnea is large. The derived indices include apnea and hypopnea indices, snore percentage, and estimated sleep efficiency (parameters that have a value between 0% and 100%, where 100% means that the subject slept for the whole night), oxygen desaturation index (the number of times the SpO_2 level drops below some predefined level per hour of sleep), and so on.

Redundant sensors are sensors added to the device to support major sensing modalities. A survey on redundant sensor arrangements in the automotive environment is given in Leonhardt et al. (2018). Several reasons for using redundant sensors mentioned in Leonhardt et al. (2018) are also applicable to wearable devices, including:

- Motion artifact detection: for example, an accelerometer is often used to detect motion artifacts, and then signal segments when these motion artifacts are detected can be excluded from further processing.
- Motion artifact correction: for example, the contribution of the motion artifact to the signal of interest can be potentially removed if one or more sensors are used only for acquiring the noise or interference signal.

- Signal enhancement: by utilizing multiple sensors measuring the same physiological event, one can make sure that the event is detected (e.g., to detect the pulse pressure waveform in arterial tonometry) and that the higher quality signal can be obtained by combining their multiple signals.

10.3.2 Frequency of measurements, storing, and communication

The classification discussed in this section is shown in Fig. 10.7. Measurements can be done at regular intervals continuously or performed intermittently. Intermittent measurements can be done anytime or when some conditions are fulfilled. For example, a subject can measure blood pressure using an oscillometric device anytime. When using an ECG patch as an event recorder, the subject can initiate measurement when he/she feels arrhythmias.

Measurements, storing, and communication might all be initiated simultaneously or separately. In the examples of oscillometric blood pressure and manual arrhythmia detection using a patch, the subject initiated measurements, storing, and communication. However, in an ECG external loop recorder (see Section 7.4.1) in which the ECG signal is processed for arrhythmias, the ECG data might not be stored and transferred unless irregular beats are detected. In this case, the measurement is continuous, but storing and transferring data depend on the current events.

10.3.3 Form factor

The analysis of design factors of wearable devices that concern patients and physicians the most was presented in Bergmann and McGregor (2011). It was concluded that the following factors are important for patients: the device is not supposed to be stigmatizing and, therefore standardized look like a smartwatch is preferable; the device is supposed to be compact and has minimal risk of detachment from the patient (e.g., loose electrode connections); it should not affect daily behavior, and it should reduce the travel to the hospital/clinic. Physicians were concerned with the accuracy, compliance of the patients, techniques for attaching the device, and whether the device was attached and used properly.

As the look and the size of the devices are very important for patients, a significant amount of work was done on reducing the size of the devices and optimizing the form factor. *The form factor* refers to the size, shape, and physical specifications of the device.

The form factor depends on where the device is placed. Based on the location where the wearable device is placed on the body, we can divide the wearable devices into those placed

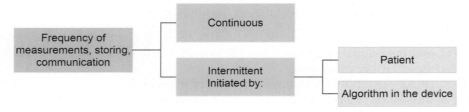

FIGURE 10.7 Classifications of measurement, storing, and communication frequency techniques.

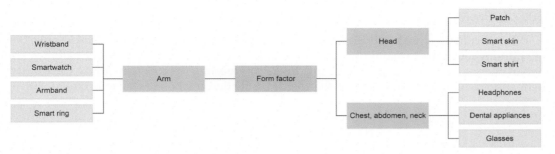

FIGURE 10.8 Classifications of different form factors of wearable devices based on the location on the body where the wearable device is placed.

on the arm, head, and the rest of the body. The form factor of the wearable devices is shown in Fig. 10.8. We will show examples of some devices with different form factors.

We will mention several interesting solutions, including smart skin, eyeglasses, headphones, and smart rings.

Smart skin devices have a goal to "noninvasively measure all aspects of human activities in daily life under conditions as natural as possible, which would enable e-skins and human skin to interactively reinforce each other and maximize their potential in several different realms, including research, wellness and fitness, and the clinical setting" (Someya and Amagai, 2019). The development of smart skin devices has been enabled because of a significant reduction in size and power consumption of electronics, the appearance of flexible electronic components, and the development of new skin-compatible materials. Almost all signals we covered in the book based on electric, mechanical, and thermal effects can be measured using smart skin devices. In addition, chemical sensors were developed to measure levels of lactic acid, potassium and sodium ions, and chloride ions (Someya and Amagai, 2019). In Sempionatto et al. (2021), a smart skin sensor was developed to measure blood pressure from the neck based on ultrasound transducers and measure lactate, alcohol, and caffeine levels. Of course, calibration is required to obtain absolute values of blood pressure.

Eyeglasses can include transducers attached to the glasses' temples or the nose pad. In Chapter 9, we mentioned a MEMS sensor attached to the nose pad that can measure the pulse wave and respiration (Nguyen and Ichiki, 2019).

Headphones can be used to measure temperature, PPG, and SpO_2 signals. For the review on *hearables*, please refer to Masè et al. (2020). It was shown in Davies et al. (2020) that SpO_2, based on a commercial PPG sensor consisting of green, red, and infrared LEDs placed in the ear canal, can be measured with low delay and high robustness to temperature changes.

The smart ring is one of the smallest form factors. There have already been several commercial ring wearable devices on the market, including Oura Ring (https://ouraring.com/) and Motiv Ring (https://www.mymotiv.com/). Oura ring includes a three-axis accelerometer, temperature, and PPG sensors and has Bluetooth connectivity. In Altini and Kinnunen (2021), the Oura ring was evaluated for sleep studies. Data were collected from 106 subjects, and machine-learning models were developed for wake—sleep detection and sleep staging.

10.3.4 Power

Wearable devices use batteries where the batteries can be rechargeable or disposable. Single lithium cells are normally utilized. Some devices can also operate without batteries and obtain energy through energy harvesting. Energy harvesting devices may utilize electromagnetic, photovoltaic, piezoelectric, or thermal transduction methods (Di Paolo Emilio, 2017).

10.4 Designing for low power

Power consumption can be reduced by applying several methods, including selecting low-power components or components with power-down modes, switching off the power or clock to the specific hardware components, applying algorithmic modifications to reduce computations, and reducing communication overhead.

Let us first understand the sources of power consumption. Power consumption depends on the application, selection of hardware components, frequency and voltage applied, and software implementation that includes implementation of the algorithms as well as power-down strategies (Oshana and Kraeling, 2013). Regarding the application, let us consider oscillometric blood pressure and beat-to-beat blood pressure monitoring devices. An oscillometric blood pressure device performs intermittent measurements. Normally, the device does not need to be small and can use batteries of larger capacity. A beat-to-beat blood pressure device measures continuously; therefore, the emphasis should be placed on reducing the algorithms' complexity and putting the device into low-power modes as often as possible.

As pointed out in Chapter 4, Section 4.7, one should reduce the supply voltage and clock frequency to lower the dynamic power. The clock frequency and the supply voltage can be reduced together with significant power savings. Therefore, modern processors have several power modes in which some parts of the chip are shut off while others work at lower frequencies and supply voltages. For example, Table 10.1 shows three operating modes for a processor TMS320C5504 by Texas Instruments. All three modes will allow for the normal operation of the processor. However, there is a trade-off between speed and power consumption—higher speed requires a larger clock frequency and higher supply voltage resulting in larger power consumption. Also, components in wearable devices operate at different voltage levels, which is why a *power management unit* is commonly used to provide the required supply voltage for different components.

TABLE 10.1　TMS320C5504 operating conditions.

Power mode	CPU operating frequency f_{clk}	Supply voltage V_{dd}
1	60 MHz	1.05 V
2	75 MHz	1.1 V
3	100 MHz	1.3 V

TABLE 10.2 Power consumption and computational times in three modes for processor TMS320C5504.

Power mode	Computational time (s)	Power consumption (mW)
1	0.83	3.9
2	0.67	5.4
3	0.5	10

Example 10.1. Let us consider the power consumption of a device running a program on a digital signal processor TMS320C5504 by Texas Instruments. The CPU operating conditions are shown in Table 10.1. Let us assume that the processor can complete its task in 0.5 s and can sleep for 0.5 s while operating in Mode 3. Also, the power consumption while operating in mode 3 is 10 mW. We ignore the power consumption in sleep mode.

Can the processor complete the task in less than 1 s if it operates in modes 1 or 2, and what is the power consumption in these cases?

Solution. In Chapter 4, we defined the dynamic power as $P = A \cdot C \cdot V_{dd}^2 \cdot f_{clk}$, where A is the switching factor which is the probability of switching from 0 to 1 or from 1 to 0 in a given clock cycle, C is the capacitive load, f_{clk} is the clock frequency and V_{dd} is the supply voltage. The total time the program runs in mode 1 (assuming nothing changes when switching between modes except the clock frequency and supply voltage) is 0.5 s $\cdot$ 100 MHz/60 MHz = 0.83 s. From the formula for dynamic power consumption in mode 3, we can find the product $A \cdot C$ and use it to compute the power consumption in modes 1 and 2 . The results are shown in Table 10.2. The processor will be able to complete tasks in all modes. Power consumption is reduced almost twice in mode 2 and even more in mode 1. Therefore, it is important to switch to low-power modes if real-time requirements can be met at lower frequencies.

10.4.1 Strategies to improve the energy efficiency

The strategies to improve energy efficiency include (Qaim et al., 2020):

- Task offloading
- Duty cycling
- Energy-aware computations and communications

Task offloading means that computationally intensive tasks are not done on the device itself. Normally there is a trade-off between power consumption due to communication and computations. If the algorithm's complexity is high and the number of data points is low, it makes sense to transfer the data and perform computations on the cloud or a personal computer.

Duty cycling conserves energy by powering off or putting hardware components that are not in use to sleep. Nowadays, CPUs and other components have multiple power modes allowing for the trade-off between performance and power consumption. For example, in idle mode, the processor cores are shut down, but some peripherals and the interrupt controller unit remain powered. Therefore, when the interrupt from a peripheral comes, the processor can be turned on. The goal is to place the processor in idle mode as often as possible to reduce power consumption.

There are several hardware techniques to reduce the power consumption of the components, including:

- *Power gating* switches off the current to specific components inside the chip or peripherals of the device,
- *Clock gating* shuts down the clock of specific components inside the chip or specific peripherals,
- *Clock and voltage scaling* means that if the clock frequency is reduced, the supply voltage can also be reduced and vice versa.

Another way to increase the duration of the idle modes is to reduce the number of events the processor needs to handle. For example, instead of sending every data sample over the network as it arrives, the data can be stored in a buffer, and then the content of the whole buffer can be sent over the network. This way, the number of events that the processor handles can be significantly reduced.

Energy-aware computation includes the design of algorithms that are energy efficient or the selection of algorithms/methods that are not computationally demanding. That means we sometimes implement algorithms that are not the most accurate but have good enough performance and low computational complexity. Having low computational complexity algorithms is important to be able to complete tasks on time and still have time to put the CPU to sleep to reduce power consumption. Energy-aware communication includes choosing a low-power communication protocol, reducing the number of messages being communicated, and reducing the transmission power. In addition, data compression is sometimes used to reduce the amount of data being transferred (Qaim et al., 2020).

10.4.2 Low-power hardware design

Integrated solutions put together multiple components on a single chip. These integrated solutions include transducers, conditioning circuitry, filters, A/D converters, and interface circuits for connecting the chip with the microcontroller and other components. They usually have several power-down modes. We will discuss the power-related design of several components, including a PPG chip, Bluetooth communication unit, a microcontroller, and voltage regulators. In addition, we will also consider batteries and the power management unit. We analyze power consumption in realistic scenarios.

10.4.2.1 Power consumption of a PPG sensor module

This section discusses the power consumption of a PPG-integrated sensor module ADP-D144RI by Analog devices. This chip includes red and infrared (IR) LEDs, photodiodes, an analog front-end (AFE) with a transimpedance and other amplifiers, an ambient light rejection circuit, as well as a 14-bit A/D converter. The output of the A/D converter is time multiplexed, so there are two outputs that correspond to red and IR LEDs. It also includes a timing circuit and LED drivers. LEDs have their own supply voltage V_{LED}, while the rest of the

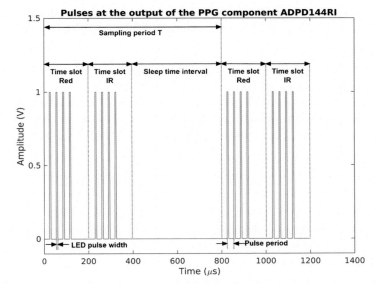

FIGURE 10.9 The timing diagram for turning on red and infrared LEDs in the PPG chip ADPD144RI by Analog devices.

circuit uses the voltage supply V_{DD}. The output signal is digital, and the communication with a microcontroller is done using a serial bus called I^2C.

To reduce the power consumption of the PPG sensor module, the LEDs are turned on during short intervals. The timing diagram is shown in Fig. 10.9. We can see that during one sampling period, the red LED is on, and then the IR LED is on, followed by the standby or sleep period. Each LED can be turned on several times in a sequence. The number of times the LED is turned on is N_{pulse}. The reason for turning on the LED more than once during the sampling period is that one can average the amplitude over N_{pulse} pulses to obtain less noisy PPG signal. However, turning on the LEDs more times during one sampling period increases the module's power consumption. The period during which one LED is turned on and off N_{pulse} times is referred to as the time slot for that LED T_{slot}. One discrete PPG sample is obtained in each time slot, and the combined time slots and the sleep time interval are considered a sampling period T_s. Sampling rate f_s is $f_s = 1/T_s$. The time during which one LED is on is referred to as the LED pulse width $T_{LEDwidth}$, while the period of turning LEDs on and off is $T_{PusePeriod}$—see Fig. 10.9.

Power consumption can be computed as the power consumed due to the operation of LEDs and the operation of the rest of the circuit. Therefore, the total power is

$$P_{ppg} = P_{circuit} + 2P_{LED} = V_{dd}I_{dd\_ave} + V_{LED} \cdot 2I_{LED\_ave}$$

Next, the power consumption of the LED is computed as the average power during one sampling period. It depends on the number of times the LED is turned on during the sampling period, the ratio of the LED pulse width and the sampling period, and the maximum LED current. The power consumption, while the LEDs are off, is zero.

$$I_{LED\_ave} = N_{pulse}I_{LED}T_{LEDwidth}/T_s$$

The power consumption of turning on the analog front-end and the rest of the circuit is proportional to the amount of time T_{slot} during which the LED is on. The current of the analog front-end (AFE) is I_{AFE} and the current for other processing is I_{proc}. During the time both LEDs are off, the total current is $I_{standby}$. Therefore, the average current for the analog front-end and processing components is

$$I_{dd\_ave} = \left(2T_{slot}\left(I_{AFE} + I_{proc}\right) + (T_s - 2T_{slot})I_{standby}\right)/T_s$$

Example 10.2. For the operating parameters of the ADPD144RI module given in Table 10.3, compute the following: (1) output power versus N_{pulse}, (2) output power versus the sampling rate. Comment on your results.

Solution. In Fig. 10.10, we showed the power consumption of the PPG chip for two cases.

(a) In the first case, the sampling rate is kept at 100 samples per second (sps), and the number of pulses N_{pulse} per one time slot is varied from 1 to 10. We can see an increase in power consumption by a factor of 2.7.
(b) In the second case, the sampling rate is changed from 50 sps to 500 sps while N_{pulse} is kept at 10. We can see a ten-fold increase in consumed power with a 10-fold increase in the sampling rate.

This example shows that setting the parameters of the sensor module can significantly affect power consumption. In addition, there is always a trade-off between accuracy (averaging more pulses during one time slot to reduce the noise in the PPG signal) and power consumption (the duration of the period the LEDs are on).

TABLE 10.3 Operating parameters of the PPG sensor module ADPD144RI by analog devices. Please note that we did not exactly follow the datasheet.

Timing parameters	Values of the timing parameters	Parameters related to current	Values of parameters related to current
T_s	10 ms	$I_{standby}$	3.5 µA
T_{slot} for $N_{pulses} = 10$	0.22 ms	I_{proc}	1.5 mA
$T_{LEDwidth}$	3 µs	I_{AFE}	9.3 mA
$T_{PusePeriod}$	19 µs	I_{LED}	100 mA

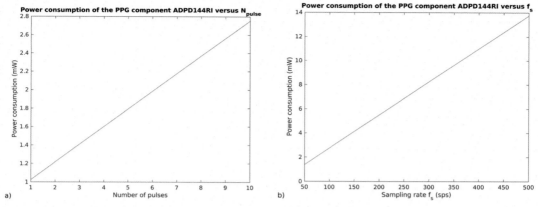

FIGURE 10.10 Power consumption of the PPG chip versus (A) the number of pulses per slot while the sampling rate is 100 sps, and (B) the sampling rate while keeping the number of pulses at 10.

10.4.2.2 Communication unit

Bluetooth Low Energy (BLE) is a common communication protocol in today's wearable devices. Bluetooth is a wireless technology standard for exchanging data over short distances in the industrial, scientific, and medical (ISM) frequency band from 2.4 to 2.48 GHz. BLE is used for sensor data, low-power, and low-bandwidth applications. BLE devices operate in one of three states: advertising, scanning, and connections. In the advertising state, a device sends packets containing useful data for others to receive and process. In the scanning state, the device is listening for the packets. Finally, in the connection state, two devices can exchange packets. To establish a connection, one device needs to start advertising while the other needs to scan for the advertisements.

The two main factors affecting power consumption in a BLE device and the transmit power and the total time the radio is active, including both transmit (TX) and receive (RX) components. The required transmit power depends on the range that needs to be reached between the two BLE devices. The range is affected by the environment and potential obstacles between the devices. The duration of time that a BLE radio is active depends on how much data needs to be transferred and how often.

The average radio current and the transmit power for the three BLE states are shown in Table 10.4. The values are obtained from the BLE Power profile calculator tool developed by NXP Semiconductors.

10.4.2.3 Microcontrollers

A microcontroller (MCU) is a device that contains a processor and several peripherals integrated into the same chip. Power-down modes have different names for different MCUs. Table 10.5 shows some power-down modes used by NXP Semiconductors microcontrollers. Depending on the requirements of the user application, a variety of modes are available that provide state retention and partial or full power down of certain logic and/or memory. The three primary modes of operation are run, wait, and stop. Table 10.5 also shows the current

TABLE 10.4   Power consumption summary based on BLE Power profile calculator by NXP Semiconductors.

BLE radio mode	Average radio current at 3.6 V at 100 ms advertising, scanning, connection intervals	Average power at 3.6 V at 100 ms advertising, scanning, connection intervals
Advertising	0.19 mA	0.68 mW
Connection at 32 bytes per packet	0.15 mA	0.54 mW
Connection at 251 bytes per packet	0.32 mA	1.15 mW
Scanning	1 mA	3.76 mW

TABLE 10.5   Partial list of power down modes of the microcontroller MKW41Z256 of Kinetis W series by NXP Semiconductor microcontrollers.

Power mode	Description	Microcontroller recovery method	Radio and other peripherals	Example current with peripherals turned off
Normal run	Allows maximum performance of the chip.	–	Peripherals and radios can be turned on if needed	6.81 mA for $f_{clk} = 48$ MHz
Wait (sleep)	Allows peripherals to function while the CPU can go to sleep	Interrupt	Peripherals and radios can be turned on if needed	3.1 mA for $f_{clk} = 21$ MHz
Stop (deep sleep)	Lowest power mode that retains all registers	Interrupt	Interfaces and radios are off	204 µA for $f_{clk} = 0$ Hz
Very low power stop	Lowest power mode with A/D converter and all pin interrupts functional.	Wakeup interrupt	Only several peripherals can still be turned on	4.3 µA for $f_{clk} = 0$ Hz
Very low leakage stop	SRAM is shut down	Wakeup interrupt	Only several peripherals can still be turned on	0.56 µA for $f_{clk} = 0$ Hz

for the case when all peripherals are off and for different system clock frequencies f_{clk} corresponding to particular power modes.

Example 10.3.   Compute the average current of the microcontroller MKW41Z256 of Kinetis W series by NXP Semiconductor in two cases:

1. The processor is in deep sleep mode. It wakes up every 1 s in the normal run mode for 10 ms to communicate, collect information from sensors as well as transfer data. The current of the used peripherals is real-time clock (RTC) 1.3 µA, counter 6 µA, I^{2}C

126 µA, and radio 12 mA. Only the RTC is used in all modes, while all the other peripherals are used in normal run mode.
2. The processor is active 50% of the time, in deep sleep mode for 48%, and transfers data 2% of the time. The same peripherals are on as above.

Solution.

1. The average current during the normal run is 18.9 mA (sum of the current of all peripherals and the current of the processor), while the current during the stop mode is 204 µA + 1.3 µA = 205.3 µA. The average current, in this case, is (18.9 mA · 0.01 s + 0.205 mA · 0.99 s)/1 s = 392 µA.
2. The average current during the normal run without communication is 6.92 mA, and with communication is 19.9 mA, while the current during the stop mode is 205 µA. The average current, in this case, is 3.9 mA. The average power for a 3 V power supply is 11.7 mW. Therefore, if we use a coin battery with a capacity of 230 mAh, the system with the microcontroller, without taking into account any other losses, will be able to run for 2 days and 11 h. The current profile, or the consumption graph, is shown in Fig. 10.11.

10.4.2.4 Batteries

We will first introduce several definitions related to batteries and then consider rechargeable coin cell batteries commonly used in wearable devices. The batteries are composed of one or more cells representing the smallest electrochemical unit. Cells can be single-use or *primary* (nonrechargeable) cells and rechargeable or *secondary* cells. The *nominal voltage* of the cell

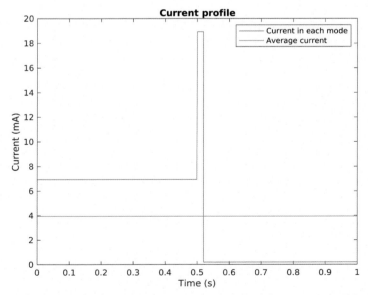

FIGURE 10.11 The current profiles for case 2 of Example 10.3 for the microcontroller MKW41Z256 of Kinetis W series by NXP semiconductor.

is fixed for a given type of cell. It is, for example, 1.2 V for NiCad cells or 3 V for lithium-based cells. The *capacity of cell Q* is given in ampere-hours and represents the charge the cell is supposed to hold. *The total energy capacity* is the capacity multiplied by nominal voltage and is given in Wh (Plett, 2015).

Example 10.4. You are asked to design a wearable device with coin cells and an average current of 3.9 mA. How long will it take until the battery is completely discharged?

Solution. Since the battery's capacity is 230 mAh and the average current is 3.9 mA, it will take about 59 h to discharge the battery.

Making a battery by connecting cells in series increases the battery's voltage but not the current so that the capacity remains the same. The capacity can be increased by connecting cells in parallel.

Example 10.5. We are building a wearable device using coin cells with a nominal voltage of 3 V and a capacity of 230 mAh. How many cells do we need to double the capacity and the battery's nominal voltage?

Solution. As we need to double the voltage, we will need to use 2 cells in series. Since we need to double the capacity, we need to use two other cells connected in parallel with the first two. So, the total number of cells is 4.

Energy density is the stored energy per unit of weight or volume. When the cell is not loaded, the voltage of the cell is the open-circuit voltage.

The state of charge (SOC) $z(t)$ of a cell is defined as (Plett, 2015)

$$z(t) = z(t_0) - \frac{1}{Q} \int_{t_0}^{t} i(\tau)d\tau,$$

where $z(t_0)$ is a state of charge at the time t_0. $i(t)$ is the current considered positive during discharging and negative during charging. If the battery were fully charged, the SOC would be 100%. Open-circuit voltage is plotted as a function of time or SOC in the battery's datasheets.

A smart battery management system monitors the status of the battery, especially its SOC.

The secondary cell commonly used in wearable devices is coin cell battery lithium/manganese dioxide (Li/MnO_2), such as CRC2032. Its nominal voltage is 3 V, while the typical voltage is 2.8 V. Typical capacity is 230 mAh, and the peak current is limited to 13 mA. Lithium-ion cells are also very popular in wearable devices requiring higher-capacity batteries. They have a higher energy density, higher voltages, and lower self-discharge rate than most secondary cells. Lithium-ion batteries come in different shapes, such as slim or cylindrical batteries. Their typical voltage is 3.6 V, and the capacity varies but can easily go up to 3000 mAh.

10.4.2.5 Regulators

Voltage regulators provide constant DC output to the other electronic components in the system even if the battery voltage on its input decreases over time. DC/DC voltage regulators convert DC voltage at the input to the DC voltage at the output. If there is a small difference between the input and output supply voltages, then linear regulators can be used. *Linear regulators* contain a voltage-controlled current source and a feedback loop that monitors the output voltage and increases or decreases the current in the voltage-controlled current sources. One representative of linear regulators is the *low dropout (LDO) regulator*. Low dropout refers to the smallest difference between the input supply voltage and the output voltages that allows the LDO circuit to still regulate the output voltage. LDOs can provide very stable voltage and the output current in the 50–500 mA range.

Switching regulators provide a much larger range of output voltage levels. If the output voltage is lower than the input voltage, the regulator is referred to as a *step-down*, or *buck converter*. If the output voltage is higher than the input voltage, the regulator is said to be a *step-up* or *boost* converter. For example, wearable devices often use a 3.6 V lithium-ion battery, while the other components of the system need smaller (1.2 V for microcontrollers) or larger voltages (5 V for sensors). Therefore, wearable devices use multiple buck and boost converters. The switching step-down regulator accepts the DC input, converts it into a pulse width modulated (PWM) signal, and produces the DC output proportional to the PWM duty cycle. The regulator controls the duty cycle depending on the value of the output voltage. As LDOs have stable output voltage, they are often placed after switch-mode regulators to ensure a noise-free voltage supply.

Several parameters are important to observe when choosing the regulators: efficiency, quiescent current, voltage accuracy, load, and line regulation.

Efficiency is the ratio of output to input power. The efficiency of switch-mode converters in transforming battery power to system power is often greater than 90%. Linear regulators can have very high efficiency only when the difference between the battery voltage and the linear regulator output voltage is small. Quiescent current I_Q refers to the current drawn by the LDO in an enabled and no-load condition. It can be said that the I_Q is the current required for LDO's internal circuitry. Power dissipation of LDOs in full operation (i.e., supplying current to the load) is computed as

$$P_D = (V_{IN} - V_{OUT})I_{OUT} + V_{IN}I_Q$$

Many applications do not require that the LDO is in full operation constantly. To preserve power, when the components the LDO is providing voltage are shut down, the LDO can be idle as well. During that time, the LDO still draws a small quiescent current to keep its internal circuits ready when a load is presented. Therefore, the current I_Q should be as small as possible. Typical linear regulators usually have an output voltage specification that guarantees the regulated output will be within some nominal voltage, such as 5%. *Load regulation* is defined as the circuit's ability to maintain a specified output voltage under varying load conditions, and it is expressed as $\Delta V_{out}/\Delta I_{out}$. *Line regulation* is defined as the circuit's ability to maintain the specified output voltage with the varying input voltage: $\Delta V_{out}/\Delta V_{in}$.

Let us take a look at the boost converter MIC2877 by Microchip. It can boost input voltage, for example, from 3 to 5 V. Of course, other combinations are possible. The typical line regulation is 0.3%, while the load regulation is 0.2%/A. The quiescent current is 125 μA, and the efficiency is up to 95%.

10.4.2.6 Integrated power management units

An example of a power management integrated circuit (PMIC) is shown in Fig. 10.12. Let us assume that a 3.6 V battery is attached to its input. PMIC contains several DC/DC

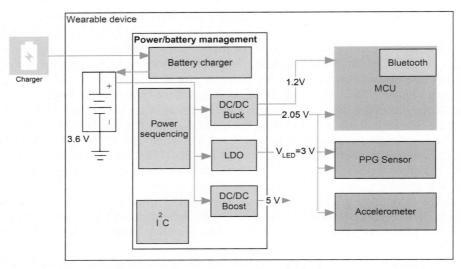

FIGURE 10.12   Integrated power management unit based on the configuration of MAX77659 by maxim integrated. Components directly related to data acquisition and processing are *blue*, while the others are *gray*.

converters that provide specific supply voltages for different components. For example, a microcontroller can be supplied with 1.2 V or 2.05 V, depending on the power mode, using a buck converter. Moreover, 5 V is generated for sensors using a boost converter. The Bluetooth communication module might be connected to 1.8 V LDO. Besides providing different voltage levels for different components, PMIC commonly includes:

1. Battery charger
2. The block for serial configuration with a microcontroller so that the microcontroller can set different configurations
3. A power sequencer that allows for entering different power modes programmatically or turning some or all the components on or off.

An example of a PMIC is MAX77659 by Maxim Integrated. It contains three switch mode regulators that can be configured as buck or boost and one LDO.

10.4.3 Modeling power consumption of the overall system

There have been many attempts to model the power consumption of wearable or Internet of Things (IoT) devices. In Oletic and Bilas (2018), a wearable device for monitoring asthmatic wheezing was developed. The system includes a microphone, A/D converter, microcontroller, and wireless communication module. Detailed analysis was performed in selecting components that consume the least amount of power. The power consumption of three implementation types was considered, as shown in Fig. 10.13, based on the task offloading introduced in Section 10.4.1. In Fig. 10.13A, only data acquisition is performed by a wearable device, and raw data is transferred to a computer. This would allow for simpler maintenance and design of embedded software on the device since the software does not need to be

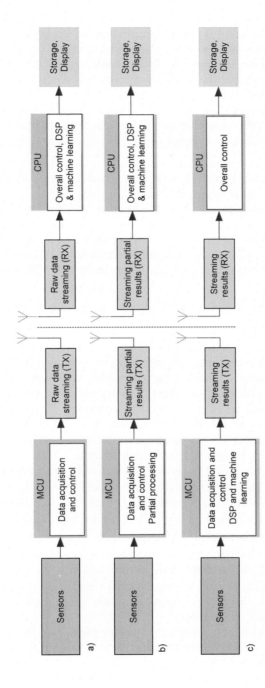

FIGURE 10.13 Different operating scenarios of wearable devices based on the trade-off between the amount of communication and computation done on a wearable device, including (A) raw signal streaming without any processing done on a device, (B) some processing done on a device to reduce communication, (C) the majority of processing is done on a device. *Modified from (Oletic and Bilas, 2018).*

optimized to work on the microcontroller. Fig. 10.13C shows the other possible implementation where the communication is minimized and the majority of computation is done on the wearable device itself. This is useful if the communication is power-hungry or when it is expected that the communication link between the device and the computer is not reliable or the computer is not near the device. Fig. 10.13B is a trade-off solution with some processing done on a device and the rest done on a computer. An example of this kind of device could be an ECG event monitor that detects specific arrhythmias. If the arrhythmias are detected, it can store and/or transmit the raw ECG signal to the computer for further processing.

Software implementation on a wearable device has a significant effect on power consumption. Regarding the implementation of signal processing algorithms, several issues need to be taken into account, including:

- Algorithm selection—out of algorithms that provide satisfactory accuracy, one should choose the one that is the least computationally complex.
- Modification of the algorithm to operate on a microcontroller—for example, the execution time can be reduced by simply rewriting the code to use less complex operations. For example, we can reduce the number of integer multiplications just by the distributive property: $a \cdot b + a \cdot c = a(b + c)$. Modifications include optimizing instruction selection, data manipulations, looking at compiler optimization levels, and so on (Oshana and Kraeling, 2013).

To evaluate the effect of the software implementation on power consumption, one can do the following.

- Counting the number of arithmetic operations and memory accesses that an algorithm of interest is going to take when implemented on a microcontroller. This number of operations can be obtained after compiling the embedded code for a processor. After the number of arithmetic operations is obtained, it is possible to give a rough estimate of how long the algorithm will take on a particular processor. However, as processor architectures are becoming more complex with multicores or several parallel processing elements in a single core, it becomes increasingly difficult to estimate the execution time based on the number and type of arithmetic instructions.
- Modeling the execution time based on knowledge of the execution time from references. For example, it is known that the execution time to perform the fast Fourier transform (FFT) is proportional to $N \log(N)$. Therefore, it is suggested that the processing current of a microcontroller executing the FFT algorithm can be estimated as $I_{prc} = kQ_{OP}N \log(N)/T_p + I_{SYS}$, where Q_{OP} is an estimate of the average charge per arithmetic operation for a given microcontroller, k is a constant factor that depends on the algorithm's implementation, T_p is the duty cycle period which is the time between two consecutive FFT runs and I_{SYS} is the average current required for other nonarithmetic operations (Martinez et al., 2015). As we can see, there are a number of components in this model that are not easy to evaluate.
- Profiling the code on the available profilers to understand how the processing elements and memory are utilized. This requires that the code is written, but it is not executed yet on a device. By analyzing profiler data such as the utilization of ALUs or other components by each software routine, code hot-spots, memory accesses, and so on, one can

get the idea about the part of the code where the processor spends the most time and therefore consumes the most power. This part of the code can be optimized then.
- Using system-level tools that allow for performing power level analysis at system-level design—for example, VisualSim by Mirabilis Design Inc. (Mirabilis Design Inc.). The tool allows for specifying the architecture (components and their connections) as well as the computational load and communication between the components. In addition, power states of the processors and other components, voltage and clock gating, and many other features can be included.

Mathematical models for power consumption of wearable and IoT devices were developed, including Dekkers et al. (2018) and Martinez et al. (2015). Both models include modeling of power consumption of communication P_{comm}, data processing P_{prc}, and system-level processing P_{sys}, as well as the power consumption of the sampling system and sensors P_{acq}: $P = P_{comm} + P_{acq} + P_{prc} + P_{sys}$. The models are very effective in exploring trade-offs between accuracy, power and execution time, but the proper assignment of model parameters is the major difficulty.

10.4.3.1 *Example of modeling power consumption*

We will estimate the power consumption of a device that continuously computes heart rate and SpO_2 based on a PPG sensor and detect if the person moves using an accelerometer. We will consider a system similar to the one shown in Fig. 10.12. We consider two examples corresponding to cases (a) and (c) in Fig. 10.13. They are described in Example 10.6 and Example 10.7.

The sensor is the PPG sensor module ADPD144RI by Analog Devices. The microcontroller is MKW41Z256 by NXP, and it operates at 32 MHz. Power management used is MAX77659 by Maxim Integrated. The battery is a lithium-ion battery with a capacity of 1000 mAh and 3.6 V. The accelerometer is ADXL345 by Analog devices, which provides the digital output and is connected to the MCU using an I^2C interface. It is supplied by 2.05 V provided by the LDO. The supply current is 50 μA in the operating mode at 100 sps and 100 nA in the standby mode. The efficiency of PMIC is 85%.

Example 10.6. Only data acquisition is performed by a wearable device, and raw data are transmitted to a computer. If we are using a 100 sps PPG module with red and IR LEDs, this would require that the data are collected from red and IR channels. Assuming 2 bytes are needed per channel, this will result in 4 bytes per sample. We will also read x, y, and z coordinates from the accelerometer at 100 sps resulting in 6 bytes per sample. Let us assume that the processor communicates over I^2C bus with the PPG and accelerometer module for 1 ms every 10 ms, stores the data, and sleeps for the remaining 9 ms. Then the processor turns on every 250 ms for 20 ms and communicates the collected data ((4 bytes + 6 bytes)·250 ms/10 ms = 250 bytes, which can fit a BLE packet) over Bluetooth with the radio current of 6.3 mA. If we assume that the number of pulses of the PPG chip $N_{pulse} = 10$, then based on Fig. 10.10A, the power consumption of the PPG module is 2.75 mW. Estimate the power consumption of the device.

Solution. The average current of the microcontroller is $T_{normal} = 1$ ms·250 ms/10 ms = 25 ms and similarly $T_{stop} = 250$ ms − 25 ms = 225 ms. The average MCU current is computed as

$$I_{ave} = \frac{I_{normal}T_{normal} + I_{stop}T_{stop} + I_{comm}T_{comm}}{T_{normal} + T_{stop} + T_{comm}}$$

$$= \frac{6.71 \text{ mA} \cdot 25 \text{ ms} + 0.2 \text{ mA} \cdot 225 \text{ ms} + (6.71 \text{ mA} + 6.3 \text{ mA}) \cdot 20 \text{ ms}}{270 \text{ ms}} = 1.8 \text{ mA}$$

The total power consumed, including the power consumption of the PPG chip and the power consumption of the MCU and the accelerator, assuming that the accelerator is always on, is:

$$P = 1.8 \text{ mA} \cdot 1.2 \text{ V} + 50 \ \mu A \cdot 2.05 \text{ V} + 2.75 \text{ mW} = 5 \text{ mW}$$

We will assume that the efficiency of all regulators is about 85%. Therefore, if we ignore the low power mode of the PMIC, the required power is 5 mW/0.85 = 5.9 mW. Therefore, for a battery that has a 3600 mWh capacity, the battery life will be 25 days and 13 h (Table 10.6).

Example 10.7. Heart rate and SpO$_2$ estimations are computed on a device every second, and only the final estimates are transferred to the computer once per second. We assume that it will take a total of 500 ms to perform computations and to read PPG and accelerometer data using I^2C communication, 8 ms to communicate data since the data is now processed and only results are communicated, and 492 ms to be in the stop mode. Estimate the power consumption of the device.

Solution. In this case, the average power is $P = 7.2$ mW computed in the same way as shown in Example 10.6. The total power needed is 8.5 mW, and the battery can last for 17 days and 16 h. Of course, the assumptions on how long processing would take are arbitrary. They would need to be obtained by using profilers and/or simulators.

Example 10.6 and Example 10.7 allow for performing the design space exploration. For example, if we increase the sample rate in Example 10.6 by 10 times, the number of packets of raw data that need to be transferred from the MCU to the computer will also increase tenfold. If we do the same in Example 10.7, the number of messages that are communicated remains the same, but the processing time of the algorithm will increase because the MCU

TABLE 10.6 Components used to estimate power consumption and their main operational characteristics.

Component type	Selected component	Vdd	Current in active mode	Current in standby mode
MCU	MKW41Z256.	1.2 V	6.84 mA without radio 13.1 mA with the radio	Stop mode: 204 μA
PPG	ADPD144RI	VLED = 3.2V, Vdd = 1.8V	See Table 10.3	See Table 10.3
Accelerometer	ADXL345	2.05V	50 μA	100 nA

needs to handle more data. In addition, increasing the sampling rate would increase the power consumption of the PPG module (see Fig. 10.10B) and accelerometer in both examples.

10.5 Examples of wearable devices

10.5.1 Wearable devices for cardiovascular and respiratory monitoring

The list of several commercial devices and their use for cardiac monitoring is shown in Table 10.7 based on the wearable device review (Sana et al., 2020). The list is not comprehensive, and we listed devices only as examples. The devices are used for ECG data acquisition, heart rate monitoring, detecting arrhythmias, SpO$_2$, estimating continuous blood pressure and heart rate variability, as well as stress diagnosis. We observe that the devices are becoming increasingly multimodal and that they estimate multiple indices. For example, the ViSi Mobile by Sotera Wireless can measure and estimate continuous blood pressure, ECG, SpO$_2$, heart rate arrhythmia, and respiration rate.

Heart failure monitoring includes monitoring blood pressure and heart rate, activity monitoring, as well as detecting thoracic fluid accumulation (Duncker et al., 2021). In Stehlik et al. (2020), a wearable patch sensor by Vital Connect, San Jose, CA, was attached to the patient's chest. Patients were monitored at home for several months. The sensor continuously collected ECG waveform, 3-axis accelerometer data, skin impedance, and temperature. The data were used to estimate heart rate, heart rate variability, arrhythmia burden, respiratory rate, gross activity, walking, sleep, body tilt, and body posture. These estimates were then used to

TABLE 10.7 Examples of wearable devices used for cardiac applications.

Application	Device type	Device name
ECG monitoring, arrhythmia detection	Patch	Zio patch by iRhythm Technologies, Nuvant by SEEQ mobile Cardiac telemetry
ECG monitoring, arrhythmia detection	Pocket-based hand-held device with external electrodes	PocketECG by MediLynx
ECG monitoring, heart rate, atrial fibrillation detection	Wristwatch	Apple iWatch
SpO$_2$, HR	Wristwatch	Samsung galaxy
HR variability, skin temperature, stress diagnosis	Patch	CorSense by Elite HRV, Polar H7/H10 HR by Polar Electro Oy
Continuous blood pressure	Wristwatch	BPro by HealthStats Inc., Aktiia by Aktiia
Continuous blood pressure, ECG, SpO$_2$, heart rate arrhythmia, respiration rate	Multiple points	The ViSi mobile by Sotera Wireless

develop models to calculate the number of days before hospitalization and predict hospital-ization due to the worsening of heart failure. The results include detecting heart failure events with 76%–88% sensitivity and 85% specificity. The median time between the initial alert and hospital readmission was 6.5 days.

In Inan et al. (2018), a wearable patch that measures SCG and ECG was proposed. Subjects were asked to perform a controlled exercise with the 6-min walk test. In healthy subjects, the exercise leads to substantial changes in the shape and timings of the SCG wave-form, while fewer changes are observed for decompensated heart failure patients. In addi-tion, a similarity score was developed to compare features extracted from the SCG signal after the exercise. The results showed that detecting subjects with decompensated heart fail-ure is possible.

Some of the devices mentioned in Table 10.7 also measure the breathing rate. The breath-ing rate is often extracted from the PPG signals. Biosensor BX100 by Philips is a chest patch that uses a bioimpedance sensor to measure breathing rate. Zephyr Performance Systems by Healthcare Medtronic is a belt for monitoring breathing and heart rate. PneumoWave devel-oped a chest warn sensor that continuously measures respiration rate and detects apneas and other conditions. It was used in several studies related to COVID-19.

10.5.1.1 Commercial platforms for wearable system development

There are several available platforms used for the development of wearable devices. They can be used for data collection and the development of software applications on top of these platforms. One of the leading devices is E4 by Empatica, which includes a PPG sensor, accel-erometer, temperature sensor, and sensor for measuring the impedance of the skin (electro-dermal activity). Raw unprocessed data can be collected from these sensors and used for developing different applications, including stress level monitoring, estimation of heart rate and SpO$_2$, etc.

Commercial wearable data acquisition systems can be used for data acquisition, including Shimmer3 by Shimmer Inc., Bitalino cardiac and respiratory data acquisition modules by PLUX—Wireless Biosignal, and so on.

Embedded development systems for designing Internet of Things applications can also be used for designing wearable devices. There are several recent solutions based on Arduino and Raspberry Pi boards for wearables with PPG, accelerometers and electro-dermal activity sensors.

10.5.2 Examples of research-based wrist-worn wearable devices

We will show several approaches to designing wrist-worn devices, including open hard-ware/open software systems and a wearable device used to monitor blood pressure.

Open-source software and hardware are very important for speeding up the development process because it allows designers to start the development with existing, partially or fully functional hardware and software. The license is such that the hardware and/or software can be freely modified. In Boateng et al. (2019), a wrist-worn device called Amulet was designed. It is based on a 16-bit microcontroller and includes a number of sensors that can measure ac-celeration, rotation, ambient sound, ambient light, and ambient temperature. A micro-SD card reader is used to allow for storing up to 2 GB of data. A lithium-ion 110 mAh battery was used.

In Ganti et al. (2021), a wrist-worn device comprises two printed circuit boards. One board is based on a microcontroller with an ARM Kortex M4 core and includes an accelerometer and gyroscope sensors. Another board includes PPG and ECG front-end electronics. Two ECG metal electrodes are placed inside the wristband, while the third one is placed outside so that it can be touched with a finger from another hand. The two electrodes placed inside the wristband act as the driven right leg electrode and the positive electrode, whereas the external electrode is the negative electrode. A 1.2 Ah lithium-ion rechargeable battery was used to power the system.

10.6 Summary

In this chapter, we started with the classification of wearable devices based on the type of cardiac and respiratory signals they sense. Then we focused on three modalities we did not cover in previous chapters: SCG, PCG, and arterial tonometry. We showed several other classifications based on multimodal monitoring and sensor fusion, form factor, frequency of measurement, storing and communication, and so on. Next, the chapter focused on low-power design. We showed how one could estimate, at a very high level, the power consumption of existing integrated circuits, including sensors and microcontrollers. In addition, we introduced the power management unit. The chapter concludes with several examples of wearable devices.

This chapter did not cover devices based on gyrocardiography, magnetocardiography (Pannetier-Lecoeur et al., 2011), ballistocardiography, etc. We hope to include these modalities in the next editions of the book.

Energy harvesting was not covered in this chapter. Instead, please refer to Romero et al. (2009), Di Paolo Emilio (2017), and Xu et al. (2021) for detailed information about energy harvesting.

Wearable devices set new challenges for standardization. Some devices do not even have their own interface, but they transfer data to an external device, such as a smartphone or a computer, where the data are stored and displayed. Therefore, regulations are different for wearable medical devices than for standard medical devices. For details about the standardization of wearable devices with an emphasis on the efforts in the European Union, please refer to Ravizza et al. (2019).

10.7 Problems

10.7.1 Simple questions

10.1. Select the valid answer
    (a) SCG signal is generated by
        a. Mechanical heart motion measured at the surface of the chest
        b. Force due to the motion of blood
        c. Sound vibrations of heart and blood circulation.

(b) Bioimpedance measurement requires current to be injected into the body
  a. True
  b. False

10.2. What is Audio CODEC and how is it used in PCG applications?

10.3. List at least three advantages of using sensor fusion when measuring physiological signals.

10.4. What are the reasons for using redundant sensors?

10.5. Explain strategies for reducing power consumption, including task offloading, duty cycling and energy-aware computations and communications.

10.7.2 Problems

10.6. Let us consider the tonometric arterial pulse sensor method, where a displacement transducer is used to measure the displacement due to the motion of the radial artery.
  (a) If the main criterion in choosing a transducer in this application is a high sensitivity, what displacement sensor would you select? Why?
  (b) Please do a Web search to find one strain gauge sensor used for tonometry application and describe its parameters.
  (c) Model the front-end circuit (bridge and amplifier) for that sensor in Simscape.
  (d) Replace the Wheatstone bridge configuration with the one presented in Section 3.3.5.1. Model changes in ΔR as a sine wave with a frequency of 1 Hz and amplitude of about 0.1 Ω. Assume that all resistors have values of 120 Ω. Compare the outputs of the differential amplifier in case a regular Wheatstone bridge is used and when the configuration shown in Fig. 3.7A is used.

10.7. In Example 10.2, let us assume that the PPG device is used for heart rate monitoring, and therefore, it would use only one LED, not two. Repeat the simulations for Fig. 10.10 for that case.

10.8. Analyze power down modes of ADS1293 component by Texas Instruments that is used for ECG signal conditioning in wearable devices. Consider the power consumption when only one channel is used and all other components are off.

10.9. *Fig. 10.1 shows several blocks that we did not cover in this chapter. Read and discuss about:
  (a) Security solutions such as secure microcontrollers (e.g., Arm SecurCore SC300 by ST Microelectronics) and secure elements (e.g., ST54H by ST Microelectronics).
  (b) Touchscreen controllers.
  (c) Battery chargers.

10.10. This example is from Texas Instruments (Fwu, 2021) and is based on Texas Instruments TPS7A05 LDO, which generates 1.8 V at the output from a 3 V coin battery. The output current in the active mode is 100 μA, and in the low-power mode, it is 1 μA. The quiescent current is 1 μA.
  (a) Estimate power dissipation on the LDO during active and low power modes.
  (b) Why is it important to reduce quiescent current?

10.11. In Example 10.6, compute the power consumption if the sampling rate increases to 400 sps. Assume that the typical current of the accelerometer ADXL345 by Analog devices at 400 sps sampling rate is 90 µA. The power consumption of the PPG module can be obtained from Fig. 10.10A. Introduce your assumptions on the increase in power consumption due to data transfer from the MCU to the computer.

10.7.3 Simulations

10.12. In Example 10.2, analyze power consumption when I_{LED} is changed from 25 to 250 mA. Comment on why it is often important to use more than 25 mA to get a good signal.

10.13. Evaluate the power consumption of a wrist-worn device by Ganti et al. (2021) by proposing your own duty cycle and using the same components as in Fig. 5 of Ganti et al. (2021).

10.8 Further reading

Additional resources related to different topics covered in this chapter are listed below. Please note that the following material is selected based on availability and the author's preferences and that many good resources might be omitted unintentionally.

The design of wearable devices and Internet of Things devices at a high level is presented in Raad (2021). An excellent book on voltage regulators and techniques for power management is Chen (2016). For a review of power management and other issues regarding energy-efficient wearable devices, please refer to Qaim et al. (2020).

To explore the power consumption of different microcontrollers, there are several available power estimation tools. These tools come as Excel spreadsheets or as stand-alone applications. We used KINETIS-PET: Kinetis Power Estimation Tool to evaluate the power consumption of different NXP microcontrollers.

An excellent book on modeling batteries is Plett (2015).

References

Altini, M., Kinnunen, H., 2021. The promise of sleep: a multi-sensor approach for accurate sleep stage detection using the Oura ring. Sensors 21 (4302).

Bentley, P., Nordehn, G., Coimbra, M., Mannor, S., 2011. The PASCAL Classifying Heart Sounds Challenge 2011 (CHSC2011) Results. http://www.peterjbentley.com/heartchallenge/index.html.

Bergmann, J.H.M., McGregor, A.H., 2011. Body-worn sensor design: what do patients and clinicians want? Annals of Biomedical Engineering 39 (9), 2299–2312.

Boateng, G., et al., 2019. Experience: Design, Development and Evaluation of a Wearable Device for mHealth Applications. International Conference on Mobile Computing and Networking.

Chen, K.-H., 2016. Power Management Techniques for Integrated Circuit Design. Wiley.

Ciaccio, E.J., Drzewiecki, G.M., 2008. Tonometric arterial pulse sensor with noise cancellation. IEEE Transactions on Biomedical Engineering 55 (10), 2388–2396.

Davies, H.J., Williams, I., Peters, N.S., Mandic, D.P., 2020. In-ear SpO₂: a tool for wearable, unobtrusive monitoring of core blood oxygen saturation. Sensors 20 (17), 4879.

Dekkers, G., Rosas, F., Lauwereins, S., et al., 2018. A Multi-Layered Energy Consumption Model for Smart Wireless Acoustic Sensor Networks, abs/1812.06672. ArXiv.

Di Paolo Emilio, M., 2017. Microelectronic Circuit Design for Energy Harvesting Systems. Springer.

Duncker, D., Ding, W.Y., Etheridge, S., Noseworthy, P.A., Veltmann, C., Yao, X., Bunch, T.J., Gupta, D., 2021. Smart wearables for cardiac monitoring—real-world use beyond atrial fibrillation. Sensors 21 (2539).

Esmaili, A., Kachuee, M., Shabany, M., Dec. 2017. Nonlinear cuffless blood pressure estimation of healthy subjects using pulse transit time and arrival time. IEEE Transactions on Instrumentation and Measurement 66 (12), 3299—3308.

Fang, J.C., T O'Gara, P., 2019. The history and physical examination: an evidence-based approach. In: Zipes, D.P., et al. (Eds.), Chapter in Book: Braunwald's Heart Disease: A Textbook of Cardiovascular Medicine, eleventh ed.

Fwu, W., 2021. Understanding the Foundations of Quiescent Current in Linear Power Systems. Technical document, SLYY206Texas Instruments.

Ganti, V., Carek, A.M., Jung, H., et al., 2021. Enabling wearable pulse transit time-based blood pressure estimation for medically underserved areas and health equity: comprehensive evaluation study. JMIR Mhealth Uhealth 9 (8), e27466.

García-González, M.A., Argelagós-Palau, A., Fernández-Chimeno, M., Ramos-Castro, J., J., 2013. A comparison of heartbeat detectors for the seismocardiogram. In: Computing in Cardiology Conference (CinC).

Inan, O.T., Pouyan, M.B., Javaid, A.Q., et al., 2018. Novel wearable seismocardiography and machine learning algorithms can assess clinical status of heart failure patients. Circulation. Heart Failure 11 (1).

King, R.C., Villeneuve, E., White, R.J., Sherratt, R.S., Holderbaum, W., Harwin, W.S., 2017. Application of data fusion techniques and technologies for wearable health monitoring. Medical Engineering and Physics 42, 1—12.

Leonhardt, S., Leicht, L., Teichmann, D., 2018. Unobtrusive vital sign monitoring in automotive environments-A review. Sensors, MDPI 18 (9), 3080.

Liu, C., et al., 2016. An open access database for the evaluation of heart sound algorithms. Physiological Measurements 12, 2181—2213.

Martinez, B., Montón, M., Vilajosana, I., Prades, J.D., Oct. 2015. The power of models: modeling power consumption for IoT devices. IEEE Sensors Journal 15 (10), 5777—5789.

Masè, M., Micarelli, A., Strapazzon, G., 2020. Hearables: new perspectives and pitfalls of in-ear devices for physiological monitoring. A scoping review. Frontiers in Physiology 11 (568886).

Mirabilis Design Inc, "Power Modeling and Estimation in Early System Design, Part 1," https://www.mirabilisdesign.com/power-modeling-and-estimation-in-early-system-design-part-1/, Last accessed on February 28, 2022.

Mondal, H., Mondal, S., Saha, K., 2018. Development of a low-cost wireless phonocardiograph with a Bluetooth headset under resource-limited conditions. Medical Sciences 6 (4).

Munck, K., Sørensen, K., Struijk, J.J., Schmidt, S., 2020a. Multichannel seismocardiography: an imaging modality for investigating heart vibrations. Physiological Measurement 41 (11).

Munck, K., Sørensen, K., Struijk, J.J., Schmidt, S., 2020b. Data for: Multichannel Seismocardiography: An Imaging Modality for Investigating Heart Vibrations. Mendeley Data, p. V2. https://doi.org/10.17632/scn464x7xd.2.

Nguyen, T.V., Ichiki, M., 2019. MEMS-based sensor for simultaneous measurement of pulse wave and respiration rate. Sensors, MDPI 19 (4942).

Oletic, D., Bilas, V., 2018. System-level power consumption analysis of the wearable asthmatic wheeze quantification. Journal of Sensors, Hindawi 2018. Article ID 6564158.

Oliveira, J., 2021. The CirCor DigiScope Dataset: From Murmur Detection to Murmur Classification. arXiv:2108.00813v1.

Omari, T., Bereksi-Reguig, F., 2019. A new approach for blood pressure estimation based on phonocardiogram. Biomedical Engineering Letters 9 (3), 395—406.

Oshana, R., 2013. Optimizing embedded software for power. In: Chapter in Book: Software Engineering for Embedded Systems. Methods, Practical Techniques, and Applications by R. Oshana and M. Kraeling, Newnes.

Pannetier-Lecoeur, M., Parkkonen, L., Sergeeva-Chollet, N., Polovy, H., Fermon, C., Fowley, C., 2011. Magnetocardiography with sensors based on giant magnetoresistance. Applied Physics and Letters 98 (153705).

Pariaszewskaa, K., Młyńczaka, M., Niewiadomskib, W., Cybulski, G., 2013. Digital stethoscope system — the feasibility of cardiac auscultation. Proceedings of SPIE - The International Society for Optical Engineering.

Pinsky, L.E., Wipf, J.E., "Learning and Teaching at the Bedside: Examination for Heart Sounds and Murmurs," https://depts.washington.edu/physdx/index.html, last accessed on December 02, 2021.

Plett, G.L., 2015. Battery management systems. In: Battery Modeling, Vol. 1. Artech House.

Qaim, W.B., et al., 2020. Towards energy efficiency in the Internet of wearable Things: a systematic review. IEEE Access 8, 175412–175435.

Raad, H., January 20, 2021. Fundamentals of IoT and Wearable Technology Design. John Wiley and Sons.

Rahmani, M.H., Berkvens, R., Weyn, M., 2021. Chest-worn inertial sensors: a survey of applications and methods. Sensors 21 (2875).

Ravizza, A., De Maria, C., Di Pietro, L., Sternini, F., Audenino, A.L., Bignardi, C., 2019. Comprehensive review on current and future regulatory requirements on wearable sensors in preclinical and clinical testing. Frontiers in Bioengineering and Biotechnology 7.

Romero, E., Warrington, R.O., Neuman, M.R., 2009. Energy scavenging sources for biomedical sensors. Physiol Meas 30 (9), 35–62.

Sana, F., Isselbacher, E.M., Singh, J.P., Heist, E.K., Pathik, B., Armoundas, A.A., 2020. Wearable devices for ambulatory cardiac monitoring: JACC state-of-the-art review. Journal of American Colloids and Cardiology 75 (13).

Sempionatto, J.R., Lin, M.Y., Yin, L., et al., 2021. An epidermal patch for the simultaneous monitoring of haemodynamic and metabolic biomarkers. Nature Biomedical Engineering 5, 737–748.

Someya, T., Amagai, M., 2019. Toward a new generation of smart skins. Nature Biotechnology 37, 382–388.

Springer, D.B., Tarassenko, L., Clifford, G.D., 2016. Logistic regression-HSMM-based heart sound segmentation. IEEE Transactions on Biomedical Engineering 63 (4), 822–832.

Stehlik, J., Schmalfuss, C., Bozkurt, B., et al., 2020. Continuous wearable monitoring analytics predict heart failure hospitalization: the LINK-HF multicenter study. Circulation: Heart Failure 13 (3).

Taebi, A., Solar, B.E., Bomar, A.J., Sandler, R.H., Mansy, H.A., 2019. Recent advances in seismocardiography. Vibration 2 (11), 64–86.

Thoms, L., Collichia, G., Girwidz, R., 2019. Real-life physics: phonocardiography, electrocardiography, and audiometry with a smartphone. Journal of Physics: Conference Series 1223 (012007).

Tseng, Y.-L., et al., 2012. Detection of the third and fourth heart sounds using Hilbert-Huang transform. Biomedical Engineering Online 11 (8).

Wahlström, J., et al., 2017. A hidden Markov model for seismocardiography. IEEE Transactions on Biomedical Engineering 64 (10), 2361–2372.

Wang, T.-W., Lin, S.-F., 2020. Wearable piezoelectric-based system for continuous beat-to-beat blood pressure measurement. Sensors, MDPI 20 (851).

Wang, Y.-J., Chen, C.-H., Sue, C.-Y., Lu, W.-H., Chiou, Y.-H., 2018. Estimation of blood pressure in the radial artery using strain-based pulse wave and photoplethysmography sensors. Micromachines 9 (556).

Xu, C., et al., 2021. Portable and wearable self-powered systems based on emerging energy harvesting technology. Microsystems and Nanoengineering 7 (25).

Conclusion and research directions: beyond wearable devices

Acronyms and explanations

3D camera	a camera that provides 3D perception in a way that provides 3D coordinates of the pixels.
BVP	blood volume pulse
CW	continuous wave
FPS	frames per second
iPPG	imaging photoplethysmography
NETD	temperature difference
RGB	red, green, blue
RoI	region of interest
rPPG	remote photoplethysmography
UWB	ultrawide band

11.1 Introduction

In this chapter, we will briefly mention topics covered in this book. Then, we focus on the research directions related to electronics used in wearable devices, types of measurements, algorithms, machine learning, and others. The focus in this chapter then shifts to contactless monitoring, including both short (up to several cm) and long-range contactless monitoring (up to several meters). We believe that future technologies and monitoring systems will combine all these technologies.

11.2 What was covered in the book

The main theme of this book is to understand issues in designing biomedical devices for use at home and to model and analyze the devices. Ideally, we would like to be able to develop continuous, noninvasive, and unobtrusive devices for monitoring physiological variables.

11.2.1 Transducers and electronics

Therefore, when describing a technology, we focused on defining and understanding the problem and issues in measurements, signals, and noise that are typically dealt with, performance metrics and how one can evaluate them, common algorithms, and hardware block diagrams. In the first part of the book, we described the performance metrics commonly used when designing biomedical devices and then introduced transducers and conditioning electronics.

Piezoelectric transducers are used as parts of a large number of sensors for measuring pressure or force or as a component of a microphone. Piezoelectric transducers are conditioned using a charge amplifier followed by other necessary amplifiers and filters. Piezoelectric transducers are used in several applications, such as oscillometric blood pressure, arterial tonometry, and respiratory belts. Microphones are used for PCG.

Capacitive transducers are used in pressure sensors, accelerometers, and microphones. As pressure sensors, they are used, for example, for blood pressure measurements and in respiratory belts. As part of accelerators, they are used for SCG measurements. Also, as they are often parts of microphones, capacitive transducers are used for PCG measurements. There are several ways one can connect capacitive sensors to a circuit, such as, for example, synchronous demodulation. Integrated circuits for capacitive sensors are capacitance to digital converters.

Photodiodes and photodetectors are used for PPG and pulse oximetry measurements. They require current-to-voltage converters or transimpedance amplifiers as conditioning circuits. Integrated current to digital converters can be used.

We discussed electrodes for ECG and bioimpedance measurements. There are several available integrated circuits for both applications. For ECG applications, they allow for connecting two or more electrodes, and they contain an instrumentation amplifier, filters, and A/D converter. Bioimpedance circuits contain a demodulator, amplifier, filter, and A/D converter.

Table 11.1 summarizes the transducers, conditioning electronics, and integrated circuits that were covered. Examples of applications of strain gauges are oscillometric blood pressure devices, nasal pressure sensors, and so on. In these applications, the strain gauge is used as a part of the sensor to measure the pressure due to airflow in the tube. In both cases, the Wheatstone bridge, followed by the instrumentation amplifier and filters, can be used. However, they are also available integrated components that can be used for that design, such as bridge

TABLE 11.1   Transducers and circuits that we covered in the book.

Transducer	Sensor made based on these transducers	Conditioning circuit	Integrated circuit with the conditioning circuit and other components	Applications
Strain gauges	Pressure	Bridge, differential amplifiers, other amplifiers and filters, A/D converter	Bridge transducer A/D converter	Blood pressure Respiratory belts, nasal pressure sensor
Piezoelectric	Microphones, pressure, force	Charge amplifier, other amplifiers and filters, A/D converter	An integrated solution that includes a charge amplifier	Blood pressure, arterial tonometry, respiratory belts, PCG
Capacitive	Pressure, accelerometers, microphones	AC bridge, synchronous demodulation	Capacitance to digital converter	Blood pressure, respiratory belts, SCG
Photodiodes	Photodetectors	Transimpedance amplifier	Current to digital converters, current to frequency converters	PPG, pulse oximetry
Electrodes	- Biopotential electrode - Bioimpedance	- Instrumentation amplifier, filters, A/D converter - Demodulator, amplifier, filter, A/D converter	- ECG integrated circuits - Bioimpedance	- ECG - Breathing belt based on bioimpedance

transducer A/D converter. These devices contain integrated instrumentation amplifiers, programmable gain amplifiers, A/D converters, digital filters, serial interface for connecting the device to a microcontroller or a microprocessor, and so on. As such, these devices represent a great solution that replaces a number of discrete components on the board.

11.2.2 Biomedical devices

Table 11.2 shows the devices and methods covered in the second part of the book. They include devices based on ECG, respiratory devices for measuring breathing rate, volume, and flow and for breathing pattern classification, PPG devices and pulse oximeters, oscillometric blood pressure, SCG, and continuous blood pressure devices. Devices are compared based on measurement techniques, main applications, whether the processing is performed online or after data are collected, waveforms or features required for processing, indices presented at the output of the device, the common placement of the device, and the form factor. In addition, they are compared based on whether machine learning or fitting is needed, whether the measurements are intermittent or continuous, and whether the device requires the application of energy to perform the measurement. As an example, let us consider an oscillometric blood pressure device. The main application is for monitoring hypertension. The processing is done after data are collected, and the measurements are performed intermittently. To do the processing, the oscillometric waveform needs to be stored. The outputs of the device are systolic and diastolic blood pressure. To estimate the output, traditional signal processing algorithms are commonly used. To perform the measurements, external mechanical energy has to be applied. It requires pushing air into the bladder to occlude the artery. The device is placed on the upper arm or the wrist and is a cuff-based device. Table 11.2 includes a similar description for each device covered in the book.

11.3 Trends and research directions for biomedical devices

Throughout the book, we discussed some trends and research directions related to electronics and devices. Some trends related to improving measurement capabilities, design issues, learning, and modeling are summarized below and in Fig. 11.1.

11.3.1 Measuring deeper

A new approach that describes several important concepts, including sensor arrays and smart skin, is shown in Wang et al. (2021). The array is based on ultrasound transducers that allow for deep penetration in the skin, and therefore it allows for obtaining arterial pulses from arteries that are not superficial.

11.3.2 Measuring unobtrusively (with smart skin)

Even though this book focuses only on physiological signal measurements, there is a great need to collect other variables besides physiological ones, including serum electrolyte or hormone levels and transepidermal water loss. Recent smart skin solutions have been developed

TABLE 11.2 Factors and parameters related to the devices we covered in the book.

Devices and methods	Techniques	Applications	Online processing or after data is collected	Waveform/Features required	Metrics	Machine learning or some kind of fitting needed	Intermittent or continuous	Require application of energy	Common placement	Form factor of wearable devices
ECG	Holter or patch long-term monitoring	Arrhythmia detection and classification	Data collection and storage—process offline	ECG signal, all waves	Classification of rhythm and beats	Yes, to detect and classify arrhythmias	Continuous	No, except for the leadoff detector	Chest	Patches, small devices with electrodes worn at the belt
	Short-term monitoring	Collect data when the subject feels arrhythmia or other symptoms	Activate measurement when needed—process online or offline	ECG signal, all waves	Detection of arrhythmia	Yes	Intermittent	No, except for the leadoff detector	Chest, wrist	Smartwatches, patches
	Heart rate	Physical well-being	Online, moving window	ECG signal, R wave	Single: beats per minute, vital sign	No	Can be both	No, except for the leadoff detector	Wrist—requires touching with another arm	Smartwatches, smart shirts, patches
	Heart rate variability	Stress, emotions	Can be both; long-term 24-hour measurements are processed offline	ECG signal, R wave	Several indices are used.	Depending on whether indices are observed only, or classification is done based on them	Can be both	No, except for the leadoff detector circuit	Chest, wrist; requires touching with another arm	Smartwatches, smart shirts, patches
Respiratory devices	Breathing rate based on volume	Sleep apnea, COVID 19	Online	Impedance, inductance, piezoelectric, accelerometers, ...	Single: breaths per minute, vital sign	No	Can be both	Yes, if the impedance is measured, otherwise no	Chest	Breathing belt or patch
	Breathing rate based on flow	Sleep apnea, COVID 19	Online	Thermal or pressure sensor	Single: breaths per minute, vital sign	No	Can be both	No	Oral or nasal	Nasal cannula

(Continued)

TABLE 11.2 Factors and parameters related to the devices we covered in the book.—cont'd

Devices and methods	Techniques	Applications	Online processing or after data is collected	Waveform/Features required	Metrics	Machine learning or some kind of fitting needed	Intermittent or continuous	Require application of energy	Common placement	Form factor of wearable devices
	Volume and flow	COPD	Offline, after the measurement procedure is done	Spirometers	Several parameters such as FEV_1 and FEV_1/FVC	No	Intermittent	No	Mouth	Mouthpiece
	Breathing pattern classification	Heart failure, sleep apnea, radiology	Can be online	Any sensor that can be used to obtain respiratory signal	Classes of interest	Yes	Can be both	No	Commonly chest	Commonly breathing belt
PPG	Heart rate	Used for SpO_2, heart rate monitors	Online	LED + photodetector	Single—beats per minute	No	Normally continuous	Yes, light	Wrist, finger	Smartwatches, PPG probes
Pulse oximeters	Oxygen level	COVID 19, sleep apnea	Online	2 LEDs + photodetector	Single—oxygen level	Yes, there is a regression formula	Normally intermittent	Yes, light	Finger	Oximeter probes
Blood pressure	Oscillometric	Hypertension	After data are collected	Oscillometric waveform	Systolic and diastolic blood pressure	No in general	Intermittent	Yes, mechanical energy for occluding the artery	Upper arm	Cuff-based device
SCG	Heart rate	Physical well-being	Online	Accelerometers	Features of the SGC waveform	Yes	Can be both	No	Chest	Smartphone or a small sensor placed on a chest
Continuous blood pressure	PAT	Hypertension	Online	ECG and PPG	Systolic and diastolic blood pressure	Yes	Continuous	No	Wrist, chest	Smartwatches, embedded devices that measure ECG and PPG
	PTT or other features	Hypertension	Online	Two PPGs at different locations, PPG and SCG, ...	Systolic and diastolic blood pressure	Yes	Continuous	No	Depending on the sensors used	Embedded devices that can measure multiple signals

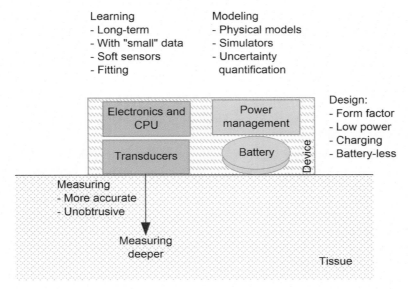

FIGURE 11.1 Several trends and research directions related to biomedical devices.

that can measure physiological and chemical variables. It was pointed out in Someya and Amagai (2019) that monitoring the health status of individuals over an extended period requires robust and reproducible measurements. In addition, the measurement duration depends on battery life and how long patients can wear or tolerate the device while performing normal daily activities. Therefore, the robustness of measurements, low power design, and making smart skin sensors almost a part of the body (through miniaturization, reduction of thickness and weight, and improvement of visual appearance) are major research trends.

11.3.3 Integrating devices into the items that people already wear

We mentioned above that an important aspect of using wearable devices is how much the subject is able to tolerate the device while performing daily activities. If the devices are embedded in clothing, then they become unobtrusive. Smart clothes are common-looking clothes with embedded physiological sensors and electronics, as well as features such as charging, displaying, etc. E-textiles are fabrics with electronics embedded in them that can be used to make smart clothes. The electronics embedded into fabric can be composed of classical electronic components such as integrated circuits, LEDs, and conventional batteries. On the other side, the electronics can be directly integrated into the textile substrate (Lund et al., 2021).

11.3.4 Energy harvesting and low-power design

Energy harvesting for wearable devices is mainly based on four energy sources: motion from walking and/or moving arms, light, electromagnetic waves, and body heat.

The amount of power generated is: about 15 μW from walking, about 100 μW from lighting in indoor conditions, about 10 μW at ultra-high frequency band, and about ~3 μW from body heat (Bhat et al., 2020). Making devices operate without interruptions requires that the consumed energy is smaller or equal to the harvested energy. This can be achieved by improving the design of energy harvesters, applying low-power design principles, and using power management techniques.

11.3.5 Battery-less systems

Passive battery-less sensors based on Radio Frequency Identification (RFID) technology can be designed to monitor the breathing rate. These tags are embedded in t-shirts or sweaters and rely on having a flexible antenna whose size changes with the expansion or contraction of the thorax. When the wearable RFID tag antenna changes its size due to inhalation or exhalation, there is a mismatch of impedance between the chip and the antenna, and therefore, the received signal strength at the reader side varies with breathing (Araujo et al., 2018). The device is powered up when it is in the proximity of an RFID reader. In this way, RFID tags can be used for monitoring the breathing rate.

11.3.6 Providing uncertainties in measurements

It is still uncommon to encounter biomedical devices that perform uncertainty quantification and provide confidence levels of the measurement quantity. Therefore, we see a great opportunity in developing systems capable of quantifying uncertainty starting from transducers, through the circuitry, A/D converters, as well as the algorithms (such as, for example, regression models). A guide on performing uncertainty quantification and sensitivity analysis in cardiovascular applications is given in Eck et al. (2016).

11.3.7 Soft and self-validating sensors

Soft (or software) sensors are commonly used in industrial processes and include one or more physical sensors and a physical or data-driven model. The soft sensors do not provide raw but processed data based on implemented models/algorithms. They can also perform operations such as imputing missing data if the proportion of missed data is small and detecting outliers (Jiang, 2021).

There have been several recent works based on the pioneering work of Henry and Clarke (1993) on extending the capabilities of sensors so that they can perform the self-validate their outputs and states and perform the following actions: (1) self-diagnose the fault or sensors failure, (2) if possible, reconstruct and replace the measurements of the faulty sensors, and (3) estimate the uncertainties of the output measurements. These sensors then provide the following outputs besides raw sensor data: validated measurements, confidence levels, and the status of measurements or the fault types. If the sensor is faulty, the raw measurements would be incorrect, and validated or reconstructed measurements should be used. If the sensor is faulty, the self-validation sensors should provide the fault type.

11.3.8 Developing better models

Models can be white-box models described using differential equations in which the parameters of equations have physical sense. Also, models can be data-driven, such as regression models whose parameters might or might not have physical sense. Significant efforts have been made in modeling physiological systems as well as devices.

11.3.9 Developing better simulators

While writing the book, we observed the lack of simulators that would allow for easy simulation of physiological systems, devices and algorithms. Some of the simulators were too complex for our use. Many simulators were not open source or were written in different languages and, therefore, were difficult to integrate. Having well-developed and maintained simulators is very important for developing biomedical devices. Significant research investment needs to be made in that direction, and we believe this book is a small step forward.

11.3.10 Learning with small data

In biomedical applications, we often deal with data collected from several or tens of subjects. This is considered "small data" in machine learning, and therefore, this data cannot be used for fitting larger models. Therefore, physics-based models or hybrid physics and machine learning models might be better alternatives for parameter estimation. Please refer to Cuomo et al. (2022) for an excellent review of physics-informed neural networks.

11.3.11 Learning longer

Most works on biomedical signals focus on processing signals in a short-time window and learning information directly from the signals obtained during that window. There is a significant difference in the type of information that can be obtained from a subject monitored for several minutes and monitored for hours or days. Therefore, significant effort needs to be made to allow for long-term monitoring and learning based on the subject's behavior, habits, and conditions. Applications include monitoring for the changes in behavior (Diraco et al., 2019), changes in the conditions of the patients (such as worsening of heart failure symptoms (Krittanawong et al., 2021)), performing early diagnosis, and analyzing treatment effects.

11.3.12 Screening large populations using cardiac or respiratory monitoring

Several attempts have been made to screen a large population using available smartwatches, including the analysis of PPG data for detecting atrial fibrillation done with Apple (Perez et al., 2019) and Huawei (Guo et al., 2019) smartphones/watches on hundreds of thousands of people. That number of subjects was impossible to obtain from any study before the appearance of smartwatches. In the Apple study, 0.52% of subjects were notified of having an irregular heart rhythm.

Smartwatch data were analyzed in several studies related to detecting COVID-19 symptoms. In Mishra et al. (2020), heart rate and activity data from more than 4000 subjects wearing smartwatches were analyzed. COVID-19-related symptoms were detected in 32 subjects.

11.4 Beyond wearable devices: touch and proximity

In this section, we will briefly discuss sensors and devices embedded in the environment so that they can sense the subject's physiological variables when the subject touches commonly used objects or the subject is near objects with sensors attached to or embedded in them. Sensors that require touch are those integrated into the computer mouse, steering wheel, etc. Proximity sensors include sensors activated only when the subject is in close proximity to them, such as pressure-sensitive mats placed on the top or under a mattress.

Challenges in monitoring physiological parameters based on touch or proximity sensors include:

- The range of the sensors is very small. For example, dry ECG electrodes are galvanic electrodes that require direct touch and good contact. Capacitive ECG electrodes do not require direct contact but can operate up to several mm. The signal quality depends on the quality of the contact and/or proximity between the object and the sensors.
- Even a small motion changes the contact area or the distance to the proximity sensor, causing motion artifacts.
- Even if the contact with the sensor is good, the signal quality is inferior to that of sensors used in wearable devices.
- It is not known who is monitored. In wearable devices, the patient is normally wearing the device. Here, multiple people can touch the object with the sensor. This might not be a problem if we are interested in measuring physiological variables only during a short activity—for example, during exercising on the treadmill. However, if we would like to monitor a patient in environments where multiple people live, work or exercise, it is important to identify the patient before the measurement is done.

11.4.1 Sensors embedded into objects where people spend much time

In some situations, attaching or putting on a wearable device is not a practical option, such as in the case of monitoring the physiological variables of drivers. As the driver is in the driver's seat and he or she is touching the steering wheel for the whole duration of the ride, it makes sense to embed sensors in the seat and the steering wheel. Some vehicle sensors are contactless and allow monitoring at distances longer than several cm. Examples of these sensors are cameras and radars, which are covered in the next section. The car seat, including the backrest and seating area, is used to embed or attach capacitive ECG electrodes and BCG sensors. The steering wheel could contain contact ECG electrodes, a pulse oximeter, and a thermometer. ECG electrodes were used for heart rate and heart rate variability estimation. The problem with placing two dry electrodes on the steering wheel is that sometimes people drive with one hand or move their hands to different parts of the steering wheel while driving. Therefore, a hybrid ECG solution was proposed where one electrode is a dry electrode covering the steering wheel while another electrode is the capacitive electrode embedded into the seat. The quality of obtained ECG signal is low, but the R wave can be easily recognized (Xu and Ta, 2013). BCG sensors are used for heart and breathing rate estimation. Piezoelectric sensors were integrated into the seatbelt and used to monitor breathing. A detailed description of different types of sensors used in a car is given in Leonhardt et al. (2018) and Wang et al. (2020).

An example of an application where the subject spends a lot of time on the same platform is a wheelchair. There have been several solutions for monitoring the vital signs of wheelchair users. In Arias et al. (2016), four electromechanical films were embedded into the seat of the wheelchair and used to obtain the BCG signal. Heart rate and breathing rate were then extracted from the BCG signal.

Pressure-sensitive mats measure pressure underneath the patient from many points placed in a 2D grid. As an output, the mats provide a 2D map or an image that changes in time. Mats are mainly used to lower the incidence of pressure ulcers in patients with very limited mobility. Regarding cardiorespiratory monitoring, they are also used to acquire BCG signals so that it is possible to obtain both breathing rate (Mokhtari et al., 2019) and heart rate from them.

11.4.2 Sensors embedded into objects that people touch frequently

Sensors can be embedded in a variety of objects that are used during normal daily activities. An example of such a sensor is a smart mouse with a temperature and PPG sensor on the top. The problem with monitoring heart rate using a PPG device embedded in the mouse is the motion artifacts since the mouse gets moved all the time, and the possibility of having good contact between the fingers or palm and the LED/photodetector (Lin et al., 2017). Multiple sensors increase the likelihood of good contact with a palm.

11.5 Beyond wearable devices: contactless monitoring

In general, wearable devices are unsuitable for monitoring that takes weeks or months because they cause discomfort after some time, and people forget to wear or recharge them. Also, some people do not want to wear devices, such as dementia patients, people not comfortable with new technologies, and inmates. Contactless monitoring can help us achieve continuous, noninvasive, and unobtrusive monitoring of the vital signs and activities of patients without wearing the devices. Contactless monitoring can be done with long-range sensors as well as with proximity sensors. Proximity sensors, including pressure mats and bed occupancy sensors, were mentioned in the previous section and will not be considered here. Long-range sensors can monitor physiological variables at distances from 50 cm to about 10 m. Contactless monitoring devices become part of the infrastructure; therefore, they do not need to be turned on and do not interfere with normal activities. In addition, during the COVID-19 pandemic, these contactless sensors were useful in situations where the risk of cross-infection is high. Disadvantages include lower accuracy compared to the accuracy of wearable devices and the ability to measure only when the subject is in their range.

In addition, many sensors cannot identify the subject being monitored unless a camera with face detection is used—however, this introduces privacy issues, and people generally do not like to be monitored using cameras.

Contactless monitoring of cardiorespiratory parameters at a distance can be done by observing mechanical and thermal effects (Fig. 11.2). Mechanical effects include organ motions and fluid displacement. They can be observed by monitoring the displacement of the body surface and superficial perfusion. Monitoring displacement of the body surface is performed using radars, sonars, lasers, RGB (red, green, blue), and 3D cameras. Superficial perfusion can be detected and monitored using cameras (Bruser et al., 2015). Thermal effects are observed by measuring skin temperature changes due to blood flow or by measuring the temperature changes in front of the nose or mouth due to airflow. These measurements can be done using thermal cameras. This is all summarized in Fig. 11.2. Please note that Fig. 11.2 does not include contactless monitoring using proximity sensors. For example, BCG and capacitive ECG can be applied in a noncontact way—however, we are interested in long-range contactless monitoring.

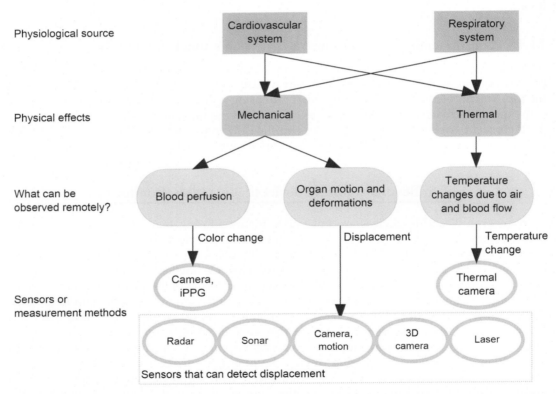

FIGURE 11.2   Physiological sources and physical variables that can be measured using contactless sensors that can detect and measure changes in color, displacement, and temperature are shown. *The figure is modified and extended from (Bruser et al., 2015).*

Challenges in contactless monitoring include:

- The signal of interest is often very small and noisy.
- The signal is extracted in uncontrolled situations when the subject is in the sensor range, often resulting in very low signal quality. Examples include the extraction of the heart rate when the subject is not oriented toward the camera or when the subject is covered by a blanket.
- Environmental factors can cause different problems for sensors. For example, environmental factors affecting video processing include darkness, a background similar to the color of the subject's face or clothes, and lighting variations. For radar, clutter, movement at frequencies that correspond to breathing frequencies, multiple people and/or animals in the room, and a working fan are all causing problems with heart rate and breathing rate measurements.
- Multiple people identification and tracking in cases cameras are not used. It is challenging to identify people using noncamera-based sensors such as radars and sonars, making it difficult to continuously monitor subjects' vital signs in environments with multiple people. In addition, cameras cannot be used in some cases due to privacy issues.
- Other issues include distance to the sensors, sensor positions, number of monitored people, and cost of devices.

11.5.1 Cameras

Camera-based methods can be used to detect both displacements of the skin due to breathing and blood flow and skin color changes due to superficial perfusion. Methods that detect color changes are referred to as imaging photoplethysmography (iPPG) or remote PPG (rPPG). Noncontact camera-based iPPG is a simple, inexpensive, and noninvasive method to detect color changes during perfusion. Normally, the region of interest (RoI) is selected, and the values of the pixels from that region are processed to get a single point. These points in each frame form a time series that can be further processed to extract the heartbeat and breathing signals. Extracted heartbeat signal is sometimes also called blood volume pulse (BVP). With appropriate filtering, it is possible to extract the breathing and heartbeat signal from this time series (Poh et al., 2011). The heart rate and breathing rate are computed in the same way described previously in this book after the heartbeat and breathing signals are obtained.

Displacement of the body surface recorded by the camera can be used for both heart rate and breathing rate estimations. Heart rate estimation based on displacements includes detecting BCG through subtle head motions caused by the Newtonian reaction to the influx of blood in each cardiac cycle (Balakrishnan et al., 2013), or minute skin motions due to pulsations. Methods for estimating breathing rate based on displacement range from detecting large movements of the abdomen, shoulders, thorax, or pit of the neck (Corres et al., 2018) to detecting small displacements on the skin of the neck and the face.

Fig. 11.3 on the top shows the processing steps needed to extract the signal from the camera and to estimate the heart rate using iPPG. Fig. 11.3 at the bottom also shows steps to extract the breathing rate using a thermal camera which is explained in the next section.

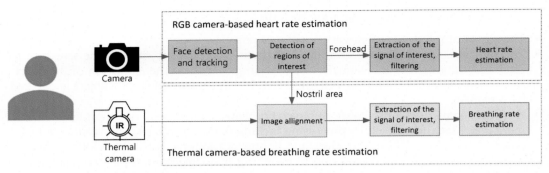

FIGURE 11.3   Top: iPPG processing steps for heart rate estimation where the input signal is the video stream from the RGB camera. Processing steps for breathing rate estimation from the video stream obtained from the thermal camera where the region of interest (nostril area) is detected by the RGB camera.

So, the first step is to point the RGB camera toward the subject's face. Normally, it is required that the subject remains still for several seconds in front of the camera. This is because even small motion will cause significant artifacts. After the video is acquired or during video acquisition, the first step is to detect and track the face. The next step is to detect the region of interest on the face. Normally, for heart rate estimation, the region of interest is the forehead. After the RoI is detected, the time series is formed where the number of points in the time series corresponds to the video duration multiplied by the frame rate. For example, for a 10 s video with 60 frames per second, we will have 10 s time series with 600 samples. Very often, only green color is selected when converting the extracted RoI images into time series. However, there are algorithms, such as (Balakrishnan et al., 2013), where all three channels (red, green, and blue) are combined and used to obtain a single point (sample) from the frame. After the heartbeat or BVP signal is extracted, it is commonly filtered in the range corresponding to cardiac frequencies—for example, between 0.7 and 2.5 Hz. After that, the heart rate can be estimated using the frequency method or by counting the number of beats in the BVP signal segment.

An experiment was performed in which a subject stood 1 m away from the webcam facing the camera. The BVP signal was extracted from the video, and it is shown in Fig. 11.4A as a blue waveform. We simultaneously collected the PPG signal from a reference PPG device. It is shown as a red waveform in Fig. 11.4A. As seen in the figure, the extracted signal based on the iPPG method is of good quality, and each beat obtained using the iPPG method corresponds to the beat obtained using the reference PPG device. The amplitudes of the waveforms are normalized to be between zero and one.

11.5.2 Thermal cameras

A thermal camera provides a video stream where pixels in each frame have values that correspond to the estimated temperature of an object related to that pixel. The thermal camera is called a thermographic camera, thermal imaging camera, or thermal imager. Compared with an RGB camera that relies on visible light between 400 and 700 nm, the thermal camera operates in an infrared radiation range, mainly between 7 and 14 μm. Thermal cameras indirectly

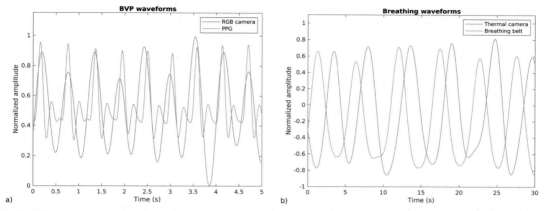

FIGURE 11.4 (A) Heartbeat waveform extracted from the RGB camera (*blue*) and the reference PPG waveform (*red*). (B) Breathing waveform extracted from the thermal camera (*blue*) and the reference waveform using a breath belt (*red*).

measure the temperature by measuring the energy emitted by an entity (human, engine) in the infrared frequency band. When looking at the image acquired using a thermal camera, the brightest spots in the image correspond to the warmest part of the image, red and yellow correspond to an intermediate temperature, while the dark parts correspond to the coolest part of the image. A scale should be provided next to an image to relate colors to temperatures.

Several important parameters determine the quality of the thermal camera, and they are listed in Table 11.3. The definitions of the parameters are presented too. The spatial resolution of thermal cameras is considerably lower than that of RGB cameras, especially for inexpensive cameras where the resolutions can be as low as 160×120 pixels. The resolution determines how fine the details are in the image. Spatial resolution is based on the number of pixels in horizontal and vertical directions and the field of view (FOV) specification. A combination of them defines the area the imager sees. The higher resolution means that each image contains more information and details. As the breathing rate is normally measured by observing the temperature changes under the nose, the resolution should be high enough to allow for detecting several pixels in both horizontal and vertical directions in the RoI. Therefore, low-resolution cameras can detect physiological signs only at very short ranges of several tens of centimeters. Thermal sensitivity or Noise Equivalent Temperature Difference (NETD) describes the smallest temperature difference the thermal camera can detect. This is important when there are only small changes in the temperature under the nose when measuring the breathing rate. Measuring heart rate requires thermal cameras with relatively high sensitivity. Thermal cameras are much more expensive than RGB cameras, which is one of the main reasons they are not used pervasively for contactless vital signs measurements. Cameras with high resolution and low NETD are an order of magnitude more expensive than RGB cameras.

Next, we discuss the use of the thermal camera for breathing rate estimation. The RGB camera is used to detect the RoI that corresponds to the nostril area. Thermal cameras are

TABLE 11.3   Parameters of a thermal camera.

Parameters	Definition	Typical values for low-end cameras	Typical values for high-end cameras	Comment
Frames per second (fps)	The number of thermal image frames acquired per second	9 fps	60 fps	Even 9 FPS cameras were used for heart rate and breathing rate estimation
Thermal resolution	The number of pixels in the horizontal and vertical direction	160×120	1280×1024	A larger resolution results in a larger distance between the camera and the subject
Temperature accuracy	The closeness between the temperature value obtained by the camera and the reference value.	$\pm 2°C$	Better than $\pm 1°C$	Normally for breathing and heart rate estimation, the absolute value of the temperature is less important.
Noise equivalent temperature difference (NETD)	Temperature sensitivity = the smallest temperature difference that the camera can detect	70 mK	20 mK	The sensitivity is a very important parameter for detecting small temperature changes, especially for heart rate measurements.
Field of view	The viewing angle of the camera in the horizontal and vertical direction	$36° \times 36°$	$80° \times 60°$	A wide field of view is important if we are not interested only in the subject standing just in front of the camera
Cost	Cost of the device in the US dollars	<250 $	>20,000 $	Please note that even inexpensive cameras can be used for estimating the breathing rate of a person several tens of cm away from the camera.

often associated with RGB cameras, and RGB image is used for RoI estimation—see Fig. 11.3. However, RGB and thermal cameras have different resolutions and fields of view and, therefore, the ROI coordinates in the RGB image need to be matched with the RoI coordinates in the image acquired by the thermal camera.

Fig. 11.4B shows the breathing rate estimation of a healthy subject standing about 1 m away from the thermal and RGB cameras. RGB and thermal videos are collected by the RGB and thermal cameras simultaneously, and the sampling rate of these two cameras is 30 fps. The reference signal is shown in red, and it is collected from a piezoresistive breathing belt. The breathing signal obtained from the thermal camera after several processing steps shown in Fig. 11.3 is presented in blue in Fig. 11.4B. As can be seen, the signals from the breathing belt and the thermal camera have the opposite phase. During expiration, the temperature in the nostril area increases while the chest volume decreases. The opposite happens during the inspiration.

A detailed review of applying thermal cameras for physiological monitoring is given in Manullang et al. (2021). An example of a work where the thermal camera is used for respiratory rate tracking based on mobile thermal imaging is Cho et al. (2017). The detection distance between the human face and the thermal camera is limited to less than 50 cm in Cho et al. (2017), and the breathing rate estimation error increases with increasing the distance between the subject and the camera.

11.5.3 Radars

Radars have been widely used for the detection of vital signs such as breathing and heart rate (Chioukh et al., 2014; Girbau et al., 2012; Massagram et al., 2009). In addition, they have been applied for fall detection, stress level detection (Wang et al., 2020), classification of human activities, detecting humans under rubble, and for suicide warning (Ashe, 2013). Several books have been published on using radars for physiological monitoring, including Boric-Lubecke et al. (2016) and Amin (2017). The operation of the radar is based on transmitting electromagnetic waves and observing the reflecting wave from objects. The reflected wave is processed to extract information about these objects. The time delay between the transmitted signal and the radar return determines the distance between the radar and the object. The shift in frequency of the reflected signal due to the Doppler effect allows us to compute the subject's velocity.

Without any movements, the radar reflections are considered clutter. We are interested in motion due to heartbeat and breathing. Small periodic movements observed by the radar are called micro-Doppler. As pointed out in Boric-Lubecke et al. (2016), peak-to-peak chest motion due to breathing is 4—12 mm, while peak-to-peak motion due to heartbeat is up to 0.5 mm. Depending on the frequency and power, the radar can detect these small periodic skin displacements due to breathing and heartbeat.

The two main radar technologies used for this purpose are the pulsed ultrawideband radar (UWB) and continuous-wave (CW) radar. The pulsed radar transmits modulated pulses of short duration and then receives the echo from reflecting objects. The frequency band of commonly used UWB radars for the research on breathing rate estimation is 6—8.5 GHz (for example, X4M03 radar chipset by Novelda). At this frequency range, the radar signal propagates through walls so that it is possible to monitor the physiological parameters of subjects without a direct line of sight between the radar and the subject. The CW radar transmits and receives a continuous electromagnetic wave. Nowadays, CW radars used for breathing and heartbeat estimation are mainly frequency-modulated CW radars (FMCW), also used for automotive applications. Their frequency range is between 77 and 81 GHz (see automotive radars by, for example, Texas Instruments).

The radar captures and down-converts the signal modulated by the physiological displacements and then identifies the human heartbeat and respiration rates by processing the baseband signal. The signal received at the radar is a superposition of signals from a movement of the torso, movement of the limbs, and expansion and contraction of the chest cavity in addition to noise and interference. Even for a stationary subject, detecting small displacement due to breathing is challenging because of clutter from the walls, other objects, and the proximity of other people. Therefore, physiological radar signal processing requires the following steps: noise and clutter suppression, estimation of the number of people, range estimation, classification of activities, and breathing signal extraction for stationary subjects (Singh et al., 2011; Yarovoy et al., 2008).

Processing stages for estimating breathing rate from radar returns are shown in Fig. 11.5. The figure shows a radar that transmits electromagnetic wave. This wave is reflected by many objects and from the subject of interest. The first step in processing is clutter removal, and preprocessing can include bandpass filtering. Next, the radar can divide the area into a number of bins called range bins. For example, the X4M03 radar chipset by Novelda divides the 10 m range into about 5 cm wide range bins. The next processing stage is determining the range bin where the subject is located. If we assume that there is no other movement in the room, then this can be done by observing the variance of the radar signal over time in each range bin. The larger variance would correspond to larger movements. Fig. 11.6B) shows normalized variance per range bin when there are two stationary subjects in the room where one subject is about 2 m away from the radar and another one is less than 3 m away from the radar. The peaks of the variance, in this case, correspond to the chest movements of both people. However, these movements could result from the subject's motion or other motions in the room (for example, the fan). Therefore, the type of motion needs to be classified. If we can detect the subject and classify her/his activities as stationary, then the next step is to extract the breathing and heartbeat signal and then apply standard processing steps to determine breathing rate and heart rate.

An example of a breathing signal extracted from the UWB radar based on X4M03 radar chipset by Novelda is shown in Fig. 11.6B. In the experiment, there were two subjects and the algorithm detected the range bin with the largest variance corresponding to the first subject. The breathing signal is extracted as a phase from the complex radar returns. As can be seen, even though the subject is stationary, the signal is quite noisy.

Some challenges when working with radars for physiological monitoring are listed in Table 11.4. In addition to challenges, we list the consequences if these challenges are not addressed as well as some ways to address them. For example, the presence of multiple

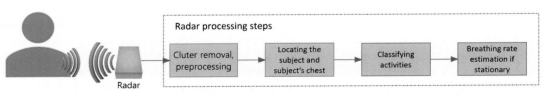

FIGURE 11.5   Processing steps for estimating breathing rate from radar returns.

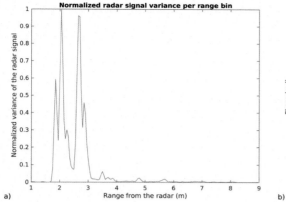 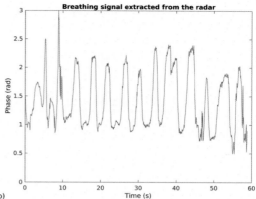

a)   b)

FIGURE 11.6 (A) Normalized variance per range bin for the experiment with two stationary subjects that were sitting in their chairs breathing normally. (B) A breathing signal of the closer subject extracted from the radar returns.

TABLE 11.4 Challenges when using radars to extract breathing and heartbeat signals.

Main challenges	Consequences	Approach
Cannot detect the heartbeat	Radar cannot be used for heart rate monitoring	Use higher frequency (for example, 80 GHz) radar. Make sure that the subject is completely stationary before trying to extract the heartbeat signal.
Occlusion of the subject	If a subject is behind a metal object or another subject, she/he will not be detected by the radar. If the subject is behind the wall and a millimeter wave FMCW radar is used, the subject will not be detected.	Use multiple radars to cover the whole room. Use UWB radar in the 3–10 GHz range for through-the-wall detection.
Presence of multiple people	Inability to distinguish among people who are close to one another	This is a problem for radars. Multiple radars of radars with steering beams could be used in this case.
Weak Doppler signature depending on the orientation of the subject	This can result in misdetection of the subject or his/her breathing signal.	Multiple radars placed at different locations can be installed.
Reading through walls if UWB radar is used	Detecting people outside the space of interest	Location capabilities and floorplan of the room where the radar is placed together with the radar learning about landmarks of the environment would allow for constraining processing within the walls.
Interferers at a similar frequency as the breathing frequency.	Interferers such as fan or water movement can be detected as the breathing signal by the radar.	More advanced processing that would take into account the morphology of the signal needs to be applied.

people close to one another might result in their breathing signals mixed in the same range bin. Therefore, their breathing signals cannot be distinguished. Multiple radars or radars with multiple antennas can be used in this case. Multiple radars need to be placed to detect peoples' chests in different range bins.

11.6 Combining all: pervasive health monitoring systems

Real pervasive sensing systems are designed for the continuous monitoring of people. This includes monitoring in their home and working environment as well as during recreation or exercises. Therefore, pervasive health monitoring systems will be based on wearables, proximity sensors as well as contactless sensors. Combining touch-based/proximity sensors with contactless sensors has been done in automotive applications. Also, the integration of capacitively coupled ECG, reflective PPG, bioimpedance sensors, and BCG sensors into an armchair was done to obtain heart rate with the goal of reducing motion artifacts (Antink et al., 2017). This represents a large number of sensors to estimate the heart rate. More advanced monitoring could be done in this case.

The major problem in pervasive monitoring using both contactless and wearable sensors is the following:

- Monitoring is performed in an uncontrolled environment, and therefore there is no guarantee that any measurement procedure is followed
- Identifying each subject to be monitored is challenging when multiple people are being monitored.

Potential research directions include:

- Improvement of sensors and the development of novel sensing modalities
- Sensor fusion at the massive scale
- Inferring actions and conditions using unsupervised learning techniques
- Inferring complex actions of subjects to understand the context in which the measurement is performed.

11.7 Conclusion

In this chapter, we summarized the electronics and devices covered in this book. Then, we briefly discussed devices that do not require direct contact to perform measurements. We believe that all of these devices have their place for future monitoring of patients at home. We hope this book will provide good training for students and practitioners to design these devices.

References

Amin, M., 2017. Radar for Indoor Monitoring: Detection, Classification, and Assessment. CRC Press.

Antink, C.H., Schulz, F., Leonhardt, S., Walter, M., 2017. Motion artifact quantification and sensor fusion for unobtrusive health monitoring. Sensors, MDPI 8 (issue 38).

Araujo, J.I.L., Morais, S.M.A., de Freitas Serres, G.K., et al., August 2018. Passive RFID Tag for Respiratory Frequency Monitoring, 3rd International Symposium on Instrumentation Systems. Circuits and Transducers.

Arias, D.E., Pino, E.J., Aqueveque, P., Curtis, D.W., 2016. Unobtrusive support system for prevention of dangerous health conditions in wheelchair users. In: Mobile Information Systems, vol 2016. Hindawi. Article ID 4568241.

Ashe, J.M., Oct. 2013. Unobtrusive Suicide Warning System. Final Technical Report, Phase III. Technical report, Document No.: 243922.

Balakrishnan, G., Durand, F., Guttag, J., 2013. Detecting pulse from head motions in video. IEEE Conference on Computer Vision and Pattern Recognition 3430–3437.

Bhat, G., et al., 2020. Self-powered wearable IoT devices for health and activity monitoring. Foundations and Trends® in Electronic Design Automation 13 (3), 145–269.

Boric-Lubecke, O., et al., 2016. Doppler Radar Physiological Sensing. Wiley.

Bruser, C., Antink, C.H., Wartzek, T., Walter, M., Leonhardt, S., 2015. Ambient and unobtrusive cardiorespiratory monitoring techniques. IEEE Reviews in Biomedical Engineering 8, 30–43.

Chioukh, L., Boutayeb, H., Deslandes, D., Wu, K., 2014. Noise and sensitivity of harmonic Radar architecture for remote sensing and detection of vital signs. IEEE Transactions on Microwave Theory and Techniques 62 (9), 1847–1855.

Cho, Y., Julier, S.J., Marquardt, N., 2017. Robust tracking of respiratory rate in high-dynamic range scenes using mobile thermal imaging. Biomedical Optics Express 8 (10), 4480–4503.

Corres, J., et al., 2018. Contactless monitoring of breathing patterns and respiratory rate at the pit of the neck: a single camera approach. Journal of Sensors 2018, 4567213.

Cuomo, S., di Cola, V.S., Giampaolo, F., Rozza, G., Raissi, M., Piccialli, F., 2022. Scientific Machine Learning through Physics-Informed Neural Networks: Where We Are and What's Next arXiv:2201.05624.

Diraco, G., Leone, A., Siciliano, P., 2019. AI-based early change detection in smart living environments. Sensors 19 (16).

Eck, V.G., Donders, W.P., Sturdy, J., et al., 2016. A guide to uncertainty quantification and sensitivity analysis for cardiovascular applications. International Journal for Numerical Methods in Biomedical Engineering 32 (8).

Girbau, D., Lázaro, A., Ramos, A., et al., 2012. Remote sensing of vital signs using a Doppler radar and diversity to overcome null detection. IEEE Sensors Journal 12 (3), 512–518.

Guo, Y., et al., 2019. Mobile photoplethysmographic technology to detect atrial fibrillation. Journal of the American College of Cardiology 74, 2365–2375.

Henry, M.P., Clarke, D.W., 1993. The self-validating sensor: rationale, definitions and examples. Control Engineering Practice 1 (4), 585–610.

Jiang, Y., Yin, S., Dong, J., Kaynak, O., 2021. A review on soft sensors for monitoring, control, and optimization of industrial processes. IEEE Sensors Journal 21 (11), 12868–12881.

Krittanawong, C., Rogers, A.J., Johnson, K.W., et al., 2021. Integration of novel monitoring devices with machine learning technology for scalable cardiovascular management. Nature Reviews Cardiology 18, 75–91.

Leonhardt, S., Leicht, L., Teichmann, D., 2018. Unobtrusive vital sign monitoring in automotive environments-A review. Sensors, MDPI 18 (9), 3080.

Lin, S.-T., Chen, W.-H., Lin, Y.-H., Lin, Y.-H., 2017. A pulse rate detection method for mouse application based on multi-PPG sensors. Sensors, MDPI 17 (7).

Lund, A., Wu, Y., Fenech-Salerno, B., Torrisi, F., Carmichael, T.B., Müller, C., 2021. Conducting materials as building blocks for electronic textiles. MRS Bulletin 46, 491–501. Springer.

Manullang, M.C.T., Lin, Y.-H., Lai, S.-J., Chou, N.-K., 2021. Implementation of thermal camera for noncontact physiological measurement: a systematic review. Sensors, MDPI 21 (7777).

Massagram, W., Lubecke, V.M., Host-Madsen, A., Boric-Lubecke, O., 2009. Assessment of heart rate variability and respiratory sinus arrhythmia via Doppler radar. IEEE Transactions on Microwave Theory and Techniques 57, 2542—2549.

Mishra, T., Wang, M., Metwally, A.A., et al., 2020. Pre-symptomatic detection of COVID-19 from smartwatch data. Nature Biomedical Engineering 4, 1208—1220.

Mokhtari, M., Aloulou, H., Abdulrazak, B., Fernanda, C.M., 2019. Smart mat for respiratory activity detection: study in a clinical setting. In: Chapter in: How AI Impacts Urban Living and Public Health. by Springer International Publishing, pp. 61—72.

Perez, M.V., et al., 2019. Large-scale assessment of a smartwatch to identify atrial fibrillation. The New England Journal of Medicine 381, 1909—1917.

Poh, M.Z., McDuff, D.J., Picard, R.W., 2011. Advancements in noncontact, multiparameter physiological measurements using a webcam. IEEE Transactions on Biomed. Eng. 58, 7—11.

Singh, S., Liang, Q., Chen, D., Sheng, L., 2011. Sense through wall human detection using UWB radar. Journal on Wireless Communications and Networking 1—11.

Someya, T., Amagai, M., 2019. Toward a new generation of smart skins. Nature Biotechnology 37, 382—388.

Wang, J., Warnecke, J.M., Haghi, M., Deserno, T.M., 2020. Unobtrusive health monitoring in private spaces: the smart vehicle. Sensors, MDPI 20 (9), 2442.

Wang, C., Qi, B., Lin, M., et al., 2021. Continuous monitoring of deep-tissue haemodynamics with stretchable ultrasonic phased arrays. Nature Biomedical Engineering 5, 749—758.

Xu, X., Ta, L., 2013. A novel driver-friendly ECG monitoring system based on capacitive-coupled electrode. Information Technology Journal 12, 4730—4734.

Yarovoy, A.G., Ligthart, L.P., Matuzas, J., Levitas, B., 2008. UWB radar for human being detection. IEEE Aeroscope and Electronic Systems Magazine 23, 36—40.

Index

Note: 'Page numbers followed by "*f*" indicate figures and "*t*" indicate tables.'

Printed in the United States
by Baker & Taylor Publisher Services